Springer Proceedings in Earth and Environmental Sciences

Series Editors

The series Springer Proceedings in Earth and Environmental Sciences publishes proceedings from scholarly meetings and workshops on all topics related to Environmental and Earth Sciences and related sciences. This series constitutes a comprehensive up-to-date source of reference on a field or subfield of relevance in Earth and Environmental Sciences. In addition to an overall evaluation of the interest, scientific quality, and timeliness of each proposal at the hands of the publisher, individual contributions are all refereed to the high quality standards of leading journals in the field. Thus, this series provides the research community with well-edited, authoritative reports on developments in the most exciting areas of environmental sciences, earth sciences and related fields.

Luan Vu · Peter Sly · Stephania A. Cormier ·
Phuong Thao Nguyen · Thi Thu Hoai Nguyen
Editors

Global Perspectives on the Climate-Health Interface

Proceedings of the 20th Pacific Basin Consortium Conference, 28–31 May 2025, Ho Chi Minh, Vietnam

Springer

Editors
Luan Vu
University of Texas at San Antonio
San Antonio, TX, USA

Stephania A. Cormier
Pennington Biomedical Research Center
Louisiana State University
Baton Rouge, LA, USA

Thi Thu Hoai Nguyen
Research Center for Infectious Diseases
International University, Vietnam National University
Ho Chi Minh City, Vietnam

Peter Sly
Child Health Research Centre
The University of Queensland
South Brisbane, QLD, Australia

Phuong Thao Nguyen
Research Center for Infectious Diseases
International University, Vietnam National University
Ho Chi Minh City, Vietnam

ISSN 2524-342X ISSN 2524-3438 (electronic)
Springer Proceedings in Earth and Environmental Sciences
ISBN 978-3-032-29578-1 ISBN 978-3-032-29579-8 (eBook)
https://doi.org/10.1007/978-3-032-29579-8

This Springer imprint is published by the registered company Springer Nature Switzerland AG
The registered company address is: Gewerbestrasse 11, 6330 Cham, Switzerland

Contents

The Impact of Prenatal Exposure Heavy Metals to Pregnancy Outcome in Makassar City

Ana Utami Zainal[1], Hasnawati Amqam[2(✉)], Agus Bintara Barawida[2], Stang[3], Rina Previna Amiruddin[4,5,5], Anna Khuzaimah[6], Isymiarni Syarif[1], Irawati[1], and Sarmaliana[1]

[1] Faculty of Public Health, Hasanuddin University, Makassar, Indonesia
[2] Departement of Environmental Health, Faculty of Public Health, Hasanuddin University, Makassar, Indonesia
hasnawati.amqam@ms.unhas.ac.id
[3] Departement of Biostatistic, Faculty of Public Health, Hasanuddin University, Makassar, Indonesia
[4] Hospital of Mother and Children Kartini, Makassar, Indonesia
[5] RSUD Labuang Baji, Makassar, Indonesia
[6] Departement of Nutrition, Faculty of Public Health, Hasanuddin University, Makassar, Indonesia

Abstract.
Background The prenatal period is a critical phase of human development during which the fetus is highly susceptible to environmental exposures, including heavy metals. Makassar City has been identified as an area with significant environmental contamination, particularly with heavy metals such as chromium (Cr), manganese (Mn), copper (Cu), and arsenic (As), which may pose health risks to both mothers and infants.

Objective This study aimed to analyze the impact of prenatal exposure to heavy metals (Cr, Mn, Cu, As) on pregnancy outcomes, specifically neonatal birth weight and length, in Makassar, Indonesia.

Methods A cross-sectional analytical study was conducted involving 70 pregnant women meeting inclusion criteria. Maternal blood concentrations of heavy metals were measured using Inductively Coupled Plasma Mass Spectrometry (ICP-MS), and neonatal anthropometric data were collected within 24 h of birth. Correlation analyses were performed to evaluate associations between maternal metal concentrations and neonatal outcomes.

Results Significant negative correlations were found between maternal blood levels of Cr and birth weight ($r = -0.331$, $p = 0.005$). Mn levels were also negatively correlated with birth weight ($r = -0.277$, $p = 0.020$). No significant associations were observed between Cu or As and neonatal birth measurements.

Conclusion Prenatal exposure to certain heavy metals, particularly Cr, was associated with reduced neonatal birth weight and length. These findings highlight the

L. Vu et al. (Eds.): PBC 2025, 2026.
https://doi.org/10.1007/978-3-032-29579-8_1

potential risks of environmental contamination on fetal development and underscore the importance of integrating environmental monitoring and protection into maternal health strategies, especially in urban areas like Makassar.

Keywords: Heavy metals · Pregnancy · Birth weight · Birth length

1 Introduction

The prenatal period is a critical and foundational stage in human development, during which the fetus is highly vulnerable to various external factors, including environmental toxicants such as heavy metals. These contaminants have been increasingly recognized for their potential to cause complications during pregnancy and interfere with fetal development [1]. In this context, prenatal exposure to heavy metals has emerged as a significant issue in environmental and maternal-child health, particularly in urban settings with considerable environmental pollution, such as Makassar City.

Heavy metals are chemical elements with high atomic mass and density, known for their toxic effects even at low concentrations. These elements are widely present in daily life through air, water, food, household dust, and industrial products [2, 3]. In Makassar City, environmental monitoring has revealed the presence of heavy metals such as Cadmium (Cd), Chromium (Cr), Manganese (Mn), Lead (Pb), and others, with some measurements exceeding environmental safety thresholds [4]. For instance, Cd levels in Makassar waters remained consistently high over a four-year observation period (0.001 $\pm$ 2.28 μg/L), and Cr concentrations rose significantly in 2020 [4]. Moreover, heavy metals such as Cr and Cd have been detected in high concentrations in fish and shellfish consumed by the public [5], as well as in soils around landfill areas such as final solid waste disposal site in Antang. These findings indicate that the public, including pregnant women, may be chronically exposed to multiple sources of heavy metal contamination.

Several heavy metals are known to cross the placental barrier and accumulate in fetal tissues, posing teratogenic and neurotoxic risks [6]. Prenatal exposure to Pb, Cd, Cr, and Hg has been associated with preterm birth, reduced birth weight and length, and impaired neurodevelopment [7]. Although some metals like Mn are essential in trace amounts, excessive exposure can also result in adverse health effects [8].

In South Sulawesi, data from the 2018 Nasional Basic Health Research show that the prevalence of short birth length was 27.1%, higher than the national average of 22.6%. The proportion of low birth weight (<2500 g) was also above the national average, reaching 4.3% [9]. In Makassar City, data from the Health Office revealed fluctuating trends in low birth weight (LBW) cases, with a significant increase to 816 cases in 2023 [9]. These statistics suggest a possible link between environmental exposure and poor pregnancy outcomes in the region.

Most previous studies have investigated the effects of individual heavy metals, while in reality, pregnant women are typically exposed to complex mixtures of multiple metals. Research from countries like China, Sweden, and the United States has shown that combined exposure to Cd, Pb, and Hg may significantly affect neonatal birth metrics [10]. However, these studies also highlight inconsistencies and sex-specific effects, underscoring the complexity and non-linearity of the associations involved.

In Indonesia, research on prenatal exposure to heavy metals remains limited and mostly focuses on a single metal such as Hg [11]. Additionally, classical statistical approaches like multiple linear regression are commonly used, which may not sufficiently account for the complex interactions and nonlinear effects of multiple exposures.To address these complexities, advanced statistical methods such as Bayesian Kernel Machine Regression (BKMR) have been developed. BKMR allows for the simultaneous analysis of correlated mixtures, capturing potential nonlinear and interactive effects among metals [12].

This study offers both theoretical and practical contributions. Scientifically, it represents the first study in Indonesia to evaluate the influence of combined prenatal exposure to Cr, Mn, Cu and As on neonatal birth weight and length. It deepens our understanding of the interactive effects of heavy metal exposure on fetal development. Practically, the study provides a foundation for public health strategies and environmental health policies aimed at improving maternal and child health. The results are expected to inform future regulatory standards and health promotion efforts targeting environmental toxicants in urban Indonesian settings.

Based on the above context, this study aims to analyze the influence of prenatal exposure to Cr, Mn, Cu and As on neonatal outcomes in Makassar City. It will also assess maternal blood concentrations of these metals, identify exposure sources, and evaluate their contribution to birth length and weight outcomes.

2 Methods

This study employed an observational analytic design with a cross-sectional approach. The research was conducted in Makassar City, South Sulawesi, Indonesia, an urban area known to be affected by environmental contamination, particularly by heavy metals. This study used data collected from a total 70 pregnant women from Health Center in Makassar City. The inclusion criteria were pregnant women aged 18–40 years and living in the area for at least one year. Exclusion criteria for this study were women with preeclampsia, diabetes, or other pregnancy-related complications that could confound birth outcomes.

A purposive sampling technique was used to select participants from 14 primary health centers in Makassar City. The minimum required sample size was determined based on power analysis to detect differences in neonatal outcomes related to exposure levels, with adjustments for possible non-response or incomplete data.

Data Collection

Venous blood specimens (10 mL) were procured from every participant utilizing tubes containing ethylenediaminetetraacetic acid (EDTA) by the technician and maintained at a temperature range of 2–8 °C prior to subsequent analysis. In summary, blood specimens were drawn from the EDTA tube employing a 0.5 mL micropipette and subsequently transferred to a test tube. Each specimen was subjected to treatment with 10 mL of 98% nitric acid (HNO_3). The specimens were placed on a hotplate at temperatures ranging from 200–300 °C for decomposition until the vapor at the periphery of the tube evaporated. Each specimen received 49 mL of distilled water (Amidis). The mixture was homogenized, transferred to a 50 mL volumetric glass, and sealed with parafilm. Prior to analysis, the specimen was filtered utilizing 0.45 μm membrane filters. Cr, Mn, Cu and As concentrations were determined using inductively coupled plasma mass spectrometry (ICP-MS) (Thermo Fisher Scientific, Bremen, Germany) and measured in μg/L. To ensure the precision of the analytical methods and confirm validation, Certified Reference Materials (CRM BCR®-636, lyophilized human blood, Joint Research Center-JRC, European Commission) were employed as quality control benchmarks. For the implementation of quality assurance and quality control (QA/QC), calibration curves were constructed utilizing multi-element standard solutions of established concentrations. Negative control samples were incorporated into each analytical batch to assess contamination levels. Recovery rates were found to be within the acceptable range (85–115%).

Statistical analysis

Selected demographic, maternal and infant characteristics, exposure concentrations, and birth outcomes (birth length and weight) were summarized among the 70 mother-infant. The median, standard deviation (SD), minimum, and maximum of each blood metal concentration were presented overall. Correlations between heavy metals and birth outcome were assessed using Spearman's rank correlation coefficients.

Ethical Considerations

Ethical approval was obtained from the Ethics Committee of the Faculty of Public Health, Hasanuddin University (Protocol Number 31724093044). Written informed consent was obtained from all participants prior to data collection. Participants were assured the right to withdraw at any time, and confidentiality was strictly maintained by anonymizing and securely storing all data. Non-monetary incentives were provided in recognition of participants' time and contributions.

Results

Table 1 summarizes the selected characteristics of the 70 mother-infant that most pregnant women in this study had a high school education background, were unemployed (housewives), and were exposed to cigarette smoke from other people around them even though pregnant women almost never smoked actively. The average birth weight and length of newborns were within the normal range (13). However, a significant proportion of pregnant women with a history of low birth weight and miscarriage likely had obstetric risk factors related to heavy metal exposure, which could negatively impact pregnancy outcomes.

The distributions of the blood metals overall was summarized in Table 2. The marked variability in blood concentrations of Cr, Mn, and As, as indicated by high standard deviations and large gaps between minimum and maximum values. Therefore, median values and interquartile ranges (IQR) were added to ensure accurate representation of central tendency and data dispersion. Log-transformed values were also calculated to address extreme right-skewness and the influence of outliers. Chromium and manganese showed a very wide range of concentrations, while arsenic, despite its low median,

Table 1. Distribution characteristics of the pregnant women.

Characteristics (n = 70)	n (%)	Mean ± SD
Infant sex	40 (57.1)	
Male	30 (42.9)	
Female		
Education	16 (22.9)	
University level		
High school		
Basic		
Occupational	40 (57.1)	
Unemployment	14 (20.0)	
Employment		
Smoking During Pregnancy	12 (17.1)	
Never	58 (82.9)	
Former		
Tobaco Exposure	69 (98.6)	
No	1 (1.4)	
Yes		
Birth Outcomes	26 (37.1)	2914.76 ± 575.87
Newborn weight (gr)	44 (62.9)	47.65 ± 2.83
Newborn Length (cm)		
History Low Birth Weight	33 (47.1)	
Yes	37 (52.9)	
No		
History Miscarriage	39 (55.7)	
Yes	31 (44.3)	
No		

Table 2. Heavy Metal Concentrations (μg/L).

	Mean	SD	Median (IQR)	Minimum	Maximum	Log^{10} Mean ± SD
Cr	14.73	84.95	3.30 (1.10–8.45)	0.01	714.00	0.42 ± 0.81
Mn	25.80	67.97	4.46 (1.22–10.78)	0.01	454.69	0.53 ± 0.76
Cu	20.77	12.66	18.49(11.33–27.66)	0.01	83.61	1.19 ± 0.31
As	1.78	13.74	0.01 (0.01–0.45)	0.01	115.00	−0.88 ± 0.52

showed a high maximum, indicating sporadic elevated exposure. In contrast, copper levels were more evenly distributed, with a narrower range and lower variability.

Table 3 presents the Spearman correlations between maternal blood heavy metal concentrations and newborn anthropometric outcomes. Chromium showed a negative correlation with newborn length and weight, with a statistically significant association with weight. Manganese levels also had a negative impact on newborn weight, reaching statistical significance, while its correlation with newborn length was not statistically significant. Copper and arsenic did not show significant correlations with either outcome. Additionally, 95% confidence intervals (CI) have been added to strengthen statistical reporting.These findings suggest that high maternal Cr and Mn levels may be associated with decreased fetal growth, especially in terms of infant birth weight (Fig. 1).

Table 3. Correlation between heavy metals and newborn length and weight.

Heavy metals	Newborn length (95% Cl)		Newborn weight (95% Cl)	
	r value	p value	r value	p value
Cr	−0.205 * (−0.39 to 0.02)	0.088*	−0.331* (−0.50 to–0.12)	0.005**
Mn	−0.122 * (−0.32 to 0.09)	0.316**	−0.277* (−0.46 to–0.06)	0.020**
Cu	−0.053* (−0.27 to 0.16)	0.664**	0.018* (0.21–0.24)	0.879**
As	−0.163* (−0.36 to 0.05)	0.177**	0.025* (−0.20–0.25)	0.837**

** spearman correlation test

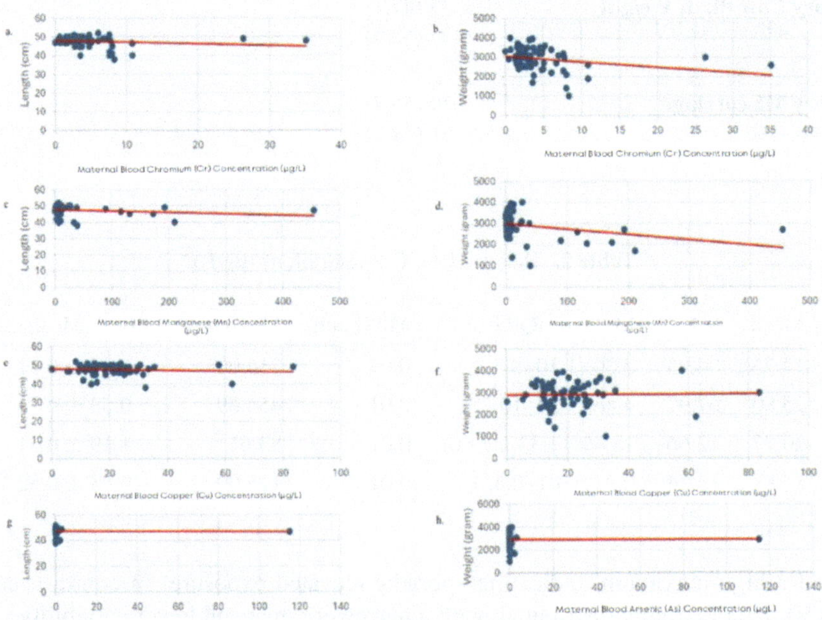

Fig. 1. Scatter plot of maternal blood (Cr ^{ab}, Mn^{cd}, Cu^{ef}, and As^{gh}) levels and newborn anthropometric (Length and Weight).

The scatterplot illustrates the relationship between maternal blood Cr levels and birth weight. Although the data show substantial clustering at lower Cr concentrations, a negative linear trend was observed. This pattern is consistent with the statistical finding that higher Cr levels were significantly associated with lower birth weight ($p = 0.005$), suggesting that increased maternal chromium exposure may adversely affect fetal growth.

3 Discussions

This study aimed to examine the relationship between prenatal exposure to heavy metals Cr, Mn, Cu, As and neonatal anthropometric outcomes, particularly birth length and weight, in Makassar, Indonesia. It should be noted that exposure biomarkers were measured only at delivery, so these results primarily reflect exposure during the later stages of pregnancy. This study did not assess exposure dynamics across the first and third trimesters. Therefore, our findings are indicative and should not be considered a comprehensive picture of prenatal exposure. The findings revealed substantial variability in maternal blood concentrations of heavy metals, along with statistically significant inverse associations between certain metals particularly Cr and Mn and neonatal outcomes.

The primary analysis used Spearman correlation due to the non-normal and highly skewed distribution of the data. This analysis did not control for confounding variables such as maternal age, parity, nutritional status, cigarette smoke exposure, and gestational age. Given the sample size ($n = 70$), multivariate regression risked producing an unstable model. Therefore, the results should be interpreted as preliminary associations, not causal relationships.

This study aimed to examine the relationship between prenatal exposure to the heavy metals Cr, Mn, Cu, and As and neonatal anthropometric outcomes. The finding of a negative correlation between Cr and Mn and birth weight should be interpreted with caution. Spearman correlation did not control for confounding variables such as maternal age, parity, nutritional status, cigarette smoke exposure, and gestational age. Given the small sample size, multivariate regression was unstable and therefore not performed. Therefore, the associations found represent preliminary associations and do not constitute causal evidence. These findings are consistent with previous epidemiological studies showing an inverse association between prenatal heavy metal exposure and fetal growth. However, the proportion of explained variance ($<11\%$) suggests that other factors likely contribute significantly to birth weight variation. Therefore, these results cannot be considered causal evidence.

The most prominent and consistent finding was the statistically significant negative correlation between maternal chromium levels and newborn birth weight ($r = -0.331$, $p = 0.005$). This suggests that higher Cr exposure during pregnancy may impair intrauterine growth, likely by interfering with placental function or through direct toxic effects on fetal tissues. Chromium, particularly in its hexavalent form (Cr(VI)), is a known environmental pollutant and recognized for its oxidative stress-inducing properties, which can lead to DNA damage and disrupted cell proliferation during fetal development [10]. Similarly, manganese exposure was found to be significantly negatively associated with birth weight ($r = -0.277$, $p = 0.020$), although its correlation with birth length was not statistically significant. While Mn is an essential trace element involved in enzyme function and antioxidant defense [8], excessive levels during pregnancy have been associated with neurotoxicity and growth restriction in animal and human studies. The elevated levels observed in this study, with a mean of 25.80 μg/L and a maximum of 454.69 μg/L, raise concerns over chronic environmental or dietary exposure in the study area. Although a significant correlation was found between Cr and Mn and the outcome, the strength of the association was relatively weak. The proportion of explained variance of <11% confirms that other factors likely contributed significantly to the variation in outcome. Therefore, these results should be interpreted as preliminary associations and cannot be considered as evidence of causality.

On the other hand, copper and arsenic did not show significant correlations with neonatal anthropometry. Copper levels were relatively stable, and although Cu plays a vital role in iron metabolism and fetal development, the lack of association may reflect adequate homeostatic regulation in the maternal-fetal interface [12]. Arsenic, despite being a well-established teratogen and endocrine disruptor, was generally found in low concentrations in this study population, with a median of 0.01 μg/L and a mean of 1.78 μg/L. However, the wide standard deviation suggests sporadic high exposures in some individuals, warranting further investigation.

These findings are consistent with previous epidemiological studies that demonstrate inverse associations between prenatal exposure to certain heavy metals and fetal growth parameters. Other study found that combined exposure to Cd, Pb, and Mn was associated with reduced birth weight in a cohort of Bangladeshi infants [8]. Similarly, each study reported that prenatal exposure to Cr and Mn in Sweden was linked to low birth weight and small-for-gestational-age outcomes [10]. In Indonesia, research on prenatal metal exposure remains limited. A study by Muniroh et al. [11] assessed mercury exposure in coastal pregnant women and its association with developmental outcomes but did not include other metals. Therefore, the present study adds significant value by providing data on less frequently studied metals like Cr and Mn and their distinct impacts on birth weight and length in a local context.

The toxicological effects of heavy metals on fetal development are well-documented. Cr has been shown to induce oxidative stress, impair placental angiogenesis, and disrupt gene expression related to growth factors [1]. Mn although necessary in small amounts, can be neurotoxic and interfere with iron transport, leading to fetal hypoxia and reduced growth. Both metals may interfere with nutrient transfer in the placenta or cause epigenetic changes that affect organogenesis.

The absence of associations with Cu and As may be due to lower exposure levels, metabolic buffering, or the influence of other unmeasured factors such as maternal diet, supplement intake, or genetic polymorphisms in metal metabolism pathways. Moreover, the cross-sectional design limits causal inference, and the temporal relationship between exposure and outcomes cannot be firmly established. The wide range and elevated maximum values of Cr and Mn observed in this study suggest potential environmental contamination sources in Makassar. These may include industrial emissions, traffic-related air pollution, improper waste disposal, or contaminated water and seafood a known issue in the region [4, 5]. Given the significant proportion of women exposed to tobacco smoke (62.9%) and the high prevalence of past miscarriages (55.7%) and low birth weight history (47.1%), these findings raise public health concerns about cumulative risk exposure in urban maternal populations but these high metal levels are a possible source, not a proven pathway but this is a possible source of heavy metal exposure, not a proven pathway. However, it must be acknowledged that both variables are confounding variables in this case. Although several contextual factors suggest that urban environmental contamination may contribute to elevated metal levels, this interpretation remains hypothetical. The study did not include environmental sampling; therefore, any suggested exposure sources should be regarded as possible rather than demonstrated routes of exposure.

Preventive strategies should include stricter regulation of industrial emissions, improved monitoring of environmental pollutants, and public health campaigns to reduce indirect exposures during pregnancy. Healthcare providers should consider incorporating environmental exposure assessments during antenatal care visits. One strength of this study lies in its measurement of actual maternal blood concentrations of multiple heavy metals, which provides a more accurate reflection of internal exposure compared to proxy measures such as dietary recall. Moreover, the inclusion of multiple anthropometric outcomes enhances the reliability of growth assessments. However, several limitations must be acknowledged. The sample size ($n = 70$) limits statistical power and generalizability. Potential confounders such as maternal nutrition, inflammation, socioeconomic status, and gestational age were not fully controlled in the correlation analysis. The use of a cross-sectional design precludes causal inferences, and the lack of longitudinal follow-up limits understanding of long-term developmental consequences.

4 Limitations

Heavy metal concentrations were measured only in maternal blood at delivery, thus not representing longitudinal prenatal exposure. These values reflect exposure at the end of pregnancy, not the entire gestational period. Therefore, the results of this study cannot be considered a complete picture of prenatal exposure.

The Spearman test in this study did not control for important confounding variables such as maternal age, parity, nutritional status, cigarette smoke exposure, and gestational age. The limited sample size ($n = 70$) made multivariate regression unstable and unsuitable. Therefore, the association between heavy metal exposure and neonatal outcomes should be understood as a preliminary association and not a causal relationship.

5 Conclusion

Prenatal exposure to heavy metals particularly chromium (Cr) and manganese (Mn) was found to be significantly associated with adverse birth outcomes in Makassar, Indonesia. The study revealed that higher maternal blood levels of Cr and Mn were negatively correlated with neonatal birth weight and, in the case of Cr, also with birth length. These findings suggest that environmental contamination in urban areas like Makassar may pose a substantial risk to fetal development. While copper (Cu) and arsenic (As) showed no significant associations, the wide variability in exposure levels underscores the need for closer environmental monitoring. The study highlights the importance of integrating environmental health into maternal care practices and recommends further research using advanced statistical methods and larger, longitudinal cohorts to better understand the long-term effects of prenatal metal exposure on child health.

Acknowledgements. We extend our gratitude and acknowledge the support and encouragement provided by Beasiswa Pendidikan Indonesia (BPI). This support has significantly contributed to the smooth execution and completion of this study. No portion of this manuscript was generated using generative AI. All the content was written and edited by the authors.

Conflicts of Interest. The authors declare no conflict of interest associated with the materials used in this study.

References

1. Mesa, D.R., et al.: Prenatal exposure to heavy metals and child development: current knowledge and future directions. Environ. Res. (2022)
2. Wang, M., et al.: Heavy metals in urban environments: Implications for public health. Environ. Pollut. (2013)
3. Tang, W., et al.: Environmental and human health risk assessment of heavy metal pollution in urban soils. Environ. Geochem Health (2017)
4. Polapa, R.A.: Kajian Kualitas Air dan Kandungan Logam Berat di Kota Makassar. Makassar, Indonesia: Balai Pengelolaan Daerah Aliran Sungai dan Hutan Lindung (2022)
5. Jais, M., et al.: Kandungan Logam Berat Kromium dan Kadmium pada Ikan dan Kerang di Sungai Tallo Kota Makassar. J. Lingkung dan Kesehat (2020)
6. Gundacker, C., et al.: Heavy metals and neurodevelopment of children: a review. Environ. Health Perspect. (2012)
7. Howe, C.G., et al.: Methods for analyzing exposure to complex environmental mixtures. Curr. Environ. Heal. Reports ((2021))
8. Rahman, A., et al.: Effects of prenatal exposure to heavy metals on birth weight and length using BKMR. Environ. Pollut. (2021)
9. Makassar, D.K.K.: Profil Kesehatan Kota Makassar Tahun 2022. Makassar, Indonesia: Dinas Kesehatan Makassa (2022)
10. Gustin, K., et al.: Prenatal exposure to cadmium and birth outcomes in a Swedish cohort: a BKMR approach. Environ. Int. (2020)
11. Muniroh, L., et al.: Exposure to mercury among pregnant women in coastal and inland regions of Indonesia. BMC Public Health (2022)
12. Chen, Z., et al.: Placental transfer and concentrations of heavy metals in maternal and cord blood. Environ. Sci. Technol. (2014)
13. Zinia, S.S., Yang, K.H., Lee, E.J., Lim, M.N., Kim, J.: Effects of heavy metal exposure during pregnancy on birth outcomes, 1–10 (2023)

The Association of Climate Variability and Incidence of Dengue Haemorrhagic Fever in South Sulawesi and East Nusa Tenggara, Indonesia

Budi Haryanto[1,2(✉)], Triarko Nurlambang[2,3], Febi Dwirahmadi[4], and Al Asyary[1]

[1] Department of Environmental Health, Faculty of Public Health, Universitas Indonesia, Kota Depok, Indonesia
{bharyanto,al.asyary}@ui.ac.id

[2] Research Center for Climate Change, Universitas Indonesia, Kota Depok, Indonesia
triarko@ui.ac.id

[3] Department of Geography, Faculty of Mathematics and Basic Sciences, Universitas Indonesia, Kota Depok, Indonesia

[4] Centre for Environment and Population Health, School of Medicine, Griffith University, Queensland, Australia
F.dwirahmadi@griffith.edu.au

Abstract.
Background Climate change has intensified the transmission dynamics of dengue haemorrhagic fever (DHF), particularly through changes in temperature, rainfall, and humidity that affect mosquito survival and virus propagation. This study examined the association between climatic variability and DHF incidence across six districts and cities in South Sulawesi and East Nusa Tenggara, Indonesia, from 1999 to 2019.

Method Using ecological time-series data from BMKG meteorological stations and local health offices, correlation and multivariate regression analyses were conducted at a one-month lag period.

Results Results showed that relative humidity had the most consistent and strongest positive relationship with DHF incidence across sites, explaining up to 24% of monthly variation in Kupang. Rainfall was moderately associated with increased DHF cases, while temperature effects were inconsistent, often showing inverse or delayed relationships. Urban areas such as Makassar and Kupang exhibited stronger climatic associations than rural regions like Alor and North Luwu, reflecting demographic and environmental heterogeneity.

Conclusion The findings highlight the significant yet location-specific influence of climatic variability on dengue transmission. Integrating climate data into surveillance systems can enhance early warning capabilities and adaptive public health responses, crucial for mitigating dengue risks under ongoing climate change.

Keywords: Dengue haemorrhagic fever · Climate variability · Indonesia · Time-series analysis · Vector-borne diseases

L. Vu et al. (Eds.): PBC 2025, 2026.
https://doi.org/10.1007/978-3-032-29579-8_2

1 Background

Climate change significantly alters environmental ecosystem dynamics, affecting multiple components, including the transmission of infectious diseases (Haryanto 2016). Vector-borne diseases—such as dengue, malaria, and leptospirosis—are particularly sensitive to climatic variables like temperature, humidity, rainfall, soil moisture, and sea-level rise (Prayudhy et al. 2023). These environmental changes influence the survival, reproduction, and distribution of disease vectors and pathogens, increasing the risk of transmission where a competent vector, pathogen, and susceptible host coexist in adequate numbers.

In Indonesia, climate variability has been linked to the growing burden of vector-borne diseases, especially DHF. Vulnerable populations, particularly those with limited access to resources, are disproportionately affected by the impacts of climate-related changes on disease transmission dynamics. Rising temperatures and water contamination may further exacerbate disease spread in these communities.

Over the last three decades, Indonesia has experienced a steady increase in DHF cases. In 2008, 136,333 DF cases and 1,170 deaths were reported (CFR = 0.86%; IR = 60.06 per 100,000 population). By 2009, cases rose to 154,855 with 1,384 deaths (CFR = 0.89%). A 2014 study showed a dramatic rise in dengue incidence—from 0.05 per 100,000 in 1968 to around 35–40 per 100,000 by 2013—with the most severe outbreak in 2010 (85.7 per 100,000 population). (Haryanto et al. 2014). Despite this rise, the case fatality rate (CFR) fell from 41% in 1968 to 0.73% in 2013, indicating improved disease management.

Age distribution trends have also shifted. Until 1998, children aged 5–14 years bore the highest burden of dengue. Since 1999, incidence among individuals aged 15 and older has risen significantly, suggesting a demographic transition in dengue susceptibility.

Several Indonesian provinces report incidence rates (IR) exceeding 100 per 100,000, including Bali (484), East Kalimantan (306), DKI Jakarta (198.7), and DI Yogyakarta (167.9). In contrast, Papua and West Kalimantan exhibit the lowest IRs at 11.8 and 12.1, respectively. Nationally, Indonesia's overall IR stands at 78.0 per 100,000, highlighting a high endemic burden (Prasetyo et al. 2019).

Urban expansion, climate variability, and population dynamics have contributed to the increasing spatial spread of DHF in key provinces, including Jakarta, East Java, Bali, and Central Kalimantan. This study aims to explore the relationship between climatic factors and DHF incidence in Indonesia, emphasizing regional differences and temporal patterns.

2 Methods

An ecological time-series study design was implemented to quantify the relationships between climatic variables—specifically temperature and rainfall—and the incidence of DHF from 1999 to 2019. This approach is suitable for examining population-level trends over time and is commonly used to assess associations between environmental exposures and disease occurrence. Ecological time-trend studies are observational in nature and aim to describe the temporal correlation between variations in climatic components and fluctuations in disease incidence. The primary focus was to determine whether patterns in climate data align with the trends in DHF cases and to evaluate the potential predictive power of these associations.

The study was conducted in six selected cities and regencies within two Indonesian provinces: South Sulawesi and East Nusa Tenggara. Each study site was chosen based on the availability of a climatological observation station operated by the Meteorology, Climatology, and Geophysics Agency (BMKG), ensuring consistency and reliability of climate data. The selection of one BMKG station per district or municipality was based on the Intergovernmental Panel on Climate Change (IPCC) Assessment Report V guidelines, which define climate coverage areas as spanning 5–20 km.

Climatic data (monthly average temperature and rainfall) were obtained from the respective BMKG stations, while monthly DHF incidence reports for the same period were collected from local city and district health offices. Time-series analyses, including correlation and regression modelling, were employed to explore and predict climate-disease associations. The dataset included monthly DHF cases of 69, 2,659, and 5,290 for Alor, Sikka, and Kupang in East Nusa Tenggara, and 1,122, 4,010, and 11,005 for Luwu Utara, Maros, and Makassar in South Sulawesi, respectively.

3 Results

3.1 Kupang City

Between 1999 and 2018, Kupang experienced substantial variability in DHF incidence. The highest was in 2004 (301.32/100,000), while the lowest was recorded in 2017 (5.57/100,000). In 2002, a strong and statistically significant inverse correlation was observed between maximum temperature and dengue incidence ($r = -0.754$, $p = 0.005$), suggesting that cooler months were followed by increased dengue transmission (Table 1).

When data from the entire 20-year period were analyzed, a significant but modest overall negative association was found ($r = -0.226$, $p = 0.001$). This suggests that while individual years may show strong trends, the long-term relationship is more nuanced.

The results of the bivariate analysis of the correlation test between the variables of maximum temperature (Tmax), relative humidity (RH_avg), and rainfall (CH_RR) with DHF cases in Kupang City in the period 1999–2018, obtained p-values for each of 0.001 (Tmax), 0.002 (RH_avg), 0.0001 (CH_RR). This shows that the variables Tmax, RH_avg, CH_RR can continue to multivariate analysis because the p-value is <0.25 (Table 2).

A multivariate model incorporating relative humidity and rainfall found both factors to be significant. Specifically: Each 1% increase in RH led to an estimated increase of 1,581 dengue cases; Each 1 mm rise in rainfall corresponded to an increase of 0.072 cases; The model accounted for 24% of the variation in monthly dengue cases.

3.2 Sikka Regency

DHF incidence in Sikka peaked dramatically in 2010 at 310.30/100,000 and reached zero in several years, including 2006. The correlation between temperature and dengue

Table 1. Association between maximum temperature and DHF in Kupang City 1999–2018 (lag 1 month).

Year	IR DHF	Tmax			
	(per 100 000 population)	n	r	p-value	Statistic test
1999	50.18	11	−0.4	0.223	Spearman
2000	26.55	4	NA	NA	NA
2001	64.69	3	NA	NA	NA
2002	10.59	12	**−0.754**	**0.005**	Spearman
2003	103.10	11	−0.421	0.197	Spearman
2004	301.32	10	−0.615	0.059	Spearman
2005	281.85	12	−0.298	0.346	Spearman
2006	111.50	11	−0.021	0.951	Spearman
2007	184.51	12	−0.281	0.377	Spearman
2008	132.19	12	−0.346	0.271	Spearman
2009	23.87	12	0.057	0.862	Spearman
2010	107.40	12	−0.026	0.936	Spearman
2011	76.04	12	−0.323	0.305	Spearman
2012	256.41	12	−0.006	0.985	Spearman
2013	27.96	4	NA	NA	NA
2014	28.02	12	0.078	0.809	Spearman
2015	49.89	12	−0.37	0.236	Spearman
2016	86.01	11	−0.16	0.637	Spearman
2017	5.57	12	−0.322	0.308	Spearman
2018	54.27	12	−0.25	0.434	Spearman
1999–2018		209	**−0.226**	**0.001**	Spearman

Table 2. Results of bivariate correlation analysis between DHF and climatic factors in Kupang City at lag 1 during the 1999–2018 period.

Variable	Correlation test results		
	n	r	p-value
Tmax	209	−0.226	0.001
RH_avg	209	0.645	0.002
CH_RR	209	0.644	0.0001

was significant only in 2011 ($r = -0.654$, $p = 0.021$), but no overall long-term significant association was found ($r = -0.142$, $p = 0.069$) (Table 3).

Table 3. Association between maximum temperature and DHF in Sikka Regency 2005–2018 (lag 1 month).

Year	IR DHF	Tmax			
	(per 100 000 population)	n	r	p-value	Statistics test
2005	7.54	11	−0.399	0.225	Spearman
2006	0.00	12	NA	NA	NA
2007	5.55	12	0.324	0.304	Spearman
2008	28.43	11	−0.376	0.254	Spearman
2009	92.06	12	−0.076	0.814	Pearson
2010	310.30	12	−0.198	0.538	Spearman
2011	20.26	12	**−0.654**	**0.021**	Spearman
2012	28.43	12	−0.066	0.839	Spearman
2013	103.08	12	−0.227	0.478	Spearman
2014	11.07	12	−0.389	0.211	Spearman
2015	23.60	12	−0.52	0.083	Spearman
2016	51.98	12	−0.227	0.478	Spearman
2017	27.10	12	−0.34	0.28	Spearman
2018	47.03	12	−0.279	0.38	Spearman
2005–2018		166	−0.142	0.069	Spearman

The results of the bivariate analysis of the correlation test between the variables of maximum temperature (Tmax), relative humidity (RH_avg), and rainfall (CH_RR) with dengue cases in Sikka Regency in the period 2005–2018, obtained a p-value of 0.0001 for the variables RH_avg and CH_RR, while Tmax is 0.069 (Table 4). This shows that the variables Tmax, RH_avg and CH_RR can continue to multivariate analysis because the p-value <0.25.

Table 4. Results of bivariate correlation analysis between DHF and climatic factors in Sikka Regency at lag 1 during the 2005–2018 period.

Variabel	Correlation test results		
	n	R	p-value
Tmax	166	−0.142	0.069
RH_avg	166	0.427	0.0001
CH_RR	166	0.382	0.0001

Multivariate analysis identified RH and rainfall as statistically significant predictors: 1% RH increase → + 1,524 dengue cases; 1 mm rainfall increase → + 0.02 dengue cases; Model explained 8.1% of variance, indicating modest explanatory power.

3.3 Alor Regency

Alor had the lowest DHF incidence overall among the study sites, with many years registering zero cases. The highest IR was 4.65/100,000 in 2010. No year showed a statistically significant correlation between maximum temperature and dengue cases. Over the 2010–2018 period, the relationship remained weak and non-significant ($r = 0.139$, $p = 0.217$) (Table 5).

Table 5. Association between maximum temperature and DHF in Alor Regency 2010–2018 (lag 1 month).

Year	IR DHF	Tmax			
	(per 100 000 population)	n	r	p-value	Statistics test
2010	4.65	11	−0.252	0.455	Spearman
2011	0.00	12	NA	NA	NA
2012	0.00	1	NA	NA	NA
2013	0.00	1	NA	NA	NA
2014	0.00	10	NA	NA	NA
2015	0.00	11	NA	NA	NA
2016	0.00	12	NA	NA	NA
2017	3.45	11	0.585	0.059	Spearman
2018	0.00	11	NA	NA	NA
2010–2018		80	0.139	0.217	Spearman

The results of the bivariate analysis of the correlation test between the variables of maximum temperature (Tmax), relative humidity (RH_avg), and rainfall (CH_RR) with DHF cases in Alor Regency in the 2010–2018 period, each p-value is 0.217 (Tmax), 0.838 (RH_avg), 0.079 (CH_RR) (Table 6). This shows that only Tmax, and CH_RR variables can proceed to multivariate analysis because p-value <0.25.

Only rainfall approached statistical relevance ($r = 0.198$, $p = 0.079$), suggesting a possible effect worth further investigation. The regression model retained only rainfall, indicating: 1 mm rainfall increase → + 0.001 dengue cases; Model explained 4.2% of monthly variation.

3.4 Maros Regency

Maros had notable dengue incidence spikes, particularly in 2008 (29.61/100,000). Several years revealed strong negative correlations between temperature and dengue cases,

Table 6. Results of bivariate correlation analysis between DHF and climatic factors in Alor Regency at lag 1 during the 2010–2018 period.

Variable	Correlation test results		
	n	r	p-value
Tmax	80	0.139	0.217
RH_avg	80	0.023	0.838
CH_RR	80	0.198	0.079

including: 2002: $r = -0.724$, $p = 0.008$; 2005: $r = -0.679$, $p = 0.015$; 2008: $r = -0.877$, $p = 0.0001$; 2009: $r = -0.622$, $p = 0.031$; and 2016: $r = -0.732$, $p = 0.007$. These indicate recurring patterns where rising temperatures preceded reductions in DHF cases one month later. Over the 20-year period, the overall correlation was statistically significant ($r = -0.24$, $p = 0.0001$) (Table 7).

The results of the bivariate analysis of the correlation test between the variables of maximum temperature (Tmax), relative humidity (RH_avg), and rainfall (CH_RR) with DHF cases in Maros Regency in the 1999–2018 period, obtained a p-value of 0.0001 (Table 8). This shows that the variables Tmax, RH_avg and CH_RR can continue to multivariate analysis because the p-value <0.25.

Multivariate modelling highlighted: 1 °C increase in Tmax $\rightarrow$ + 6.365 cases; 1% RH increase $\rightarrow$ + 0.717 cases; 1 mm rainfall increase $\rightarrow$ + 0.054 cases; Combined variables explained 21.6% of dengue incidence variation.

3.5 North Luwu Regency

North Luwu saw its peak DHF incidence in 2016 (115.92/100,000). Significant correlations were detected in: 2006 ($r = -0.804$, $p = 0.002$); 2007 ($r = 0.618$, $p = 0.032$); 2009 ($r = -0.804$, $p = 0.002$). These mixed results suggest a context-dependent effect of climate on dengue dynamics. Over the full 2000–2018 period, the relationship was weak and not statistically significant ($r = 0.08$, $p = 0.248$) (Table 9).

The results of the bivariate analysis of the correlation test between the variables of maximum temperature (Tmax), relative humidity (RH_avg), and rainfall (CH_RR) with DHF cases in North Luwu Regency in the period 2000–2018, obtained p-value of 0.248 (Tmax), 0.064 (RH_avg), 0.359 (CH_RR) (Table 10). This shows that only Tmax and RH_avg variables can proceed to multivariate analysis because p-value <0.25.

Regression analysis included Tmax and RH: 1 °C Tmax increase $\rightarrow$ + 2,159 cases; 1% RH increase $\rightarrow$ + 0.002 cases; Explained only 3.7% of DHF variation. This low predictive value may reflect underreporting or non-climatic drivers dominating in this rural district.

3.6 Makassar City

As a dense urban center, Makassar showed consistent annual dengue incidence, peaking at 27.35/100,000 in 2016. Significant inverse correlations with temperature were found

Table 7. Association between maximum temperature and DHF in Maros Regency 1999–2018 (lag 1 month).

Year	IR DBD	Tmax			Statistics
	(per 100.000 population)	n	r	p-value	Test
1999	0.00	11	−0.498	0.119	Spearman
2000	0.00	12	0.083	0.799	Spearman
2001	0.00	12	−0.21	0.512	Spearman
2002	0.00	12	**−0.724**	**0.008**	Spearman
2003	0.00	12	−0.439	0.153	Spearman
2004	1.57	12	−0.43	0.163	Spearman
2005	1.14	12	**−0.679**	**0.015**	Spearman
2006	0.00	12	−0.202	0.529	Spearman
2007	0.00	12	−0.336	0.286	Pearson
2008	29.61	12	**−0.877**	**0.0001**	Spearman
2009	0.00	12	**−0.622**	**0.031**	Spearman
2010	4.65	12	0.195	0.544	Spearman
2011	0.00	12	**−0.655**	**0.021**	Pearson
2012	0.00	12	−0.522	0.082	Spearman
2013	0.00	1	NA	NA	NA
2014	0.00	9	−0.564	0.113	Pearson
2015	0.00	12	−0.429	0.164	Pearson
2016	0.00	12	**−0.732**	**0.007**	Spearman
2017	3.45	12	−0.358	0.254	Pearson
2018	0.00	12	−0.138	0.668	Pearson
1999–2018		225	**−0.24**	**0.0001**	Spearman

Table 8. Results of bivariate correlation analysis between DHF and climatic factors in Luwu Utara Regency at lag 1 during the 1999–2018 period.

Variable	Correlation Test Results		
	n	R	p-value
Tmax	225	−0.24	0.0001
RH_avg	225	0.435	0.0001
CH_RR	225	0.415	0.0001

in: 2012 ($r = -0.767$, $p = 0.01$); 2015 ($r = -0.649$, $p = 0.02$); 2016 ($r = -0.608$, $p = 0.04$); 2017 ($r = -0.594$, $p = 0.04$). Across the 9-year study period (2010–2018), the

Table 9. Association between maximum temperature and DHF in Luwu Utara Regency 2000–2018 (lag 1 month).

Year	IR DHF	Tmax			Statistics
	(per 100.000 population)	n	r	nilai-p	Test
2000	0.24	12	−0.307	0.331	Spearman
2001	2.65	12	−0.184	0.567	Spearman
2002	-	12	0.148	0.647	Spearman
2003	-	12	0.073	0.822	Spearman
2004	-	12	−0.04	0.901	Spearman
2005	3.92	12	0.246	0.442	Pearson
2006	1.98	12	**−0.806**	**0.002**	Spearman
2007	0.66	12	**0.618**	**0.032**	Spearman
2008	2.36	12	0.463	0.13	Spearman
2009	22.56	12	**−0.804**	**0.002**	Spearman
2010	5.46	12	0.562	0.057	Spearman
2011	8.36	12	0.071	0.826	Pearson
2012	24.75	12	−0.482	0.112	Pearson
2013	79.72	NA	NA	NA	NA
2014	15.31	4	NA	NA	NA
2015	55.50	12	0.337	0.285	Pearson
2016	115.92	12	0.484	0.111	Spearman
2017	5.84	12	−0.27	0.396	Pearson
2018	7.41	12	−0.014	0.965	Spearman
2000–2018		209	0.08	0.248	Spearman

Table 10. Results of bivariate correlation analysis between DHF and climatic factors in Luwu Utara Regency at lag 1 during the 2000–2018 period.

Variable	Correlation test results		
	n	r	*p-value*
Tmax	209	0.08	0.248
RH_avg	209	−0.128	0.064
CH_RR	209	−0.064	0.359

long-term correlation between Tmax and DHF remained significant ($r = -0.423$, $p = 0.001$) (Table 11).

Table 11. Association between maximum temperature and DHF in Makassar City 2010–2018 (lag 1 month).

Year	IR DHF	Tmax			Statistics
	(per 100.000 population)	n	r	p-population	Test
2010	11.77	11	−0.576	0.06	Pearson
2011	6.03	10	−0.526	0.12	Pearson
2012	6.10	10	**−0.767**	**0.01**	Pearson
2013	18.78	3	NA	NA	NA
2014	9.94	11	−0.397	0.23	Pearson
2015	9.94	12	**−0.649**	**0.02**	Pearson
2016	27.35	12	**−0.608**	**0.04**	Pearson
2017	9.07	12	**−0.594**	**0.04**	Pearson
2018	16.97	12	−0.380	0.22	Pearson
2010–2018		93	**−0.423**	**0.001**	Spearman

The results of the bivariate analysis of the correlation test between the variables of maximum temperature (Tmax), relative humidity (RH_avg), and rainfall (CH_RR) with DHF cases in Makassar City in the 2010–2018 period, obtained p-value = 0.0001 (Table 12). This shows that the variables Tmax, RH_avg, CH_RR can continue to multivariate analysis because p-value <0.25.

Table 12. Results of bivariate correlation analysis between dengue fever (DF) and climatic factors in Makassar City at lag 1 during the 2010–2018 period.

Variabel	Hasil Uji Korelasi *Lag* 1 bulan		
	n	r	*p-value*
Tmax	93	−0.423	0.0001
RH_avg	93	0.569	0.0001
CH_RR	93	0.470	0.0001

Multivariate analysis showed: 1 °C Tmax increase → + 6.308 cases; 1% RH increase → + 1,128 cases; 1 mm rainfall increase → + 0.019 cases; Model explained 18.3% of monthly case variation.

Makassar's consistent data allowed for more robust modelling, with higher predictive power compared to rural districts.

3.7 Cross-Site Comparisons

- **Relative Humidity (RH)** showed a positive correlation with dengue incidence in nearly all study sites, particularly Kupang, Sikka, and Makassar, where RH increases explained large changes in case counts.
- **Rainfall** had a modest but consistent association across locations. Although the impact per mm was small, the cumulative effect during the wet season could be substantial.
- **Maximum Temperature** correlations were inconsistent. While short-term (yearly) correlations were often significant, especially negative, the long-term role was diminished or confounded by RH and rainfall in multivariate models.

3.8 Model Fit and Variability Explained:

- Best-performing models were in Kupang (24%), Maros (21.6%), and Makassar (18.3%).
- Weakest models were in North Luwu (3.7%) and Alor (4.2%).

These differences reflect geographic, demographic, and environmental diversity—urban density, mosquito control efforts, or healthcare access may have influenced dengue dynamics more than climate alone in some regions.

4 Discussion

This study provides comprehensive evidence that climatic variability—particularly temperature, rainfall, and relative humidity—plays a critical role in influencing the temporal dynamics of Dengue Haemorrhagic Fever (DHF) across diverse ecological and demographic contexts in South Sulawesi and East Nusa Tenggara. The varying strength and direction of these associations highlight the complex, location-specific nature of climate–dengue interactions in Indonesia.

4.1 Temperature Effects

Temperature emerged as a key but context-dependent determinant of DHF incidence. In several urban sites, such as Makassar and Kupang, higher temperatures showed significant inverse correlations with dengue incidence at a one-month lag. This finding aligns with studies suggesting that while moderate increases in temperature accelerate *Aedes aegypti* development and viral replication, excessively high or fluctuating temperatures can reduce mosquito survival and viral infectivity (Carrington et al. 2013a; Carrington et al. 2013b). Laboratory studies have shown that *Aedes aegypti* transmission efficiency peaks between 26 °C and 30 °C, declining sharply beyond 35 °C (Morin et al. 2013; Liu-Helmersson et al. 2014). In tropical urban environments such as Makassar, where mean temperatures already exceed this threshold, further warming may suppress mosquito fitness and oviposition rates, explaining the negative correlations observed.

Conversely, in some districts such as Maros, lower or fluctuating temperatures in preceding months were followed by increased dengue cases, consistent with findings from Dhaka (Hashizume et al. 2012) and Singapore (Hii et al. 2009) showing delayed temperature effects due to vector maturation and extrinsic incubation period dynamics. Such lagged responses reinforce the importance of temporal modeling to capture nonlinear temperature–transmission relationships (Thai et al. 2010).

4.2 Rainfall and Vector Breeding Dynamics

Rainfall showed positive correlations with DHF incidence across most sites, particularly in Kupang and Sikka. Rain provides breeding habitats through the accumulation of stagnant water in containers, discarded materials, and household reservoirs (Teurlai et al. 2015; Banu et al. 2014). However, the strength of this relationship varied geographically. In Alor and North Luwu, where rainfall is heavy and terrain is steep, excessive precipitation likely disrupted mosquito breeding habitats by flushing larvae from containers or creating unsuitable conditions, thereby weakening the correlation.

The observed one-month lag between rainfall and DHF incidence corresponds with the approximate duration of the *Aedes aegypti* life cycle, from egg to adult, and the dengue virus's intrinsic and extrinsic incubation periods (Patz et al. 1998; Descloux et al. 2012). This lag pattern indicates that rainfall-driven vector population surges may lead to increased transmission within 4–6 weeks, supporting the integration of lagged rainfall indices into early warning systems for dengue outbreaks (Chakravarti and Kumaria 2005; Banu et al. 2014).

4.3 Relative Humidity and Vector Survival

Relative humidity (RH) exhibited the most consistent positive association with DHF incidence, especially in Kupang, Sikka, and Makassar. Sustained high humidity levels ($>70\%$) are known to enhance *Aedes aegypti* survival, adult longevity, and biting frequency, thereby extending the window for virus transmission (Hii et al. 2009; Akter et al. 2017). Our multivariate models showed that a 1% increase in RH could lead to a substantial rise in DHF cases, emphasizing the synergistic effect of humid and warm conditions on vector competence.

Moreover, RH often interacts with temperature to modulate mosquito physiology. For instance, high temperature under low humidity accelerates desiccation, while the same temperature under high humidity enhances survival and oviposition rates (Morin et al. 2013). This may explain why cities like Makassar, with both high humidity and moderate rainfall, sustain continuous DHF transmission compared to drier regions such as Alor.

4.4 Spatial Heterogeneity and Urban–Rural Gradients

The marked differences in model explanatory power—ranging from 24% in Kupang to less than 5% in Alor—reflect environmental, demographic, and infrastructural heterogeneity across sites. Urban centers with higher population density, water storage practices, and waste accumulation offer more stable breeding sites, amplifying the effects of climatic fluctuations (Teurlai et al. 2015; Lowe et al. 2018). Conversely, rural districts may exhibit weaker climatic associations due to greater ecological variability, underreporting, or dominance of non-climatic factors such as mobility patterns, sanitation, and vector control coverage (Akter et al. 2017; Banu et al. 2014).

These findings echo multi-country studies showing that climate effects on dengue are moderated by urbanization level, housing conditions, and public health interventions (Thai and Anders 2011; Kovats et al. 2016). Therefore, localized adaptive strategies—combining climate-based surveillance with community vector management—are essential for effective control.

4.5 Public Health Implications and Early Warning Potential

Given that climate variables explained up to one-quarter of dengue incidence variability, their predictive utility in public health surveillance is evident. Developing location-specific early warning systems that incorporate lagged temperature, humidity, and rainfall data could improve outbreak preparedness. For example, predictive models using El Niño–Southern Oscillation (ENSO) indicators have been successfully implemented in Southeast Asia to anticipate seasonal dengue surges (Descloux et al. 2012; Lowe et al. 2018). In Indonesia, integrating BMKG climate forecasts with Ministry of Health surveillance data could form the basis of a national climate-informed dengue early warning framework.

4.6 Limitations and Future Directions

Despite robust temporal coverage, several limitations should be acknowledged. The use of aggregate district-level data may mask intra-urban microclimatic variations and local breeding hotspots (Morin et al. 2013; Thai et al. 2010). Socioeconomic, behavioral, and intervention factors were not included, although these are known to modulate vector density and disease risk (Teurlai et al. 2015). Future studies should apply spatiotemporal modeling with higher-resolution data and include entomological and demographic covariates to capture the multifactorial nature of dengue transmission. Additionally, exploring non-linear and threshold-based models (e.g., distributed lag non-linear models) may better describe complex climate–disease interactions under changing climatic regimes.

4.7 Synthesis

Overall, this study reinforces that climate variability—especially in humidity and rainfall—significantly influences dengue transmission patterns, with spatial heterogeneity reflecting local ecological and socio-environmental contexts. The findings are consistent with global evidence indicating that dengue risk will expand under ongoing climate change, particularly in urban and coastal Indonesia (Patz et al. 1998; Banu et al. 2014; Akter et al. 2017). Thus, incorporating climate information into dengue control programs is no longer optional but essential for adaptive public health resilience.

5 Conclusion

This study concludes that climatic variability—particularly temperature, rainfall, and relative humidity—significantly influences dengue hemorrhagic fever incidence across South Sulawesi and East Nusa Tenggara. The strength and direction of associations vary spatially, with humidity showing the most consistent positive relationship. These findings highlight the need for localized, climate-informed surveillance and control strategies. Integrating meteorological data into public health systems can enhance early warning, preparedness, and adaptive responses to dengue under ongoing climate change.

Acknowledgements. This research was supported by the International Indexed Publication Grant (PUTI) for Kolaborasi Internasional (2Q2) Fiscal Year 2020 Number: NKB-769/UN2.RST/HKP.05.00/2020 University of Indonesia.

References

Akter, R., Hu, W., Naish, S., Banu, S., Tong, S.: Joint effects of climate variability and socioecological factors on dengue transmission: epidemiological evidence. Trop. Med. Int. Health **22**(6), 656–669 (2017)

Banu, S., Hu, W., Guo, Y., Hurst, C., Tong, S.: Projecting the impact of climate change on dengue transmission in Dhaka. Bangladesh. Environ Int. **63**, 137–142 (2014)

Carrington, L.B., Seifert, S.N., Willits, N.H., Lambrechts, L., Scott, T.W.: Large diurnal temperature fluctuations negatively influence Aedes aegypti life-history traits. J. Med. Entomol. **50**(1), 43–51 (2013)

Carrington, L.B., Seifert, S.N., Armijos, M.V., Lambrechts, L., Scott, T.W.: Reduction of Aedes aegypti vector competence for dengue virus under large temperature fluctuations. Am. J. Trop. Med. Hyg. **88**(4), 689–697 (2013)

Carrington, L.B., Seifert, S.N., Armijos, M.V., Lambrechts, L., Scott, T.W.: Fluctuations at a low mean temperature accelerate dengue virus transmission by Aedes aegypti. PLoS Negl. Trop. Dis. **7**(4), e2190 (2013)

Chakravarti, A., Kumaria, R.: Eco-epidemiological analysis of dengue infection during an outbreak of dengue fever. India. Virol J. **2**, 32 (2005)

Descloux, E., Mangeas, M., Menkes, C.E., Lengaigne, M., Leroy, A., Tehei, T., et al.: Climate-based models for understanding and forecasting dengue epidemics. PLoS Negl. Trop. Dis. **6**(2), e1470 (2012)

Haryanto, B.: Health adaptation scenario and dengue fever vulnerability assessment in Indonesia. Book chapter 13. R. Akhtar (ed.), Climate Change and Human Health Scenario in Southand Southeast Asia, Advances in Asian Human-Environmental Research (2016). https://doi.org/10.1007/978-3-319-23684-1_13

Haryanto, S., et al.: The changing incidence of dengue haemorrhagic fever in Indonesia: a 45-year registry-based analysis. BMC Infect. Dis. **14**, 412 (2014). https://doi.org/10.1186/1471-2334-14-412

Hashizume, M., Dewan, A.M., Sunahara, T., Rahman, M.Z., Yamamoto, T.: Hydroclimatological variability and dengue transmission in Dhaka, Bangladesh: a time series study. BMC Infect. Dis. **12**, 98 (2012)

Hii, Y.L., Rocklöv, J., Ng, N., Tang, C.S., Pang, F.Y., Sauerborn, R.: Climate variability and increase in intensity and magnitude of dengue incidence in Singapore. Glob. Health Action **2**, 2036 (2009)

Kovats, R.S., Campbell-Lendrum, D., McMichael, A.J., et al.: Climate Change and Human Health: Risks and Responses. WHO (2016)

Liu-Helmersson, J., Stenlund, H., Wilder-Smith, A., Rocklöv, J.: Vectorial capacity of *Aedes aegypti*: effects of temperature and implications for global dengue epidemic potential. PLoS ONE **9**(3), e89783 (2014)

Lowe, R., Gasparrini, A., Van Meerbeeck, C.J., et al.: Nonlinear and delayed impacts of climate on dengue risk in Barbados: a modeling study. PLoS Med. **15**(7), e1002613 (2018)

Morin, C.W., Comrie, A.C., Ernst, K.: Climate and dengue transmission: evidence and implications. Environ. Health Perspect. **121**(11–12), 1264–1272 (2013)

Patz, J.A., Martens, W.J., Focks, D.A., Jetten, T.H.: Dengue fever epidemic potential as projected by general circulation models of global climate change. Environ. Health Perspect. **106**(3), 147–153 (1998)

Prasetyo et al.: Epidemiology of dengue haemorrhagic fever in Indonesia: analysis of five decades data from the National Disease Surveillance. BMC Research Notes (2019)

Prayudhy, Y., Setiawan, A., Tugiyono, T.: Variasi Iklim dan Dinamika Kasus DBD di Indonesia: Systematic Review. Jurnal Kesehatan. Found climate variables (temperature, rainfall, humidity) are linked with DHF incidence in Indonesia (2023)

Teurlai, M., Menkès, C.E., Cavarero, V., Degallier, N., Descloux, E., Grangeon, J.P., et al.: Socioeconomic and climate factors associated with dengue fever spatial heterogeneity: a worked example in New Caledonia. PLoS Negl. Trop. Dis. **9**(12), e0004211 (2015)

Thai, K.T.D., Anders, K.L.: The role of climate variability and change in the transmission dynamics and geographic distribution of dengue. Exp. Biol. Med. **236**(8), 944–954 (2011)

Thai, K.T., Cazelles, B., Nguyen, N.V., Vo, L.T., Boni, M.F., Farrar, J., et al.: Dengue dynamics in Binh Thuan province, southern Vietnam: periodicity, synchronicity and climate variability. PLoS Negl. Trop. Dis. **4**(7), e747 (2010)

Thai, K.T.D., Anders, K.L.: The role of climate variability and change in the transmission dynamics and geographic distribution of dengue. Exp. Biol. Med. (Maywood) **236**(8), 944–954 (2011)

Integrating Honey DNA Meta-Barcoding and Portable Long-Read Sequencing Platform for Assessing Plant Biodiversity

Thi Phuong Anh Phi[1,3,4], Nhat-Thong Le[2,4], Hoang Dai Phuc Tran[3,4], and Minh-Thong Le[3,4](✉)

[1] Applied Biomedical Science, School of Applied Sciences, West of England University, International University Campus, Ho Chi Minh City, Vietnam
[2] Research Center for Infectious Diseases, International University, Ho Chi Minh City, Vietnam
[3] School of Biotechnology, International University, Ho Chi Minh City, Vietnam
lmthong@hcmiu.edu.vn
[4] Vietnam National University, Ho Chi Minh City, Vietnam

Abstract. Bees play a pivotal role in agriculture and ecosystems by serving as the principal pollinators of both wild and cultivated plants. The physicochemical and biological properties of honey are intrinsically linked to its floral origin, rendering honey a valuable analytical tool for identifying the host plants visited by bees. However, prior to DNA sequencing, DNA extraction from honey presented considerable difficulties due to its high viscosity, low water content, the presence of inhibiting chemical constituents, and tough, stable pollen grains. This study sought to assess plant biodiversity by integrating honey DNA metabarcoding and a portable long-read sequencing platform. DNA was extracted from honey collected from Nam Tra My (Quang Nam) using a modified sample preparation and extraction method designed to minimize inhibitory components and enhance DNA purification. The *rbcL* gene was effectively amplified via conventional PCR, followed by long-read sequencing utilizing Oxford Nanopore Technology (ONT). Through custom bioinformatic analysis, we successfully identified 64 plant species, covering 59 genera and 11 families. Additionally, six other family-level taxa were detected in the honey sample. Among species-level plants, a Shannon diversity index of 3.12 was observed, suggesting a highly diverse plant community represented in the honey. The data obtained from this study contributes to supplementing knowledge, enriching our understanding of local biodiversity, especially within Quang Nam forests. Furthermore, the methodology developed in this study offers an efficient approach for honey botanical origin assessment, supporting future strategies to monitor the effects of climate change and agricultural practices on apiculture and wild ecosystems.

Keywords: Honey DNA metabarcoding · Long-read sequencing (ONT) · Plant biodiversity · Pollination ecology · Quang Nam (Vietnam) · *rbcL*

Thi Phuong Anh Phi and Nhat-Thong Le these authors contributed equally to this work.

L. Vu et al. (Eds.): PBC 2025, 2026.
https://doi.org/10.1007/978-3-032-29579-8_3

1 Introduction

Bees are indispensable to ecosystems and agriculture, acting as the most dominant pollinators across a wide range of plant species [1]. Honey, a golden elixir produced by honeybees from the nectar of flowers, has been revered for its culinary and medicinal properties for thousands of years. Owing to their extensive environmental exploration and foraging activities, honey bees are also considered valuable large-scale monitoring tools [2]. The spectrum of pollen varieties found in honey captures the breadth of the plant species visited by bees during honey production, highlighting the long-recognized potential of using honey to trace the botanical biodiversity [3]. However, the traditional pollen morphological analysis (melissopalynology) is time-consuming, requires expertized taxonomic knowledge, while yielding a limited taxonomic resolution [4–8]. In this context, DNA extracted from pollen in honey serves as a valuable resource for identifying host plants [9]. DNA-based methods have proven effective for species identification from environmental materials, overcoming the limitations of traditional, low-throughput morphological approaches [10]. Notably, DNA metabarcoding achieved a 64% taxa match, significantly overwhelming melissopalynology's 28%, highlighting its superior reliability and reproducibility for honey's botanical origin identification [11]. Building on this advancement, this study employed DNA metabarcoding integrated with Oxford Nanopore Technology (ONT) to further enhance identification efficiency.

Nevertheless, the DNA analysis of plants and plant by-products such as honey faces numerous challenges due to the presence of multiple cell components, including polysaccharides, proteins, and DNA polymerase inhibitors like alkaloids and polyphenols [12]. Honey typically comprises at least 80% carbohydrates and phenolic compounds [13–15], which are particularly problematic by serving as critical factors that inhibit high-quality nucleic acid extraction and PCR [16–18]. Furthermore, the structural complexity of plant cells, especially pollen grains, poses additional barriers to DNA retrieval due to their tough double walls, surface waxes, and proteins, which resist cell wall disruption and release PCR inhibitors [19, 20]. Thus, an optimal DNA extraction protocol must ensure the acquisition of a sufficient, high-quality DNA free from inhibitors prior to PCR and sequencing.

The *rbcL* gene is widely favored in phylogenetic and molecular studies of plants due to its slow evolutionary rate and minimal divergence within chloroplast DNA (cpDNA) among flowering plants [21, 22]. Specifically, its great universality, availability of universal primers [23], high rate of successful amplification and discrimination power [24], and well-established database [25] make it a reliable marker for plant identification. Hence, *rbcL* analysis has promising applications in the molecular investigations of plants, including tracing the origin of plants [26, 27]. Metabarcoding approaches, with the integration of innovative sequencing technology, especially long-read sequencing platforms, allow large-scale biodiversity assessments with improved resolution and efficiency [28, 29]. Oxford Nanopore Technology (ONT), a third-generation sequencing platform, serves as a powerful tool in metagenomic studies due to its high throughput, portability, and long-read capability, which enables more accurate taxonomic classification of mixed-species samples [30]. ONT utilization offers a more comprehensive, time-efficient, higher resolution, and accuracy for honey's botanical origin exploration.

Investigating honey's floral origin is crucial for verifying authenticity, enhancing its medicinal and economic value, and promoting sustainable apiculture. In parallel, the accelerating impacts of climate change and the intensification of agricultural activities have detrimental effects, contributing to global biodiversity loss [31]. Within this scenario, monitoring plant biodiversity through environmental materials, such as honey, using high-throughput, DNA-based approaches, emerges as a promising strategy for assessing ecosystem changes and informing conservation efforts. Thus, identifying honey host-plant species not only provides insights into bee nutrition, habitat use, forage availability, and plant-pollinator interaction, but also supports biodiversity dynamics assessment, healthy colonies, sustainable beekeeping, and resilience against climate change [28, 32, 33].

This study aims to apply a DNA metabarcoding approach integrated with ONT to trace the floral components of honey collected from the Nam Tra My forest in Quang Nam, Vietnam, providing initial data and preliminary insights into the plant diversity represented in honey products. To achieve this, a DNA extraction protocol was established and optimised by combining a silica column-based method with glass bead-assisted disruption of pollen and magnetic bead purification. The chloroplast *rbcL* gene was subsequently amplified using conventional PCR and subjected to long-read sequencing via ONT. The sequencing output was directed to a custom bioinformatics pipeline for metabarcoding analysis. The methodology developed in this study offers valuable resources for in-field ecological studies using portable experimental devices.

2 Materials and Methods

2.1 Sample Preparation

The natural, verified forest honey was purchased from Nam Tra My (Quang Nam) and stored at room temperature (RT). We employed a combination of the protocols proposed previously [16, 18], with certain modifications. Specifically, 300 μL of honey was diluted with sterile water at a ratio of 1:3 and incubated at 60 °C for 30 min, followed by centrifuging at 13,200 × g for 10 min. Subsequent to the removal of the supernatant, the pellet was resuspended in 500 μL of 70% ethanol by vortexing for 5 min, followed by centrifugation at 13,200 × g for 10 min. Then, the pellet was allowed to air-dry at RT for 5 min. After processing, the pellet would be utilized in subsequent steps for light microscopy assessment and DNA extraction.

2.2 Microscopic Visualisation of Pollen Grains

The pollen-containing pellet was mixed in 50μL of 80% glycerol solution for observing under the light microscope (LM). The pollen micrographs were taken at 400X magnification.

2.3 Honey DNA Extraction and Purification

The extraction procedure was performed using the Salagene™ Unigene Extraction kit (Salagene, Vietnam), following the manufacturer's instructions with key modifications.

Specifically, the integration of chemical and mechanical disruption of the pollen cell wall was achieved by grinding the pellet in 1X PBS and lysis buffer with sterilised 0.5 mm glass beads (Fisher Scientific), followed by enzymatic treatment and incubation at 60 °C. To minimize inhibitors and optimize DNA quality for PCR and sequencing, the extracted DNA was purified using KAPA Pure Beads (Roche, Switzerland) according to the manufacturer's protocol and stored at -20 °C for further analysis. DNA quality and quantity were assessed using Nanodrop spectrophotometer and 1% agarose gel electrophoresis at 100V for 30 min in 1X SB buffer.

2.4 *RbcL* Amplification and Long-Read Sequencing

The amplification targeted a fragment of the plant cpDNA *rbcL* gene using previously proposed primers *rbcL*-aF 5'-ATGTCACCACAAACAGAGACTAAAGC-3' [34] and *rbcL*-aR 5'-GTAAAATCAAGTCCACCRCG-3' [22]. PCR was performed in a 20 μl reaction mixture under the following conditions: an initial denaturation at 95 °C for 5 min; 40 cycles of 95 °C for 30 s, 56 °C for 30 s, and 72 °C for 30 s; followed by a final extension at 72 °C for 5 min.

The PCR products were purified using KAPA Pure Beads (Roche, Switzerland). The library for long-read sequencing was prepared using Native Barcoding Kit 24 V14 (Oxford Nanopore Technologies), adhering to the manufacturer's guidelines. Then, the prepared library was loaded onto the Flongel flow cell (R10.4). Sequencing was performed on a MinION-Mk1b device using MinKNOW 24.11 software with standard settings. High-accuracy base-calling (SUP mode) was performed using Dorado 0.9.1.

2.5 Bioinformatic Analysis

Reads were trimmed for the presence of adapter sequence using Porechop-0.2.4 [35]. NanoFilt-2.8.0 [36] was employed for filtering low-quality reads, in which reads with a quality of at least 10 and length ranging from 200 to 800 bp were retained. High quality FASTQ file was processed using VSEARCH-2.30.0 [37] to dereplicate sequences and cluster them at the default 97% identity threshold, generating Operational taxonomic unit (OTU) representative sequences. The BLASTN algorithm (blast-2.15.0) [38] was used to align representative sequences with the reference *rbcL* database [25, 39]. A custom R language script applied the taxonomy assignment for OTU sequences according to the following rules: (1) Only sequences that matched the best in BLASTN with at least 80% similarity will be kept; (2) a species-level OTU will be assigned if there is only one species in the BLASTN result; (3) in cases of multiple hits, taxonomic classification was resolved at the most specific possible level, prioritizing genus, followed by family; and (4) if further classification is not possible, it will be labeled as unclassified. Raw reads per assigned taxa were counted and used for further analysis.

3 Results

3.1 Microscopic Visualisation of Pollen Grains

The observation of pollen using LM at 100 × magnification revealed the presence of multiple types of pollen in the honey sample. The micrographs of representative pollen grains are illustrated in Fig. 1. Generally, based on morphology observation, the majority of pollen grains are found to possess a circular or slightly circular view; spheroidal or prolate-spheroidal shape; thick and tectate exine layer; echinate, psilate, or scabrate sculpture, whereas the less-frequent pollens appeared to be morphologically varied with triangular, circular, or rounded view; spheroidal, oblate, prolate-spheroidal, or bean-like shape; thick and tectate or semitectate exine layer; echinate, psilate, or scabrate sculpture. However, the actual shape might differ depending on the microscopic viewing angle and degree of compression under the coverslip, requiring a higher-resolution imaging (e.g., SEM) for accurate characterization. Most observed pollen grains retained original morphology and intact cell wall structure, suggesting that cell wall disruption is essential prior to DNA extraction. The presence of diverse pollen grains in samples offers material for a DNA-based approach to assess plant biodiversity.

3.2 Honey DNA Extraction; *RbcL* Amplification

Before implementing the modified methodology, initial attempts to extract pollen DNA from honey using a conventional phenol-chloroform method produced inconsistent results, yielding low DNA purity. The A260/A230 ratios were particularly poor, with a value of 0.383 without glass-bead grinding, and improved to 0.601 when glass beads were used (Table 1). The data suggest contamination by organic compounds, primarily attributed to phenolic compounds found in honey.

Thus, the silica column-based integrated glass-bead grinding technique was employed, resulting in significantly improved DNA purity. Incorporation of a PBS washing step was found to be critical for purification, increasing A260/A280 and A260/A230 ratios from 1.874 and 1.380 (without PBS wash) to 1.978 and 1.989, respectively (Table 1). Using this optimized protocol, DNA was successfully extracted from a single Quang Nam honey sample in technical duplicates (S1, S2) to ensure the consistency and reproducibility of the method. The resulting DNA concentrations were 21.1 ng/μL (S1) and 41.9 ng/μL (S2), with the purity ratios A260/A280 and A260/A230 ranging from 2.107–2.201 and 2.006–2.322, respectively, indicating good DNA quality with minimal protein and polysaccharide contamination for facilitating the PCR procedure (Table 1). As shown in Fig. 2a, the presence of distinct bands confirmed successful DNA extraction and demonstrated the effectiveness of the method applied to the honey sample. As the DNA concentration is relatively low, two replicates were merged to increase the rate of successful amplification while maintaining the floral diversity. Thus, PCR amplification of the *rbcL* cpDNA yielded distinct bands ~600bp (Fig. 2b), corresponding to the expected product size. The obtained clear product bands demonstrated the integrity and quality of the extracted DNA, and the PCR product was sufficient for downstream analysis. Additional quality assessments were conducted to ensure DNA quality for the sequencing process.

3.3 Long-Read Sequencing and Metabarcoding Analysis of *RbcL* CpDNA

From the total raw data, post-processing successfully retained 1,348 high-quality *rbcL* cpDNA reads (60.9% yield). These reads had a mean length of 350 bp (range: 200–749 bp) and exhibited a mean Q-score of 16.19 ± 3.67. Among the 1,231 classified reads, 64 putative species-level plants (hereafter species) were identified, covering 59 genera and 11 families. (Fig. 3). These data account for an abundance of 234 and a Shannon and Simpson diversity indices of 3.125 and 0.894, respectively. Additionally, along with species level, five taxa were identified at the family level (Fig. 4), reflecting a relatively high level of plant biodiversity detected through honey metabarcoding.

Moreover, at the genus level, *Actinokentia* emerged as the most dominant genus, representing 25.88% of the total abundance, followed by *Strychnos* (9.02%) and *Allagoptera* (7.84%). These findings suggest the preferences and selective foraging behaviour of honeybees, underscoring their role in shaping local ecosystems while providing vital insights into plant-pollinator interactions.

4 Discussion

In this study, plant biodiversity was characterized through DNA metabarcoding of pollen derived from honey, resulting in the successful identification of 64 plant species visited by honeybees in the Nam Tra My forest, Quang Nam, Vietnam.

Conventional techniques for identifying plants frequented by pollinators include direct field observations of plant visitations or indirect approaches, such as examining pollen collected via traps or observing pollen extracted from honey. In natural settings, pollen is either actively harvested by bees to supply larvae nourishment and sustain the colony during winter or passively gathered during nectar collection [6, 40, 41], providing a potential source for biodiversity investigation. Although traditional melissopalynology is useful for geographical origin determination, it suffers from critical limitations in assessing botanical origins. Particularly, this method is time-consuming, additionally requires taxonomic expertise and knowledge, and yields limited taxonomic resolution [4, 42]. Moreover, pollen analyses struggle to resolve species-level diversity due to the pollen morphology variation and similarity in closely related taxa, thus often resulting in high taxonomic levels, including family or genus. In this study, light microscopy was used to observe pollen isolated from honey (Fig. 1), revealing a variety of pollen types and suggesting the high possibility of being a multifloral honey of Quang Nam. However, morphological observation alone could not provide accurate species counts or identification. In contrast, the metabarcoding approach identified 64 plant species and at least 13 families (Figs. 3 and 4), demonstrating its superior resolution. Furthermore, the small portion of pollen grains detected microscopically is insufficient to comprehensively reflect the full botanical diversity present in the sample.

DNA-based technique, particularly sequencing-based metabarcoding, was able to overcome the limitations of traditional microscopy by generating higher accuracy, comprehensive coverage, and more precise species determination [43–45]. The pollen metabarcoding has been technically validated and proven to provide higher taxonomic resolution at high accuracy [43, 46–49] without any requirement for taxonomic knowledge [11, 43, 49–51]. However, successful DNA extraction from pollen requires efficient disruption of the exine, the outer wall of pollen grains, which is highly resistant to degradation and can persist in the fossil record for millions of years. Additionally, components inherent to honey, such as carbohydrates and phenolic compounds [13–15], can further inhibit PCR reactions, necessitating careful optimization of extraction protocols.

In this study, the high-quality DNA was effectively isolated from pollen in honey, subsequently achieving successful amplification of *rbcL cpDNA* for biodiversity assessment. Key factors contributing to the success of the optimized extraction protocol included the process of diluting and incubating honey with distilled water and ethanol washing for proper sugar and polyphenolic compounds removal, and mechanical disruption of pollen walls using glass beads. Moreover, the combination of implementing silica column-based extraction with a magnetic bead purification step allowed generating high-quality DNA, minimising PCR inhibitors concentration while saving time and avoiding the use of hazardous chemicals. Previous studies have shown that the physicochemical properties of honey, influenced by its floral origin, significantly affect its composition [52, 53]. Our investigation revealed that chemical-based extraction methods exhibited significant variability with extremely low A260/A230 purity ratios and predominantly did not succeed in extracting pollen DNA from honey. On the other hand, the silica column-based approach, integrated with the chemical and physical disruption of pollen cell walls and subsequent purification, has shown the objective of efficiently isolating DNA and removing PCR inhibitors. The results demonstrate that this methodology exhibits technical stability in both DNA extraction (Fig. 2A) and amplification (Fig. 2B), yielding high-quality DNA for downstream analysis.

Third-generation sequencing platforms, particularly Oxford Nanopore Technologies (ONT), offer several advantages for metagenome studies, including long read lengths, simplified setup, reduced sequencing time [54], and portability (exemplified by the pocket-sized MinION device) [54, 55]. These features make ONT a suitable approach for field-based research when centralized laboratories were unavailable [56, 57]. Additionally, ONT-based DNA barcoding circumvents the morphological classification biases and allows the detection of low-abundance rare medicinal taxa with trace amounts of plant DNA (e.g., *Cyclanthus bipartitus*) that were indistinguishable via microscopy, highlighting its sensitivity [58]. In this study, we mainly focused on developing and optimizing the metabarcoding and sequencing procedures using eDNA material from honey. Although Nam Tra My district, where our sample was collected, has 76.8 kha of natural forest covering over 93% of its land area, formal forest surveys are still lacking. Moreover, conducting comprehensive surveys across such an extensive protective forest is inherently challenging. Although only one honey sample has been investigated in our study, the high Shannon (3.12) and Simpson (0.894) indices were shown to be comparable to or exceeding the typical range, indicating remarkably high floral diversity and evenness. These results demonstrate the potential of ONT-based honey metabarcoding as a practical proxy for biodiversity assessment in underexplored ecosystems. Nevertheless, ONT also possesses certain drawbacks, such as high error rates, sample preparation challenges, and database limitations. In this study, we utilized a Flongle flow cell (R10.4), which is currently reported to achieve a modal read accuracy of over 99.1%, corresponding to an error rate of approximately 1% [59]. Moreover, the low cost of the initial setup platform and the portability of the MinION sequencer also make it a useful tool for in-field applications. Altogether, the integration of metabarcoding with ONT sequencing represents a feasible and effective approach for biodiversity surveys, particularly in monitoring the effect of climate change on the reduction of forest biodiversity.

Our primary objective was to specifically validate the ONT portable long-read sequencing platform for an end-to-end, field-deployable *rbcL* honey metabarcoding workflow. Integrating a full Illumina comparison would have necessitated extensive re-optimization, significantly increasing project scope, duration, and cost. Furthermore, a comparison of ONT relative read counts with microscopy abundance is notoriously problematic and yields limited comparative value. This is due to well-established factors like PCR primer bias and variable gene copy numbers across different plant taxa, meaning DNA read counts are not expected to directly correlate with pollen grain counts. In this study, microscopic observation of honey pollen was performed to confirm the presence of diverse pollen grains, supporting the conclusion that the samples contained mixed floral sources, consistent with ONT-based plant species detection. However, detailed morphological identification and quantitative pollen counting were not feasible because pollen grains often exhibit overlapping or similar morphologies, and the microscopic differentiation of closely related plant taxa requires specialized palynological expertise and reference collections. Furthermore, microscopy-based pollen quantification does not reliably reflect species abundance due to variation in pollen size, morphology, and deposition. Larger or sticky pollen grains are more easily observed and trapped, whereas smaller or smooth ones are often underrepresented. Additionally, deposition bias occurs when bees collect or deposit certain pollen types more efficiently than others, leading to disproportionate representation that does not accurately reflect the true floral composition. Thus, microscopy-derived pollen counts are not directly comparable to sequencing-derived relative read counts. Therefore, while ecological validation is valuable, the omission of these external comparisons was a conscious choice to maintain focus on the technical validation and feasibility of the core ONT methodology.

The application of ONT sequencing has revolutionized honey analysis and represents a transformative step forward in modern apicultural research. Moving beyond conventional identification methods, the further integration of DNA metabarcoding and metabolomic analysis holds great promise for enhancing honey traceability, pharmacological effects, and conservation efforts. Looking ahead, harnessing and leveraging Vietnam's floral wealth for sustainable apiculture is not only advancing the economic and medicinal value of its honey industry but also contributing to global efforts in Eco-Economic synergy and food security. The Quang Nam honey can be served as a symbol for biodiversity, where its certified authenticity enhances market value while promoting conservation. ONT assists in adapting beekeeping to climate change by tracking shifts in floral resources, ensuring nutritional stability, hive sustainability, and resilience to environmental changes. Moreover, the honey's floral source can be used as an indicator for the diversity of flowering plants in Quang Nam forest, contributing to the discovery of new valuable medicinal species. Additionally, integrating genetics, ecology, and pharmacology into an interdisciplinary approach is crucial for maximizing honey's nutritional and medicinal value while leveraging Vietnam's botanical prosperity for apiculture and healthcare advancements.

5 Conclusion

In summary, this study demonstrated the application of the DNA metabarcoding method in identifying the types of flowers in honey from Quang Nam, Vietnam, providing initial information about the variety of plants in honey products. Moreover, by integrating portable sequencing platforms, we proposed time-effective in-field procedures in facilitating biodiversity surveying. Data from this study enhances understanding of local biodiversity and offers an innovative approach for tracking the impacts of climate change and agricultural practices (Figs. 1, 2, 3 and 4).

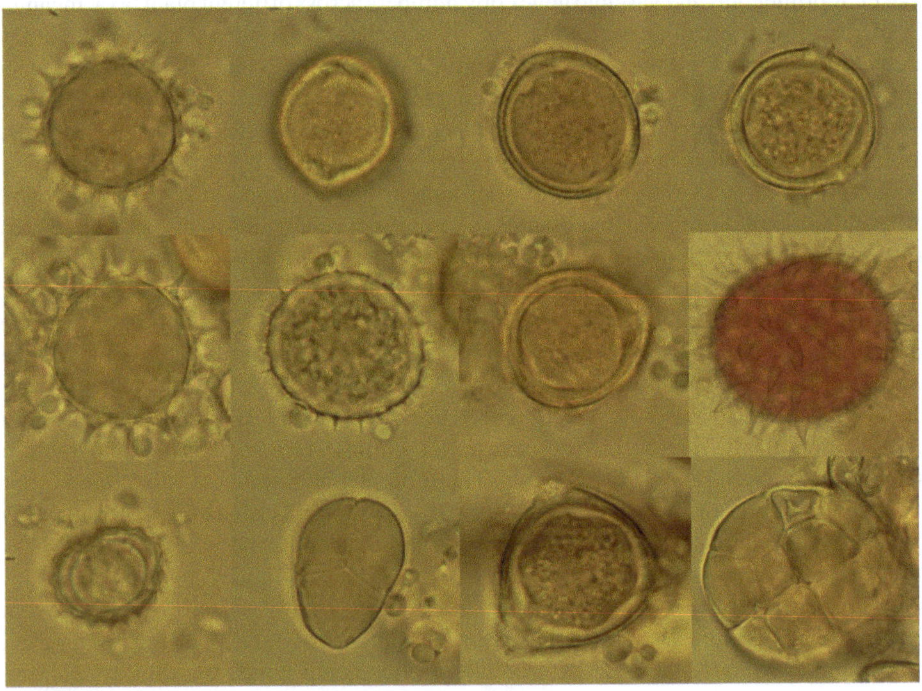

Fig. 1. Photomicrographs of representative pollen grains recovered from the honey sample.

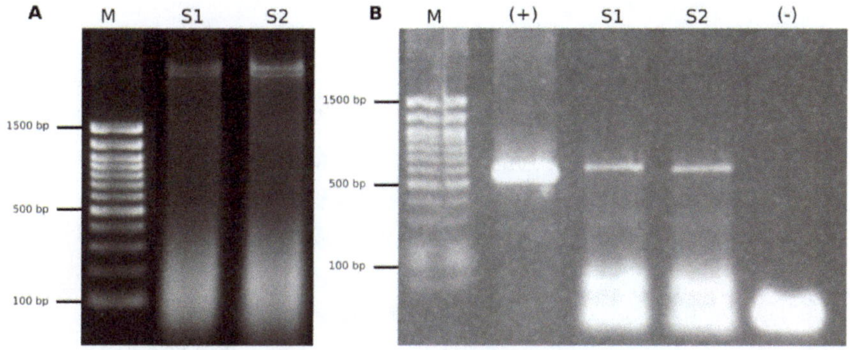

Fig. 2. **a** Extracted DNA from the honey sample. M, 100-bp marker; sample S1, S2 **b** PCR-amplified products of the cpDNA *rbcL* (~600 bp) region from honey DNA. M, 100-bp marker; (+) positive control, (S1, S2) PCR product of sample S1, S2 and (-) negative control.

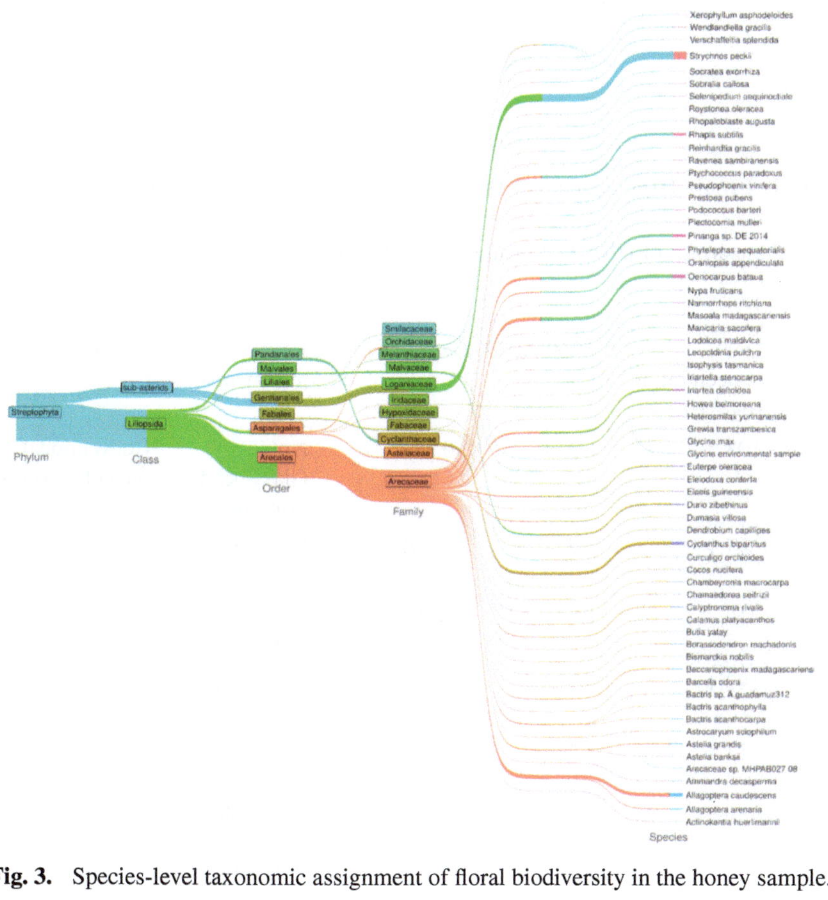

Fig. 3. Species-level taxonomic assignment of floral biodiversity in the honey sample.

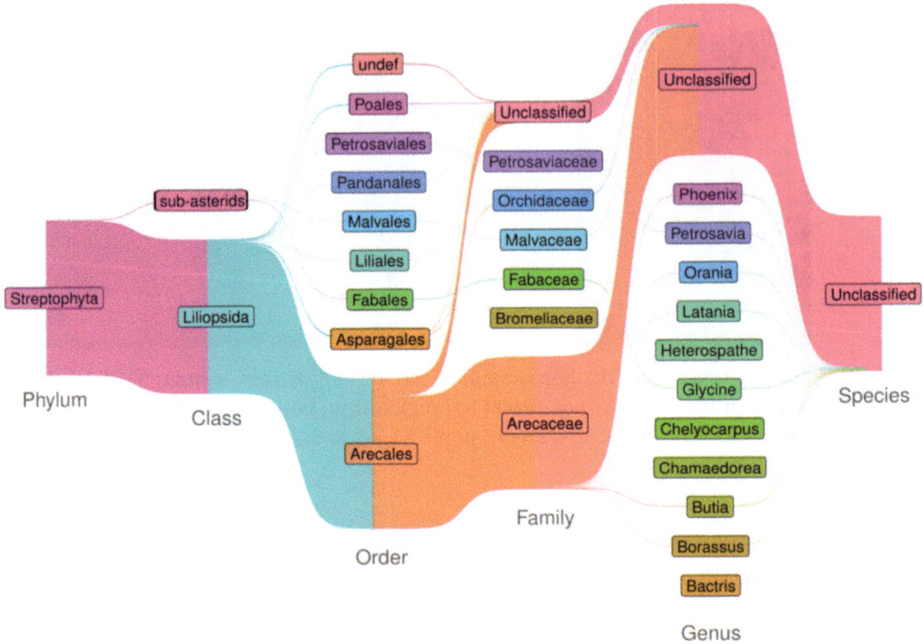

Fig. 4. Family-level taxonomic assignment of floral biodiversity in the honey sample

Table 1. Comparison of DNA yield and Purity obtained using different extraction methods and treatments.

Samples	DNA concentration (ng/μL)	A260/A280	A260/A230	Key treatments	Method
PC1	103.8	1.828	0.383	No Glass-bead grinding	Conventional phenol-chloroform
PC2	97.9	1.923	0.601	Glass-bead grinding	
PS1	256.2	1.874	1.380	No PBS washing	Silica column-based
PS2	192.1	1.978	1.989	PBS washing included	
S1	21.1	2.201	2.322	Integrating Glass-bead grinding, PBS washing, and Magnetic bead purification	Silica column-based

(*continued*)

Table 1. *(continued)*

Samples	DNA concentration (ng/μL)	A260/A280	A260/A230	Key treatments	Method
S2	41.9	2.107	2.006		

PC1 and PC2 were extracted from Quang Nam honey using the conventional phenol-chloroform method to assess the effect of glass-bead grinding on pollen disruption. PS1 and PS2 were obtained using a silica column-based method to evaluate the effect of PBS washing on sample purification. S1 and S2 represent two replicates of optimized silica column-based extraction combining glass-bead grinding, PBS washing, and magnetic bead purification. DNA concentration and purity (A260/A280 and A260/A230 ratios) were measured using a NanoDrop spectrophotometer with 1 μL of each extracted sample.

References

1. Ballantyne, G., Baldock, K.C.R., Rendell, L., Willmer, P.G.: Pollinator importance networks illustrate the crucial value of bees in a highly speciose plant community. Sci. Rep. **7**(1), 8389 (2017)
2. Bovo, S., Ribani, A., Utzeri, V.J., Schiavo, G., Bertolini, F., Fontanesi, L.: Shotgun metagenomics of honey DNA: evaluation of a methodological approach to describe a multi-kingdom honey bee derived environmental DNA signature. PLoS ONE **13**(10), e0205575 (2018)
3. Louveaux, J., Anna, Maurizio, Vorwohl, G.: Methods of Melissopalynology. Bee World. **59**(4), 139–57 (1978)
4. Özkök, A., Akel Bilgiç, H., Kosukcu, C., Arık, Gokhan, Canlı, D., Yet. Idil, et al.: Comparing the melissopalynological and next generation sequencing (NGS) methods for the determining of botanical origin of honey. Food Control. **148**, 109630 (2023)
5. Bell, K.L., Burgess, K.S., Okamoto, K.C., Aranda, R., Brosi, B.J.: Review and future prospects for DNA barcoding methods in forensic palynology. Forensic Sci. Int. Genet. **21**, 110–116 (2016)
6. Concepci6n, E.D., Moretti, M., Altermatt, F., Nobis, M.P., Obrist, M.K.: Impacts of urbanisation on biodiversity: the role of species mobility, degree of specialisation and spatial scale. Oikos **124**(12), 1571–1582 (2015)
7. Lososová, Z., Chytrý, M., Tichý, L., Danihelka, J., Fajmon, K., Hájek, O., et al.: Native and alien floras in urban habitats: a comparison across 32 cities of central Europe. Glob. Ecol. Biogeogr. **21**(5), 545–555 (2012)
8. Sniderman, J.M.K., Matley, K.A., Haberle, S.G., Cantrill, D.J.: Pollen analysis of Australian honey. PLoS ONE **13**(5), e0197545 (2018)
9. Cheng, H., Jin, W., Wu, H., Wang, F., You, C., Peng, Y., et al.: Isolation and PCR detection of foreign DNA sequences in bee honey raised on genetically modified bt (Cry 1 Ac) cotton. Food Bioprod. Process. **85**(2), 141–145 (2007)
10. CBOL Plant Working Group. A DNA barcode for land plants. Proc Natl Acad Sci U S A. **106**(31), 12794–7 (2009)
11. Hawkins, J., Vere, N., Griffith, A., Ford, C.R., Allainguillaume, J., Hegarty, M.J., et al.: Using DNA metabarcoding to identify the floral composition of honey: a new tool for investigating honey bee foraging preferences. PLoS ONE **10**(8), e0134735 (2015)

12. Abdel-Latif, A., Osman, G.: Comparison of three genomic DNA extraction methods to obtain high DNA quality from maize. Plant Methods **13**(1), 1 (2017)
13. Bogdanov, S., Jurendic, T., Sieber, R., Gallmann, P.: Honey for nutrition and health: a review. J. Am. Coll. Nutr. **27**(6), 677–689 (2008)
14. da Silva, P.M., Gauche, C., Gonzaga, L.V., Costa, A.C.O., Fett, R.: Honey: chemical composition, stability and authenticity. Food Chem. **1**(196), 309–323 (2016)
15. Miguel, M., Antunes, M., Faleiro, M.: Honey as a complementary medicine. Integr Med Insights. **1**(12), 117863371770286 (2017)
16. Lalhmangaihi, R., Ghatak, S., Laha, R., Gurusubramanian, G., Kumar, N.S.: Protocol for optimal quality and quantity pollen DNA isolation from honey samples. J. Biomol. Tech. JBT. jbt.14–2504–001 (2014)
17. Michiels, A., Van den Ende, W., Tucker, M., Van Riet, L., Van Laere, A. e.: Extraction of high-quality genomic DNA from latex-containing plants. Anal Biochem. **315**(1), 85–9 (2003)
18. Prosser, S.W.J., Hebert, P.D.N.: Rapid identification of the botanical and entomological sources of honey using DNA metabarcoding. Food Chem. **214**, 183–191 (2017)
19. Baksay, S., Pornon, A.e., Burrus, M., Mariette, J. erome, Andalo, C., Escaravage, N.: Experimental quantification of pollen with DNA metabarcoding using ITS1 and trnL. Sci. Rep. **10**(1), 4202 (2020)
20. Sonia, S., Rodrigues, F., Delerue-Matos, C.: Towards DNA-based methods analysis for honey: an update. molecules. **28**(5), 2106 (2023)
21. Ahmed, I., Biggs, P.J., Matthews, P.J., Collins, L.J., Hendy, M.D., Lockhart, P.J.: Mutational dynamics of aroid chloroplast genomes. Genome Biol. Evol. **4**(12), 1316–1323 (2012)
22. Kress, W.J., Erickson, D.L., Jones, F.A., Swenson, N.G., Perez, R., Sanjur, O., et al.: Plant DNA barcodes and a community phylogeny of a tropical forest dynamics plot in Panama. Proc. Natl. Acad. Sci. **106**(44), 18621–18626 (2009)
23. Mohamadzade Namin, S., Kim, M.J., Son, M., Jung, C.: Honey DNA metabarcoding revealed foraging resource partitioning between Korean native and introduced honey bees (Hymenoptera: Apidae). Sci. Rep. **12**(1), 14394 (2022)
24. Zhu, S., Liu, Q., Qiu, S., Dai, J., Gao, X.: DNA barcoding: an efficient technology to authenticate plant species of traditional Chinese medicine and recent advances. Chin Med. **17**(1), 112 (2022)
25. Bell, K.L., Loeffler, V.M., Brosi, B.J.: An rbcL reference library to aid in the identification of plant species mixtures by DNA metabarcoding1. Appl. Plant. Sci. **5**(3), apps.1600110 (2017)
26. Soltis, D., Soltis, P., Endress, P., Manchester, S., Judd, W., Majure, L., et al.: Phylogeny and evolution of the angiosperms: revised and updated edition [Internet]. Chicago, IL: University of Chicago Press; 2018 [cited 2025 Apr 25]. 590 p. https://press.uchicago.edu/ucp/books/book/chicago/P/bo25468735.html
27. Wolf, M.A., Sfriso, A., Andreoli, C., Moro, I.: The presence of exotic Hypnea flexicaulis (Rhodophyta) in the Mediterranean Sea as indicated by morphology, rbcL and cox1 analyses. Aquat. Bot. **95**(1), 55–58 (2011)
28. Deiner, K., Bik, H.M., Mächler, E., Seymour, M., Lacoursière-Roussel, A., Altermatt, F., et al.: Environmental DNA metabarcoding: transforming how we survey animal and plant communities. Mol. Ecol. **26**(21), 5872–5895 (2017)
29. Geldmann, J., González-Varo, J.P.: Conserving honey bees does not help wildlife. Science **359**(6374), 392–393 (2018)
30. Liu, L., Yang, Y., Deng, Y., Zhang, T.: Nanopore long-read-only metagenomics enables complete and high-quality genome reconstruction from mock and complex metagenomes. Microbiome. **10**(1), 209 (2022)
31. Mooney, H., Larigauderie, A., Cesario, M., Elmquist, T., Hoegh-Guldberg, O., Lavorel, S., et al.: Biodiversity, climate change, and ecosystem services. Curr. Opin. Environ. Sustain. **1**(1), 46–54 (2009)

32. Galliot, J.N. el, Brunel, D, B'erard, A, Chauveau, A., Blanchet, ete A e, Lanore L, et al.: Investigating a flower-insect forager network in a mountain grassland community using pollen DNA barcoding. J. Insect. Conserv. **21**(5), 827–37 (2017)
33. Oliver, A.E., Newbold, L.K., Gweon, H.S., Read, D.S., Woodcock, B.A., Pywell, R.F.: Integration of DNA extraction, metabarcoding and an informatics pipeline to underpin a national citizen science honey monitoring scheme. MethodsX. **11**(8), 101303 (2021)
34. Hasebe, M., Omori, T., Nakazawa, M., Sano, T., Kato, M., Iwatsuki, K.: RbcL gene sequences provide evidence for the evolutionary lineages of leptosporangiate ferns. Proc. Natl. Acad. Sci. **91**(12), 5730–5734 (1994)
35. Wick, R.R., Judd, L.M., Gorrie, C.L., Holt, K.E.: Completing bacterial genome assemblies with multiplex MinION sequencing. Microb Genomics. **3**(10), e000132 (2017)
36. De Coster, W., D'Hert, S., Schultz, D.T., Cruts, M., Van Broeckhoven, C.: NanoPack: visualizing and processing long-read sequencing data. Bioinformatics **34**(15), 2666–2669 (2018)
37. Rognes, T., Flouri, T., Nichols, B., Quince, C., Mahé, F.: VSEARCH: a versatile open source tool for metagenomics. PeerJ **18**(4), e2584 (2016)
38. Altschul, S.F., Madden, T.L., Schäffer, A.A., Zhang, J., Zhang, Z., Miller, W., et al.: Gapped BLAST and PSI-BLAST: a new generation of protein database search programs. Nucleic Acids Res. **25**(17), 3389–3402 (1997)
39. Bell, K.L.: rbcL July 2021 [Internet]. figshare; 2021 [cited 2025 Apr 23]. https://figshare.com/articles/dataset/rbcL_July_2021/14936007/1
40. Grimm, N.B., Faeth, S.H., Golubiewski, N.E., Redman, C.L., Wu, J., Bai, X., et al.: Global change and the ecology of cities. Science **319**(5864), 756–760 (2008)
41. Thorp, R.W.: The collection of pollen by bees. Plant Syst. Evol. **222**(1), 211–223 (2000)
42. Bruni, I., Galimberti, A., Caridi, L., Scaccabarozzi, D., De Mattia, F., Casiraghi, M., et al.: A DNA barcoding approach to identify plant species in multiflower honey. Food Chem. **170**, 308–315 (2015)
43. Bell, K.L., de Vere, N., Keller, A., Richardson, R.T., Gous, A., Burgess, K.S., et al.: Pollen DNA barcoding: current applications and future prospects. Genome **59**(9), 629–640 (2016)
44. Cheptou, P.O., Avendaño, V.L.G.: Pollination processes and the Allee effect in highly fragmented populations: consequences for the mating system in urban environments. New Phytol. **172**(4), 774–783 (2006)
45. Geslin, B., Gauzens, B., Thébault, E., Dajoz, I.: Plant pollinator networks along a gradient of urbanisation. PLoS ONE **8**(5), e63421 (2013)
46. Baldock, K.C.R., Goddard, M.A., Hicks, D.M., Kunin, W.E., Mitschunas, N., Osgathorpe, L.M., et al.: Where is the UK's pollinator biodiversity? The importance of urban areas for flower-visiting insects. Proc. R. Soc. B. Biol. Sci. **282**(1803), 20142849 (2015)
47. Keller, A., Danner, N., Grimmer, G., Ankenbrand, M., von der Ohe, K., von der Ohe, W., et al.: Evaluating multiplexed next-generation sequencing as a method in palynology for mixed pollen samples. Plant Biol. **17**(2), 558–566 (2015)
48. Sickel, W., Ankenbrand, M.J., Grimmer, G., Holzschuh, A., Härtel, S., Lanzen, J., et al.: Increased efficiency in identifying mixed pollen samples by meta-barcoding with a dual-indexing approach. BMC Ecol. **15**(1), 20 (2015)
49. Smart, M.D., Cornman, R.S., Iwanowicz, D.D., McDermott-Kubeczko, M., Pettis, J.S., Spivak, M.S., et al.: A comparison of honey bee-collected pollen from working agricultural lands using light microscopy and ITS metabarcoding. Environ. Entomol. **46**(1), 38–49 (2017)
50. Richardson, R.T., Lin, C.H., Sponsler, D.B., Quijia, J.O., Goodell, K., Johnson, R.M.: Application of ITS2 metabarcoding to determine the provenance of pollen collected by honey bees in an agroecosystem. Appl Plant Sci. **3**(1), 1400066 (2015)

51. Wilson, E.E., Sidhu, C.S., LeVAN, K.E., Holway, D.A.: Pollen foraging behaviour of solitary Hawaiian bees revealed through molecular pollen analysis. Mol. Ecol. **19**(21), 4823–4829 (2010)
52. Chakir, A., Romane, A., Marcazzan, G.L., Ferrazzi, P.: Physicochemical properties of some honeys produced from different plants in Morocco. Arab. J. Chem. **1**(9), S946–S954 (2016)
53. Ciucure, C.T., Geană, E.I.: Phenolic compounds profile and biochemical properties of honeys in relationship to the honey floral sources. Phytochem. Anal. **30**(4), 481–492 (2019)
54. Jain, M., Olsen, H.E., Paten, B., Akeson, M.: The Oxford Nanopore MinION: delivery of nanopore sequencing to the genomics community. Genome Biol. **17**(1), 239 (2016)
55. Laver, T., Harrison, J., O'Neill, P.A., Moore, K., Farbos, A., Paszkiewicz, K., et al.: Assessing the performance of the Oxford Nanopore Technologies MinION. Biomol Detect Quantif. **3**, 1–8 (2015)
56. Castro-Wallace, S.L., Chiu, C.Y., John, K.K., Stahl, S.E., Rubins, K.H., McIntyre, A.B.R., et al.: Nanopore DNA Sequencing and Genome Assembly on the International Space Station. Sci. Rep. **7**(1), 18022 (2017)
57. Egeter, B., Veríssimo, J., Lopes-Lima, M., Chaves, C., Pinto, J., Riccardi, N., et al.: Speeding up the detection of invasive bivalve species using environmental DNA: a Nanopore and Illumina sequencing comparison. Mol. Ecol. Resour. **22**(6), 2232–2247 (2022)
58. Dumschott, K., Schmidt, M.H.W., Chawla, H.S., Snowdon, R., Usadel, B.: Oxford Nanopore sequencing: new opportunities for plant genomics? J. Exp. Bot. **71**(18), 5313–22 (2020)
59. Ni, Y., Liu, X., Simeneh, Z.M., Yang, M., Li, R.: Benchmarking of Nanopore R10.4 and R9.4.1 flow cells in single-cell whole-genome amplification and whole-genome shotgun sequencing. Comput. Struct. Biotechnol. J. **21**, 2352 (2023)

The Effect of Maternal Blood Heavy Metals on Fetal Weight Gain Rate in Indonesian Population

Irawati[1], Hasnawati Amqam[2](✉), Rahayu Indriasari[3], Masni[4], Agus Bintara Birawida[2], Shinta Werorilangi[5], Isymiarni Syarif[1], and Ana Utami Zainal[1]

[1] Doctoral Student of Public Health, Faculty of Public Health, Hasanuddin University, Makassar 90425, Indonesia

[2] Enviromental Health Departement, Faculty of Public Health, Hasanuddin University, Makassar 90425, Indonesia
hasnawati.amqam@ms.unhas.ac.id

[3] Nutrition Departement, Facult[y of Public Health, Hasanuddin University, Makassar 90425, Indonesia

[4] Biostatistic Departement, Faculty of Public Health, Hasanuddin University, Makassar 90425, Indonesia

[5] Marine Science Departement, Hasanuddin University, Makassar 90425, Indonesia

Abstract. Exposure to heavy metals during pregnancy has raised growing concerns due to its potential impact on fetal development. Substances such as lead (Pb), cadmium (Cd), and mercury (Hg) can cross the placental barrier and interfere with intrauterine growth processes, including fetal weight gain. This study aimed to evaluate the effect of maternal blood concentrations of cadmium, lead, and mercury on the fetal weight gain rate during the second and third trimesters. This prospective cohort study investigated whether maternal blood concentrations of Pb, Cd, and Hg are associated with the rate of fetal weight gain. The analysis found no statistically significant correlation between any of the heavy metals and fetal growth: Pb ($\rho = 0.065$; $p = 0.289$), Cd ($\rho = 0.084$; $p = 0.168$), and Hg ($\rho = -0.015$; $p = 0.805$). The median metal concentrations remained relatively consistent across different growth categories. Interestingly, Pb and Cd levels showed a strong positive correlation ($\rho = 0.89$), suggesting common sources or shared exposure pathways among the participants. While the findings did not support a direct link between metal exposure and fetal weight gain, they emphasize the need to monitor environmental exposure in pregnant populations, given their potential cumulative effects on maternal–fetal health.

Keywords: Heavy metals · Pregnancy · Fetal weight gain · Lead · Cadmium · Mercury

L. Vu et al. (Eds.): PBC 2025, 2026.
https://doi.org/10.1007/978-3-032-29579-8_4

1 Introduction

Fetal growth is a critical indicator of prenatal health and a strong predictor of perinatal outcomes, including neonatal morbidity and long-term health conditions such as obesity, diabetes, and cardiovascular diseases [1, 2]. Suboptimal fetal growth, especially during key developmental stages, can have profound implications for the future health of the child. While much attention has been paid to birth weight as a marker of fetal growth, the rate of fetal growth during the second and third trimesters is an equally important metric that remains underexplored [3]. Monitoring fetal growth rate across these trimesters provides more nuanced insights into how environmental and biological factors, including exposure to heavy metals, may influence fetal development [4, 5].

Exposure to heavy metals such as lead (Pb), mercury (Hg), and cadmium (Cd) has raised significant concern due to their potential adverse effects on fetal development [6]. These metals can cross the placental barrier and interfere with vital biological processes essential for fetal growth [7], such as cell division, nutrient transport, and hormone regulation [8, 9]. Despite growing recognition of their toxic effects, the evidence linking maternal heavy metal exposure to fetal growth is still limited, particularly in low- and middle-income countries, including Indonesia.

The few studies conducted in Southeast Asia have shown inconsistent results, with some linking heavy metal exposure to impaired fetal development [3] and others showing no significant associations [10, 11]. This disparity underscores the complexity of the relationship between heavy metal exposure and fetal growth, which may be influenced by various factors such as exposure levels [12], maternal nutritional status, and genetic predispositions [13, 14].

Evidence regarding heavy metal exposure among pregnant women in Indonesia remfains severely limited, particularly in the Eastern Indonesian region. Maternal exposure data from Makassar are rarely reported, this study makes an important contribution to addressing this knowledge gap. Furthermore, most studies conducted in Indonesia and Southeast Asia have predominantly focused on birth outcomes such as birth weight and prematurity, rather than on the dynamics of fetal growth in utero. By assessing changes in Estimated Fetal Weight (EFW) between the second and third trimesters, this study provides novel insights into fetal growth patterns during pregnancy—a critical developmental window that has been largely overlooked in developmental toxicology research in this region.

This study aims to evaluate the effect of maternal blood concentrations of Pb, Cd, and Hg on the rate of fetal weight gain during the second and third trimesters of pregnancy in a cohort of 270 pregnant women in Makassar, Indonesia. By assessing fetal weight gain rate, this study provides valuable insights into how low-to-moderate levels of heavy metal exposure may affect fetal growth, an area that has been under-researched in the Southeast Asian context. Our findings are expected to contribute to the growing body of evidence on the potential risks of environmental pollutants on maternal–fetal health, particularly in regions where environmental contamination is poorly monitored.

2 Methods

2.1 Design, Setting, and Participants

This quantitative study employed a prospective cohort design and was conducted between June and November 2024 across 20 health centers in the city of Makassar, Indonesia. A total of 270 pregnant women were recruited using purposive sampling based on the following inclusion criteria: (1) currently in the second trimester of pregnancy and (2) carrying a singleton pregnancy.

Participants were followed up from the second to third trimester during routine antenatal care. Maternal blood samples were collected during the second trimester and analyzed for Pb, Cd, and Hg concentrations using Inductively Coupled Plasma Mass Spectrometry (ICP-MS). Fetal weight was estimated using ultrasonography (EFW, Estimated Fetal Weight) in both the second and third trimesters. The fetal weight gain rate was then calculated by subtracting the second trimester EFW from the third trimester EFW, divided by the number of weeks between the two ultrasound assessments.

Fetal weight gain rate was calculated using the formula:

$$\text{Fetal Weight gain rate} = \frac{\text{EFW}_2 - \text{EFW}_1}{\text{Weeks between scans}}$$

where EFW_1 is the estimated fetal weight (EFW) measured during the second trimester, and EFW_2 is the EFW measured during the third trimester. The EFW was estimated using the Hadlock formula, which is a commonly used formula in ultrasound-based fetal weight calculations. The Hadlock formula incorporates measurements of biparietal diameter (BPD), head circumference (HC), femur length (FL), and abdominal circumference (AC) to estimate fetal weight [15]. The interval between the two ultrasound measurements was expressed in weeks. Gestational age was not adjusted for in the EFW calculations, as both measurements were taken within the same cohort.

2.2 Data Collection

Additional data were collected using a structured questionnaire that had previously undergone validity and reliability tests. The questionnaire covered key variables, such as maternal exposure to cigarette smoke, environmental factors, lifestyle factors, nutrition, and dietary practices. These variables were measured to assess potential confounders, although they were not included in the adjusted regression model due to the exploratory nature of this analysis. This limitation is acknowledged, as controlling for these confounders could potentially improve the precision of the model.

2.3 Statistic Analysis

All data were tabulated and coded using Microsoft Excel and then imported into the SPSS statistical software version 29 for statistical analysis. Descriptive statistics were used to summarize maternal characteristics and heavy metal concentrations. Given the non-normal distribution of the data and the presence of extreme outliers, the relationship between heavy metal concentrations and fetal weight gain rate was analyzed using Spearman's rank correlation test. Median and IQR (Interquartile Range) were used as measures of central tendency and data dispersion, considering the non-normal distribution of the main variables. Boxplots and histograms were also used to visualize data distribution and detect the presence of outliers.

2.4 Ethical Considerations

Ethical approval for this study was granted by the Faculty of Public Health Ethics Committee, Hasanuddin University, Makassar, Indonesia, under the protocol number 26824093055. Written informed consent was obtained from all participants prior to data collection. Participants were informed about the study's objectives, and their rights were emphasized, including the right to withdraw from the study at any point without any consequences. All data were anonymized and securely stored to ensure confidentiality. Non-monetary incentives were provided to acknowledge the participants' time and contribution, as the study involved minimal risk.

3 Results

3.1 Respondent Characteristic

A total of 270 pregnant women participated in this study, which was conducted as part of a broader effort to understand maternal health in the local communities. During data collection, we found that the majority of participants (79.3%) were aged between 20 and 34 years, a range that is often associated with optimal reproductive health. A smaller proportion (16.3%) were over 34 years of age, while a few (4.4%) were under 20 years of age, a group that may face unique maternal health risks due to early pregnancy. In terms of education, more than half of the women (55.6%) had completed high school, reflecting moderate access to formal education in the study area. Meanwhile, 24.1% had only completed primary school and 19.6% had pursued higher education. A small proportion (0.7%) had never received formal schooling. These findings highlight the varying levels of educational attainment among pregnant women, which may influence health literacy and pregnancy outcomes. When we looked at occupational status, we found that a large majority of the participants (84.4%) were housewives. Only 15.6% of the participants were employed outside the home. This pattern was consistent with the observations from our field visits, where many women shared their experiences of focusing primarily on household responsibilities during pregnancy.

Most participants were Makassar (59.3%), followed by Bugis (28.5%), with smaller proportions from Toraja (2.6%), Javanese (2.2%), and other ethnic backgrounds (7.4%). These figures reflect the cultural diversity in the study region. An important aspect that we assessed was passive exposure to cigarette smoke. Alarmingly, many women reported being exposed to cigarette smoke daily (40.4%), while 27.4% reported occasional exposure. Only 32.2% of the participants stated that they had never been exposed. These findings suggest a high prevalence of household or environmental smoke exposure, an issue discussed by many respondents as part of their daily living conditions. Taken together, these descriptive findings provide an important context for understanding the environmental and social factors that shape pregnancy in this population. In particular, the combination of limited employment opportunities, varying levels of education, and widespread exposure to cigarette smoke may compound the potential risks posed by environmental pollutants such as heavy metals, which we explore further in subsequent analyses in Table 1.

Table 1. Distribution characteristic of the pregnant women (N = 270)

Variables	n	(%)
Age		
<20	12	4,4
20–34	214	79,3
>34	44	16,3
Education		
No School	2	7
Elementary school	65	24,1
High School	150	55,6
Higher School	53	19,6
Job		
Housewife	228	84,4
Working	42	15,6
Passive smoke exposure		
Everyday	109	40,4
Often	74	27,4
Never	87	32,2
Parity		
Nullipari	73	27,0
Multipari	194	71,9

3.2 Maternal Blood Levels

Maternal blood levels of Pb, Cd, and Hg showed highly skewed distributions with extreme values. No significant correlations were found between the levels of these heavy metals and the rate of fetal weight gain (Spearman's rho: Pb = 0.065, Cd = 0.084, Hg = −0.015; all $p > 0.05$). Descriptive analyses of the maternal characteristics and exposure variables are presented. The additional details are provided in Table 2.

Table 2. Concentrate Pb, Cd, Hg, and weight gain rate

Heavy Metals (μg/L)	Mean (SD)	Min	10	25	Median	60	75	90	Max	IQR
Pb	6.48 (3.61)	0.05	0.067	0.468	1.334	1.869	2.630	13.592	128.15	2.563
Hg	1.77 (7.65)	0.05	0.057	0.057	0.057	0.057	1.256	4.069	87.67	1.199
Cd	0.53 (3.61)	0.0003	0.006	0.006	0.006	0.006	0.012	0.036	47.26	0.006
Weight Gain rate	142.10 (3.61)	0.0003-	76.00	102.00	143.00	152.60	173.25	200.90	576	96.85

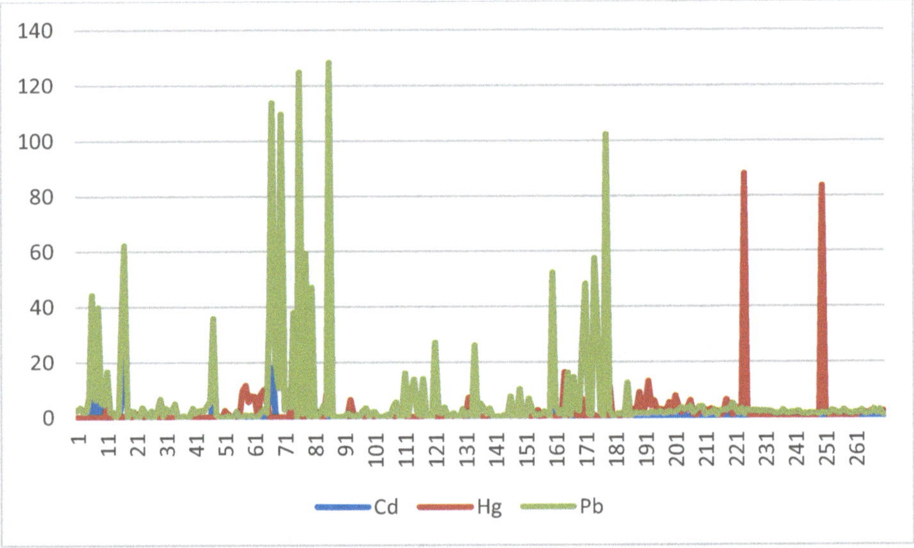

Fig. 1. Distribution of heavy metal levels in maternal blood (Cd, Hg, Pb)

The distribution of heavy metal concentrations in the maternal blood showed significant variation. The mean concentration of Pb in maternal blood was 6.49 ± 18.18 μg/L, with a minimum value of 0.0515 μg/L and a maximum value of 128.15 μg/L. The mean concentration of Hg was 1.77 ± 7.66 μg/L, with a minimum of 0.0574 μg/L and a maximum of 87.67 μg/L. Meanwhile, the mean concentration of Cd was 0.53 ± 3.61 μg/L, with a minimum of 0.0003 μg/L and a maximum of 47.26 μg/L. The histogram used to visualize the data distribution showed the presence of extreme outliers in the concentrations of Pb and Hg in Fig. 1. The majority of heavy metal concentrations were distributed in lower ranges, with several points showing very high peaks, indicating the influence of extreme outliers. To address the non-normal distribution, median and IQR (Interquartile Range) were used as measures of central tendency and data dispersion. Additionally, the relationship between heavy metal concentrations and fetal weight gain rate was analyzed using Spearman's rank correlation.

Table 3. Results of Spearman's correlation between maternal blood heavy metal concentrations and fetal weight gain rate

Variable	Koefisien Korelasi (ρ)	Nilai p	Keterangan
Pb Concentration (μg/L)	0.065	0.241	Not Significant
Hg Concentration (μg/L)	0.084	0.221	Not Significant
Cd Concentration (μg/L)	−0.015	0.853	Not Significant

The average fetal weight gain rate during the study period was 142.10 ± 59.50 g, with a minimum of 27 g and a maximum of 576 g. The fetal weight gain rate exhibited a more normal distribution compared to the heavy metal concentrations. This data showed more moderate variation without the presence of extreme outliers.

Analysis using Spearman's rank correlation in Table 3 revealed no significant association between Pb ($\rho = 0.065$, $p = 0.241$), Hg ($\rho = 0.084$, $p = 0.221$), and Cd ($\rho = -0.015$, $p = 0.853$) concentrations with fetal weight gain rate. This suggests that although there was variation in heavy metal concentrations in maternal blood, heavy metal exposure within the detected range did not have a significant impact on fetal growth in this sample. The histogram showed that the distribution of heavy metal concentrations was non-normal, with very high values (extreme outliers). The scatter plot was used to show the relationship between heavy metal concentrations and fetal weight gain rate, which did not reveal any significant correlation.

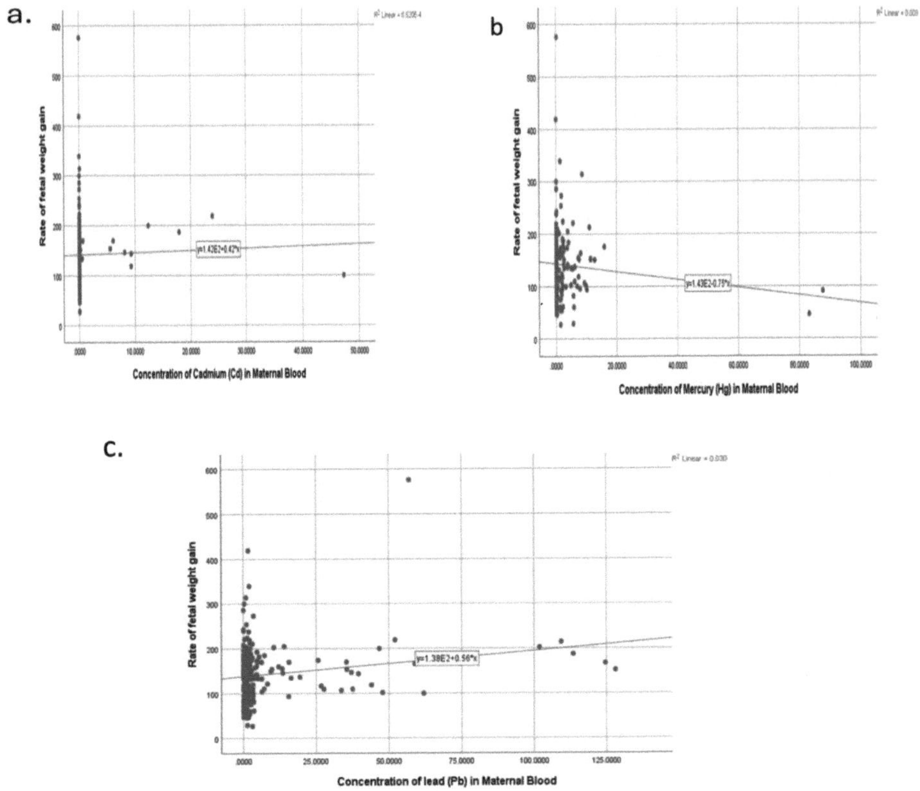

Fig. 2. Scatter plot concentration heavy metal and rate of fetal weight Gain 2a.Cadmium 2b.Mercury 2c.Lead

Although all three scatter plots in Fig. 2 show a trend in the relationship between heavy metal concentrations and fetal weight gain, the very low coefficient of determination (R2) for all three metals (Cd, Hg, and Pb) indicates that maternal blood levels of heavy metals do not have a significant effect on fetal weight gain. These scatter plots support the findings from the Spearman's rank correlation analysis, which revealed no significant relationship between heavy metals and fetal weight gain, reinforcing the idea that factors other than heavy metal exposure may play a more prominent role in influencing fetal growth.

Several points with extreme values in Pb and Hg concentrations were identified as extreme outliers that could potentially influence the analysis results. To mitigate the impact of these outliers, median and IQR were used as measures of central tendency and data dispersion, while the identified outliers were still considered in the descriptive analysis. However, since no significant relationship was found, the impact of outliers on the statistical analysis can be considered minimal in this study.

4 Discussions

The findings of this study indicate no significant association between maternal blood concentrations of Pb, Cd, and Hg and fetal weight gain rate. This result is consistent with several previous studies reporting that the effects of heavy metals on fetal growth tend to manifest only at higher exposure levels. For example, Johnsson et al. [16] in Sweden found that maternal Cd exposure affected fetal growth only among women in the highest exposure quartile [16]. Similarly, Cowell et al. [17] reported that the association between maternal Pb and reduced EFW was significant only in mothers with high Pb concentrations [17]. A comparable pattern was observed by Zhou et al. [18] in China, showing that low-level Hg exposure was not associated with significant changes in fetal size [18].

Other studies have also demonstrated that low-level exposure to heavy metals does not consistently produce measurable effects on fetal growth. Dahiri et al. [19], in a cohort study in the Spain, found that low prenatal Pb exposure was not directly associated with differences in fetal weight trajectories [19]. Kabamba et al. [12] reported that although Cd and Pb were detected in maternal blood, their relationship with fetal growth parameters was not significant at low exposure levels [12]. Likewise, a regional Southeast Asian study by Kim et al. [20] showed that Pb and Cd detected in maternal urine had very weak or non-significant associations with fetal biometric measurements, particularly during the second trimester [20].

A biological explanation for the absence of significant associations in this study involves the protective role of the placenta [21]. The placenta acts as a critical barrier limiting the transfer of heavy metals into the fetal circulation, as described by Zhang et al. [22]. Maternal nutritional status may also contribute; adequate levels of essential minerals such as iron can reduce the gastrointestinal absorption of metals such as Cd and Pb, as highlighted by Woldeamanuel et al. [23].

Conversely, studies that reported significant associations were typically conducted in populations with substantially higher levels of heavy metal exposure. For instance, Sabra et al. [6] demonstrated that elevated Pb concentrations (>10 μg/dL) were associated with impaired fetal growth [6]. Similarly, Li A et al. [24] in China found that high Cd exposure in pregnant women was linked to reductions in both birth length and birth weight [24]. These contrasting findings suggest that the toxic effects of heavy metals are highly dose-dependent and strongly influenced by maternal nutrition and environmental conditions.

Nevertheless, the interpretation of the current findings must consider several limitations. The Spearman correlation analysis employed in this study did not adjust for important potential confounders, such as maternal age, parity, education, nutritional status, and passive smoking exposure. These factors may influence both heavy metal concentrations and fetal growth trajectories. Additionally, the fetal growth indicator used was limited to weight gain rate, and more sensitive biometric parameters were not evaluated. The presence of extreme outliers in Pb and Hg may also affect interpretation, although the use of median and interquartile range mitigates some of this impact.

Despite the absence of significant associations, the long-term health risks linked to prenatal exposure to Pb, Cd, and Hg remain a major concern. Previous studies have shown that prenatal exposure to these heavy metals can contribute to neurological, metabolic, and immunological impairments in children, even when no immediate effects on fetal growth are observed. Therefore, monitoring environmental exposures among pregnant women remains essential, particularly in urban and coastal regions with potential environmental contamination.

Research on heavy metal exposure and fetal growth in Southeast Asia remains limited, making regional comparisons scarce. This study highlights the need to expand research on maternal heavy metal exposure in Indonesia and the broader Southeast Asian region, including the application of multivariate analytical models, more comprehensive fetal biometric measurements, and longitudinal study designs to assess long-term developmental outcomes.

5 Conclusions

This study used data from a prospective cohort of pregnant women to examine whether maternal blood levels of lead, cadmium, and mercury affect the rate of fetal weight gain during the second and third trimesters. The results showed no significant link between these heavy metals and fetal growth, suggesting that at the exposure levels observed in this population, maternal heavy metal levels are unlikely to be a major factor influencing fetal weight gain.

However, considering the known toxicity of these substances and the fact that some individuals have relatively high exposure levels, ongoing monitoring and preventive efforts are important to ensure the health of both mothers and their babies. Future research should consider other environmental and maternal factors that may play a role in fetal development.

Acknowledgements. This research was funded by the Directorate of Research, Technology, and Community Service of the Ministry of Education, Culture, Research and Technology, Indonesia (number 02035/UN4.22.2/PT.01.03/2024).

Conflicts of Interest. The authors declare no conflict of interest associated with the materials used in this study.

References

1. Grzesik-Gąsior, J., Sawicki, J., Pieczykolan, A., Bień, A.: Content of selected heavy metals in the umbilical cord blood and anthropometric data of mothers and newborns in Poland: preliminary data. Sci. Rep. [Internet]. **13**(1), 1–10 (2023). Available from: https://doi.org/10.1038/s41598-023-41249-4
2. Maekawa, R., Ito, R., Iwasaki, Y., Saito, K., Akutsu, K., Takatori, S., et al.: Evidence of exposure to chemicals and heavy metals during pregnancy in Japanese women. Reprod Med Biol. **16**(4), 337–348 (2017)
3. Stone, J., Sutrave, P., Gascoigne, E., Givens, M.B., Fry, R.C., Manuck, T.A.: Exposure to toxic metals and per- and polyfluoroalkyl substances and the risk of preeclampsia and preterm birth in the United States: a review. Am. J. Obstet. Gynecol. MFM [Internet]. **3**(3), 100308 (2021). Available from: https://doi.org/10.1016/j.ajogmf.2021.100308

4. García-Esquinas, E., Pérez-Gómez, B., Fernández-Navarro, P., Fernández, M.A., De Paz, C., Pérez-Meixeira, A.M., et al.: Lead, mercury and cadmium in umbilical cord blood and its association with parental epidemiological variables and birth factors. BMC Public Health. **13**(1) (2013)
5. Kim, J.H., Jeong, K.S., Ha, E.H., Park, H., Ha, M., Hong, Y.C., et al.: Association between prenatal exposure to cadmium and atopic dermatitis in infancy. J. Korean Med. Sci. **28**(4), 516–521 (2013)
6. Sabra, S., Malmqvist, E., Saborit, A., Gratacós, E., Gomez Roig, M.D.: Heavy metals exposure levels and their correlation with different clinical forms of fetal growth restriction. PLoS ONE **12**(10), 1–19 (2017)
7. Johnston, J.E., Valentiner, E., Maxson, P., Miranda, M.L., Fry, R.C.: Maternal cadmium levels during pregnancy associated with lower birth weight in infants in a North Carolina cohort. PLoS ONE **9**(10), e109661 (2014)
8. Student, M.T., Kumar, R.R., Omments, R.E.C., Prajapati, A., Blockchain, T.-A., Ml, A.I., et al.: Hubungan teman sebaya dengan penggunaan rokok elektrik. Front. Neurosci. **14**(1), 1–13 (2021)
9. Motawei, S.M., Gouda, H.E.: Screening of blood levels of mercury, cadmium, and copper in pregnant women in Dakahlia, Egypt: new attention to an old problem. Biol. Trace Elem. Res. **171**(2), 308–314 (2016)
10. Ding, G., Cui, C., Chen, L., Gao, Y., Zhou, Y., Shi, R., et al.: Prenatal low-level mercury exposure and neonatal anthropometry in rural northern China. Chemosphere [Internet]. **92**(9), 1085–9 (2013). Available from: https://doi.org/10.1016/j.chemosphere.2013.01.045
11. Jin, L., Liu, J., Ye, B., Ren, A.: Concentrations of selected heavy metals in maternal blood and associated factors in rural areas in Shanxi Province, China. Environ. Int. [Internet]. **66**, 157–64 (2014). Available from: https://doi.org/10.1016/j.envint.2014.01.016
12. Kabamba, M., Tuakuila, J.: Toxic metal (Cd, Hg, Mn, Pb) partition in the maternal/foetal unit: a systematic mini—review of recent epidemiological studies. Toxicol. Lett. **2020**(332), 20–26 (2019)
13. Molina-Mesa, S., Martínez-Cendán, J.P., Moyano-Rubiales, D., Cubillas-Rodríguez, I., Molina-García, J., González-Mesa, E.: Detection of relevant heavy metal concentrations in human placental tissue: relationship between the concentrations of Hg, As, Pb and Cd and the diet of the pregnant woman. Int. J. Environ. Res. Public Health. **19**(22) (2022)
14. Kim, H., Kim, K.N., Hwang, J.Y., Ha, E.H., Park, H., Ha, M., et al.: Relation between serum folate status and blood mercury concentrations in pregnant women. Nutrition [Internet] **29**(3), 514–8 (2013). Available from: https://doi.org/10.1016/j.nut.2012.08.012
15. Stefanie, Q., Hinkle, N., Johns, A.M., Albert, P.S., Kim, S., Grantz, K.L.: European Journal of Obstetrics & Gynecology and Reproductive Biology Longitudinal changes in gestational weight gain and the association with intrauterine fetal growth §. Eur. J. Obstet. Gynecol. [Internet] (2015). Available from: https://doi.org/10.1016/j.ejogrb.2015.04.006
16. Persson, M., Johansson, S., Villamor, E., Cnattingius, S.: Maternal overweight and obesity and risks of severe birth-asphyxia-related complications in term infants: A population-based cohort study in Sweden. **11**(5), 1–9 (2014)
17. Cowell, W., Colicino, E., Tanner, E., Amarasiriwardena, C., Andra. S.S., Bollati, V., et al.: Prenatal toxic metal mixture exposure and newborn telomere length: Modification by maternal antioxidant intake. Environ. Res. [Internet]. **190**, 110009 (2020). Available from: https://doi.org/10.1016/j.envres.2020.110009
18. Zhou, Y., Yan, L., Li, H., Li, X., Liu, Y., Liu, J.: Patterns and determinants of essential and toxic elements in Chinese women at mid-pregnancy, late pregnancy, and lactation. Nutrients **13**(2), 1–15 (2021)

19. Dahiri, B., Martín-Carrasco, I., Carbonero-Aguilar, P., Cerrillos, L., Ostos, R., Fernández-Palacín, A., et al.: Monitoring of metals and metalloids from maternal and cord blood samples in a population from Seville (Spain). Sci. Total Environ. **854**(2022)
20. Kim, S., Xu, X., Zhang, Y., Zheng, X., Liu, R., Dietrich, K., et al.: Metal concentrations in pregnant women and neonates from informal electronic waste recycling. J. Expo. Sci. Environ. Epidemiol. [Internet]. **29**(3), 406–15 (2019). Available from: https://doi.org/10.1038/s41370-018-0054-9
21. Lozano, M., Murcia, M., Soler-Blasco, R., Casas, M., Zubero, B., Riutort-Mayol, G., et al.: Exposure to metals and metalloids among pregnant women from Spain: Levels and associated factors. Chemosphere, 286 (2022)
22. Zhang, T., Wang, X., Luo, Z.C., Liu, J., Chen, Y., Fan, P., et al.: Maternal blood concentrations of toxic metal(loid)s and trace elements from preconception to pregnancy and transplacental passage to fetuses. Ecotoxicol. Environ. Saf., 264 (2023)
23. Woldeamanuel, G.G., Geta, T.G., Mohammed, T.P., Shuba, M.B., Bafa, T.A.: Effect of nutritional status of pregnant women on birth weight of newborns at Butajira Referral Hospital, Butajira, Ethiopia. SAGE Open Med., 7 (2019)
24. Li, A., Zhuang, T., Shi, J., Liang, Y., Song, M.: Heavy metals in maternal and cord blood in Beijing and their efficiency of placental transfer. J. Environ. Sci. (China) [Internet] **80**, 99–106 (2019). Available from: https://doi.org/10.1016/j.jes.2018.11.004

Distribution Characteristics of Host Animals and Disease Transmission Agents in Some Regions of Vietnam

Le Thi Lan Anh[1], Zh. A. Kasyan[2], E. A. Chumachkova[2], I. N. Sharova[2], A. M. Porshakova[2], Vo Viet Cuong[1], Trinh Van Toan[1], Le Van Quang[1], and Le Thi Van Anh[3,4](✉)

[1] Institute of Tropical Medicine, Joint Vietnam-Russia Tropical Science and Technology Research Center, Nghia Do Award, 63 Nguyen Van Huyen Street, Hanoi, Vietnam

[2] FKUN Russian Anti-Plague Institute "Microbe" Rospotrebnadzor, No. 46 Building, Universitetskaya Street, 410005 Saratov City, Russia

[3] Research Center for Infectious Diseases, International University, VNU-HCM, Ho Chi Minh City, Vietnam

ltvanh@hcmiu.edu.vn

[4] Vietnam National University, Ho Chi Minh City, Vietnam

Abstract. The aim of this study is to determine the prevalence of zoonotic pathogens including *Y. pestis*, *Leptospira*, and *Hantavirus* in wild small mammals across various provinces in the North, Central, and South regions of Vietnam from 2019 to 2023. The wild small mammal hosts (rodents and bats) were collected and the presence of pathogens of natural focal infectious diseases was tested using real-time PCR and ELISA. Of 2488 collected host species, the host species with the highest rates were *Rattus norvegicus* (3.9%) and *Rattus nitidus* (2.41%), in which the highest pathogens rate were *Leptospira* spp. (10.49% by real-time PCR) and antibodies against F1 of *Yersinia pestis* (2.13% by ELISA). These host species were mainly distributed in artificial ecosystems (6.39%) and mountainous areas (2.01%). Animal samples carrying antibodies against F1 of *Y. pestis* and DNA of *Leptospira* spp. were concentrated in the North and in artificial ecosystems with 3.09%, and 4.74%, respectively. The results of the study show that in the northern region of Vietnam there is a high risk of *Leptospira* spp., and plague infection in human. The greatest danger is posed by rodents of the genus *Rattus*, the areas with the highest risk of infection for humans are artificial ecosystems.

Keywords: *Rattus nitidus* · *Rattus norvegicus* · *Leptospira* spp. · Anti-F1 of *Yersinia pestis* antibodies · Vietnam · Ha Giang · Lai Chau

1 Introduction

In human development, pandemics caused by bacteria, viruses, and parasites have had significant impacts on human health and development. Some well-known pandemics include plague, cholera, smallpox, measles, tuberculosis,…[1], and the most recent pandemic is COVID-19. There are many pathogens are transmitted through vectors, such

L. Vu et al. (Eds.): PBC 2025, 2026.
https://doi.org/10.1007/978-3-032-29579-8_5

as ticks, mites, fleas, and lice. Small mammals, particularly bats and mice, are considered the primary hosts or reservoirs of many pathogens for humans, including viruses and bacteria. Rodents are important reservoirs of agents such as *hantavirus, Y. pestis, Leptospira, Borrelia burgdorferi, Yersinia enterocolitica*, and *Yersinia pseudotuberculosis.* More recently, bats have also been identified as important reservoirs of *Leptospira* [2]. Research on small mammals like bats and rodents not only provides information about the diversity of animal species but also plays a crucial role in understanding the presence of zoonotic pathogens. This research is essential for epidemiology, as it helps assess the risk of disease transmission and enables the development of strategies to deal with potential pandemics, especially for emerging and re-emerging diseases.

Vietnam's tropical climate and diverse natural habitats create favorable conditions for the growth and development of small mammal populations. The mountainous areas with hills, rivers, and streams, coupled with poor economic conditions, contribute to an increased risk of disease transmission from animals to humans. Prevalence of pathogens including *Rickettsia, Leptospira, Bartonella*, and *Borrelia* in rodents was reported in the northern mountainous provinces of Vietnam, such as Ha Giang, Lao Cai, and Cao Bang [3]. Serological prevalence of antibodies against *Rickettsia, Borrelia*, and *hantavirus* have also demonstrated in human blood samples collected from two northern mountainous provinces of Vietnam including Lai Chau and Son La. These studies have shown that contact with hosts and vectors increases the risk of exposure to these pathogens [4].

In this study, the biological samples collected from small mammals, including bats and rodents and their ectoparasites during 2019–2023 were used to detect the genetic material of pathogens such as *Y. pestis*, *Leptospira* spp., *hantavirus*, and *Borrelia burgdorferi* using real-time PCR. Immunological testing of blood samples from these animals was also performed to detect sero-prevalence of agents, including *Y. pestis* (antibodies), *chikungunya virus* (antigens), and *orthohantavirus* (antigens). The research aimed to evaluate the overall diversity of hosts or reservoirs carrying pathogens and the prevalence of these pathogens in some provinces in the North, Central, and South regions of Vietnam. The research results are important for assessing the risk of disease spread and outbreak in Vietnam and for helping the Vietnamese health sector develop proactive strategies for dealing with potential epidemics.

2 Materials and Methods

2.1 Study Design

Cross-sectional analysis of the molecular and serological prevalence of *Leptospira* spp., *Y. pestis, hantavirus, chikungunya virus* in mammals as well as *Borrelia burgdorferi, Y. pseudotuberculosis*, and *Y. enterocolitis* in their ectoparasites in Vietnam.

2.2 Research Subjects, Time and Location

– Research subjects:

  - Small mammals: 2488 rodent and bat samples were collected in the survey areas during the research period.

- Research period: from October 2019 to May 2023. Samples are collected twice a year, the first in April-May and the second in October-November.
- Research locations:

  - Northern provinces: Cao Bang, Ha Giang, Lang Son, Lao Cai, Lai Chau, Quang Ninh, Dien Bien;
  - Southern provinces: Binh Phuoc, An Giang, Dong Nai;
  - Central provinces: Dak Lak, Khanh Hoa, Lam Dong.

2.3 Research Methods

Classification of small mammals. Small mammals were live trapped. Trapping sites were located in densely populated areas (residential buildings, outbuildings, and their surroundings), natural habitats, and agricultural lands (artificial ecosystems). The choice of survey sites in natural habitats and artificial ecosystems was made depending on the potential risk of infection in humans. Bats were trapped in caves and natural habitats using specialized nets. Each animal specimen was collected and placed in a separate bag to capture ectoparasites on them.

The captured bats were described morphologically, weighed, and measured for parameters including head and body length, forearm, tail, tibia, hindlimb, ear, and tragus length. We identified bat species based on morphology [5].

Rodents were identified based on morphological characteristics (weight, body and tail length, paw size, ear height, fur characteristics, number of nipples, tooth structure, etc.) [6].

Collecting bats and rodents. To collect rodents, we used live traps measuring 24 × 14 × 14 cm. Sweet potatoes, cassava, or fresh corn were used as bait. After trapping, the rodents were placed in special cloth bags and taken to the laboratory. After euthanasia, the rodents were examined for the presence of ectoparasites, and then internal organs were collected including: the brain (for testing for viruses), liver, spleen, lungs, kidneys, lymph nodes (for testing for viruses and bacteria), and blood for ELISA.

To collect bats, we used 3 nets of different sizes (7 × 3 m, 9 × 3 m, and 12 × 3 m) placed on trees. The traps were checked every 30 min until 10:00 p.m. The captured bats were placed in cotton bags and traps (buckets or plastic containers). After transfer to the laboratory, as well as after euthanasia, bats were examined for the presence of ectoparasites, skin membranes were also taken, and blood, internal organs, a section of the large intestine, and rectum with feces were taken (for coronavirus detection).

Detection of *chikungunya virus* and *orthohantavirus* antigens and anti F1 of *Y. pestis* antibodies by ELISA. To detect antigens/antibodies in small mammals using ELISA, we utilized dried blood samples on 2x2 cm filter paper treated with a 0.1% sodium Merthiolate solution. The blood samples were obtained from the heart and dried on the paper. The process involved opening the thoracic cavity of the rodent, dissecting the heart and large vessels, and soaking the paper in 0.2 ml of blood before drying. Prior to the study, the paper was stored in a Petri dish in a refrigerator or freezer at minus 16 to minus 20 °C. Before the study, the paper in the test tube was soaked in 1 ml of 0.9% sodium chloride solution, resulting in a 1:10 serum dilution. After 40 min, the supernatant was used for serological testing. For live animals, 200 μl of whole blood was collected from the heart in an eppendorf tube, and the serum was separated for ELISA testing. We used the Hantagnost ELISA kit (Poliomyelitis Institute, Russia) to detect *orthohantavirus* antigens, the BioScreen-Chikungunya (Ag kit) (Bioservice, Russia) to detect *chikungunya virus* antigens, and the Plague-AT-F1 (Microbe, Russia) to detect antibodies against F1 of *Y. pestis*. The procedure was performed according to the manufacturer's instructions.

Detection of *Leptospira* spp., *B. burgdorferi, Y. pseudotuberculosis, Y. enterocolitica* and *Orthohantavirus Seoul, coronavirus* by real-time PCR. The mammal biological samples (mixed liver, spleen, lung, and kidney) were collected. For bats, a part of the rectum with feces was placed in an additional test tube for coronavirus (SARS-CoV-2). Before testing, those samples were frozen and stored at −70 °C or in liquid nitrogen under field conditions.

Real-time PCR kits used in the study: "LPS", "TBEV, *B. burgdorferi, A. phagocytophilum*, E.chaffeensis/E.muris-FL", "Yersinia enterocolitica-pseudotuberculosis-FL" (Amplisens, Russia), "OM-Screen-HFRS-RV" (Syntol, Russia) for the detection of *Leptospira* spp., *B. burgdorferi, Y. enterocolitica, Y. pseudotuberculosis*, and *orthohantavirus*, respectively. The *coronavirus* (SARS-CoV-2) detection process was performed according to WHO guidelines.

2.4 Data Analysis

Descriptive statistics were calculated for all variables. Positivity rates were expressed as percentages with exact denominators for each species, habitat, and province. Odds ratios (OR) and 95% confidence intervals (CI) were computed using univariate logistic regression. Differences in pathogen prevalence between habitats and regions were tested using Chi-square or Fisher's exact test where appropriate. The strength of association between sampling intensity and positivity was assessed using Spearman's rank correlation. All analyses were performed using IBM SPSS Statistics version 22, and a p-value <0.05 was considered statistically significant.

3 Results and Discusion

3.1 Characteristics of Host Species in Some Areas of Vietnam

Zoonotic diseases are increasingly common in Vietnam and around the world. As a result, there is growing interest in epidemiological surveillance research on zoonotic infectious diseases. Small mammals such as rodents and bats have been identified as important reservoirs for many pathogens including *Yersenia pestis, Leptospira, hantavirus, Borrelia burgdorferi*. Studying the prevalence of plague in Vietnam is an important research project between the FKUN Russian Anti-Plague Institute "Microbe", Russian Federation and Joint Vietnam-Russia Tropical Science and Technology Research Center. The sampling strategy in this study was risk-based rather than random. Provinces and habitats were selected based on historical reports of zoonotic pathogens, human rodent contact intensity, population density, ecological characteristics, and accessibility. The selection of sampling sites plays an important role in the search for the circulation of the *Yersinia pestis* (plague agent). According to the study by Liyuan Shi et al., 2018, a case of plague was diagnosed in Wuhan, China in June 2016 after many years of no cases. The patient had a history of contact with dead rat of the species *R. flavipectus.* Natural movement of rodents (rats) can carry the pathogen and spread the disease from one place to another. We hypothesize that, with the detection of the plague case in Wuhan Province, China, the risk of cross-border transmission of plague through natural movement of animals (rats, bats) may occur. In the period 2019–2023, we focused mainly on the northern provinces bordering China (including Ha Giang, Dien Bien, Lang Son and Lai Chau) to clarify our hypothesis. In addition, the Central Highlands provinces - where the plague used to occur - were also studied to control the risk of the epidemic returning. The remaining provinces are exploratory research. We have conducted research in the northern border provinces more than other provinces. Ha Giang was sampled annually from 2019-2023 and Lai Chau, Lang Son were sampled in 2019, 2020, 2021, 2023 and Dien Bien was sampled from 2019, 2021, 2022, 2023. Therefore, the provinces of Ha Giang, Lai Chau, Dien Bien, and Lang Son have more samples for analysis than other provinces (Table 1). Between 2019 and 2023, we collected 2488 samples from bats and rodents. These animal samples were identified based on their morphology, and biological samples including blood, liver, kidney, and spleen were tested for DNA/RNA of *Leptospira, orthohantavirus* and tick samples collected from rodents were tested for *B. burgdorferi, Yersinia pseudotuberculosis*, and *Yersinia enterocolitica* using real-time PCR. Additionally, antibodies against F1 of *Y. pestis*, *orthohantavirus* and *chikungunya virus* antigens were tested using ELISA. Animals or their ticks that tested positive for any of these agents by real-time PCR or ELISA were considered as disease carriers.

In Table 1 shows that, a total of 2488 samples from small mammals were collected. Out of these, 345 samples (13.87%) were showed positive for tested pathogens, while 2143 samples (86.13%) tested negative. The prevalence of pathogens in the Southern regions of Vietnam is relatively low. For instance, Dong Nai accounts for 0.6%, An Giang for 1%, Binh Phuoc for only 0.04%, and Lam Dong for only 0.08%. The prevalence of pathogens carrying animals is mainly concentrated in the Northern region of Vietnam, such as Dien Bien (2.33%), Lai Chau (3.18%), and Ha Giang, which has the highest rate of 3.5%. The prevalence in Ha Giang is 87.5 times higher than in Binh Phuoc, and this difference is statistically significant (OR = 87.5, 95%CI: 31.7-238.4, $p < 0.05$).

Research by Le Thi Lan Anh et al., 2021, showed the prevalence of *Leptospira, Bartonella*, and *Rickettsia* agents in the Northern provinces of Vietnam, including Ha Giang, Lao Cai, and Cao Bang, the co-infection of these agents was also mentioned [3]. In addition to rodents, *Leptospira* infection in bats collected in Lai Chau, Son La, and Dong Thap provinces has been reported [2]. Serological studies on blood samples of healthy people collected in Son La and Lai Chau showed a high prevalence of antibodies against *Borrelia* and *hantavirus* [4]. Finally, in a study involving 22 host animal samples collected in Quang Ninh, no prevalence of infectious agents was detected.

Table 1. Characteristics of the prevalence of host species in some areas of Vietnam.

City	Positive (Number; %)	Negative (Number, %)	Total (Number, %)
Dak Lak	4; 0.16	46; 1.85	50; 2.01
Dong Nai	15; 0.6	23; 0.92	38; 1.53
Dien Bien	58; 2.33	256; 10.29	314; 12.63
An Giang	25; 1	67; 2.69	92; 3.7
Binh Phuoc	1; 0.04	29; 1.17	30; 1.21
Cao Bang	6; 0.24	77; 3.09	83; 3.34
Ha Giang	87; 3.5	666; 26.77	753; 30.27
Khanh Hoa	10; 0.4	38; 1.53	48; 1.93
Lang Son	46; 1.85	305; 12.26	351; 14.11
Lai Chau	79; 3.18	534; 21.46	613; 24.64
Lam Dong	2; 0.08	10; 0.4	12; 0.48
Lao Cai	12; 0.48	70; 2.81	82; 3.3
Quang Ninh	0; 0	22; 0.88	22; 0.88
Total	**345; 13.87**	**2143; 86.13**	**2488; 100**

The sampling intensity varies between provinces, which may affect the detected positive rate, for example, in Ha Giang and Lai Chau, the highest number of tested samples corresponds to the highest positive rate of 3.5% and 3.18% respectively. However, not all provinces with a higher number of tested samples have a higher positive rate, for example, in Lang Son, the number of tested samples is 351, higher than Dien Bien (314 samples), however, the positive rate in Lang Son is 1.85%, lower than in Dien Bien (2.3%), in Dak Lak, the number of tested samples is 50, higher than Dong Nai (38), but the positive rate in Dong Nai is 0.6%, which is 3.75 times higher than in Dak Lak.

3.2 Proportion of Host Species Harboring Pathogens

The results of the analysis of the distribution of host species carrying the pathogen showed that 345 positive samples belonged to 27 species, mainly concentrated in the genus *Rattus* (Fig. 1). Analyzing the number of positive samples by species shows that *Menetes berdmorei, Rhinolophus sinicus, Niviventer confucianus, Hylomys suillus, Niviventer fulvescens* has very high positive rates of 100; 100; 66.67; 40 and 33% respectively. However, analyzing by the total number of collected animal samples indicates that *Rattus norvegicus* has the highest positive rate accounted for 3.9%, *Rattus nitidus* for 2.41%, and *Rattus tanezumi* for 1.17%. Other species with lower proportions included *Bandicota Indica* at 0.68%, *Bandicota savilei* at 0.24%, and *Leopoldemys edwadsi* at 0.28%. The remaining species accounted for very low proportions or none at all, such as *Crocidura fuliginosa* (0.04%), *Cynopterus sphinx* (0.04%), *Cricetidae* sp (0%), *Taphozous melanopogon* (0%), and *Tupaia belangeri* (0%). *R. norvegicus* and *R. tanezumi* are commonly distributed in Vietnam, found not only in the Northern mountainous provinces [3, 7] but also in large cities such as Hanoi and Hai Phong [8]. Research on rodent populations using real-time PCR from samples collected in some Northern provinces of Vietnam, including Ha Giang, Lao Cai, and Cao Bang, indicated that *R. tanezumi* had a high positive rate for *Rickettsia Spotted Fever Group (SFG)* and *Rickettsia typhi*. *R. norvegicus, R. nitidus, R. tanezumi, B. indica*, and *B. savilei* also showed high infection rates with *Rickettsia, Bartonella*, and *Leptospira* spp. [3, 9]. Among rodents, *R. norvegicus* and *R. tanezumi* are reported to be the primary hosts of many pathogens such as *Leptospira, hantavirus*, and *Rickettsia* [10–12]. A study on 461 *R. norvegicus* and 64 *R. tanezumi* mice collected in an urban area in southern China showed that the positive rate for *Seoul hantavirus* was 7.43%, in which 8.24% in *R. norvegicus* and 1.56% in *R. tanezumi* [13].

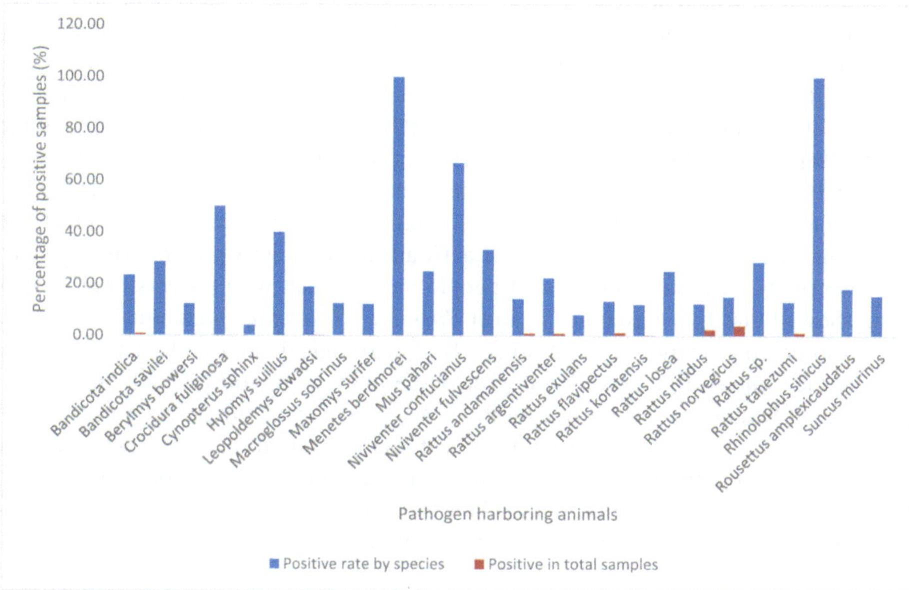

Fig. 1. Proportion of host species harboring pathogens.

3.3 Habitat Distribution of Host Species

The habitat of small mammals is quite diverse. They can be found in mountainous areas, caves, around bushes, along rivers and streams, and in artificial ecosystems such as rice fields, corn fields, cassava fields, and potatoes (collectively called agrocenoces). To determine the habitat distribution of host species, representing an epidemic danger 345 host samples were found positive for one of the research agents by real-time PCR or ELISA were analyzed. The results in Fig. 2 showed that out of the 345 small mammals were positive with investigated pathogens, 6,39% of them live in artificial ecosystems, accounting for the highest proportion and 79.88 times higher than in cave areas (0.08%)

with a statistically significant difference (OR = 79.88; 95%CI: 37.3–169.5, $p < 0.05$). Follwed by not determined group which accounts for 2.05% and mountainous areas accounting for 2.01%. Additionally, other areas account for a lower proportion such as: bushes accounting for 1.41%, and around the house accounting for 1.93%. The Not determined group also accounted for a relatively high positive rate (2.05%), which was the group of animals collected with the support of local people. However, information about the collection location was not precisely determined, possibly in the high mountains or on the hills. During the process of trapping animals, at each area we setted 100 traps/day for 7–10 days and evenly distributed in different ecological areas including artificial ecosystems, mountainous areas, bushes, around houses and caves (for bats). Traps were usually setted in the late afternoon and samples were collected early next morning. All trapped animals were used for research purposes without bias towards any particular ecosystem. The artificial ecological area includes corn, rice, potato, cassava, and wheat fields, which are rich in food, so they attract more animals than other locations. Therefore, the number of rodents trapped in this area is greater than in other areas. However, at each research site we only setted traps for 7–10 days and each day we setted them in different locations to ensure that sampling was not intentional. The number of animals collected in the artificial ecosystem was 970 samples, accounting for 39% of the total number of animal samples collected. Analyzing the positive sample rate by sampling site, the artificial ecosystem also had the highest positive rate of 16.37% (159/970), indicating a high potential risk of infection for humans. A survey of the serological rate of *Borrelia, Rickettsia*, and *hantavirus* in Lai Chau and Son La (two Northern provinces of Vietnam), showed that over 90% of people have contact with rodents, and ELISA testing showed that IgG antibodies against *Borrelia, Rickettsia*, and hantaviruses were detected in healthy people in these areas [4]. A serological survey of 21,630 healthy people in the Central Highlands, Northwest, and Southwest regions of Vietnam showed that 7.36% of healthy people carried antibodies against *Leptospira*, concentrated in the group of people working in the agricultural sector [14]. Therefore, in the research work on monitoring zoonotic diseases, there should be recommendations for people in the research area about the possible risks when working and contacting with rodents, as well as ways to avoid infection from rodents in the wild.

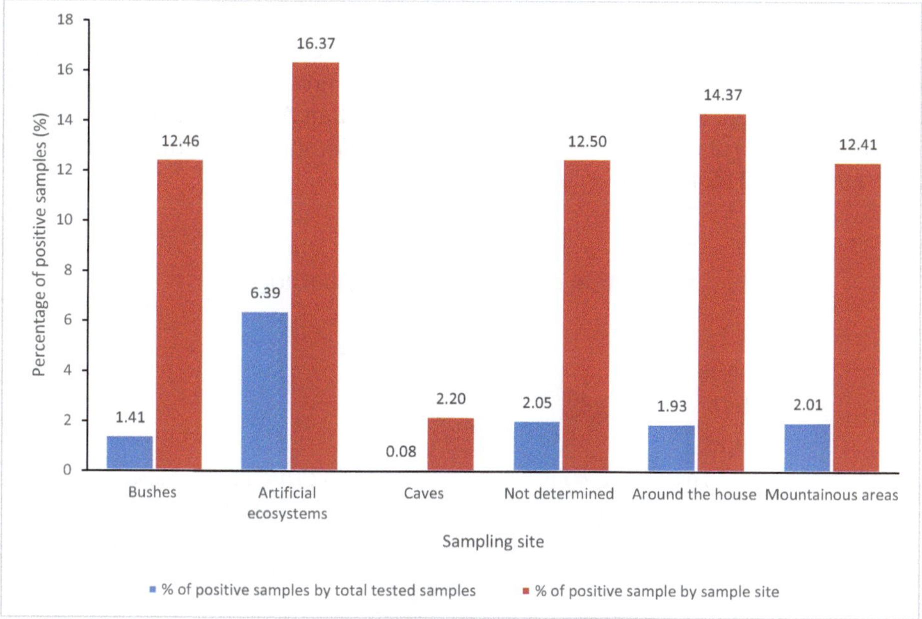

Fig. 2. Habitat distribution of host species.

3.4 Infection Rates by Pathogen

The analysis of the positive sample rate by pathogen in Fig. 3 reveals that among the surveyed pathogens, DNA of *Leptospira* spp. Has the highest rate, accounting for 10.49%, with a statistically significant difference compared to other pathogens with $p < 0.05$. Next is F1 antibodies against the *Yersinia pestis*, with a serological positive rate of 2.13%. Other pathogens, such as antigens of *chikungunya virus*, *orthohantavirus*, or DNA of *Borrelia burgdorferi, Y. enterocolitica*, and *Y. pseudotuberculosis*, have very low rates. Specifically, the ticks from rodents samples which were positive for *Y. pestis* antibody by ELISA were tested for *B. burgdorferi*, *Y. enterocolitica*, and *Y. pseudotuberculosis* by real-time PCR. We found 17 tick samples were positive for *B. burgdorferi*, of which, 10 samples collected in Lai Chau, 6 samples in Ha Giang, and 1 sample in Dong Nai. In addition, among them, 4 tick samples from *Cynopterus sphinx, Crocidura fuliginosa, R. flavipectus*, and *R. nitidus* were positive for *Y. pseudotuberculosis* DNA. Two tick samples from *R. nitidus* and *R. tanezumi* were positive for *Y. enterocolitica*. Five of 2488 collected animal samples were positive for *orthohantavirus*, including 4 samples from *R. norvegicus* and 1 sample from *Hylomys suillus*. Notably, 1 sample of *R. norvegicus* collected in Ha Giang was co-infected with *hantavirus* and *Leptospira* spp., and 1 sample of *R. argentiventer* collected in Lai Chau was simultaneously positive for *Leptospira* spp. And their tick was positive with *Borrelia burgdorferi*. A study on serological circulation in some healthy people serum samples in Phong Tho district, Lai Chau province, showed a relatively high serological prevalence of *B. burgdorferi*, with 25/220 and 53/220 people testing positive for IgM and IgG antibodies against *B. burgdorferi*, respectively [4]. ELISA results detecting antigens of *orthohantavirus* and *chikungunya virus* showed that

2 samples (*R. norvegicus* and *R. losea*) were positive for *orthohantavirus*, and 1 sample of *R. norvegicus* was positive for *chikungunya virus*. Research results on the *Leptospira* spp. Infection rate in rodents captured in some Northern provinces of Vietnam, including Ha Giang, Lao Cai, and Cao Bang, by author Le Thi Lan Anh and colleagues showed that 18% (24/133) of rodents were positive for *Leptospira* spp. by real-time PCR, mainly on *R. tanezumi* (11/24) and *R. cf. Rattus* (7/24) [3].

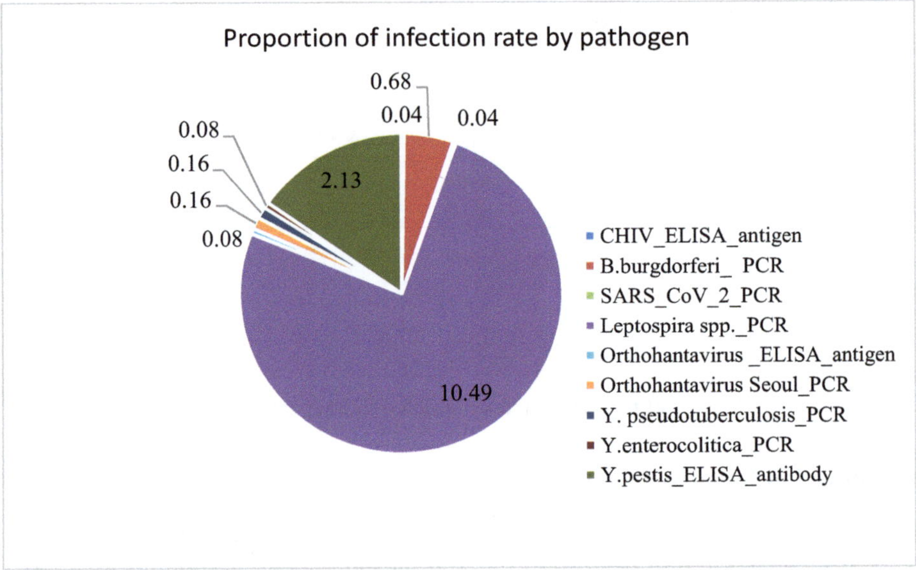

Fig. 3. Proportion of infection rate by pathogens.

In the Mekong Delta region of Vietnam, 18.3% of rodents were positive for *Leptospira* by MAT, with the highest rates in *Rattus norvegicus* (33.0%), followed by *Bandicota indica* and *Rattus tanezumi* with positive rates of 26.5% and 24.6%, respectively [15]. This study showed that among the collected bat samples, four *Rousettus amplexicaudatus* bats collected in Lai Chau were positive for *Leptospira. Leptospira* infection in bats have been reported in many countries around the world [16]; however, research on the circulation of *Leptospira* in bats in Vietnam is still limited. Recently, *Leptospira* has been isolated from *Rousettus amplexicaudatus, Rousettus leschenaultii, Myotis hasseltii, Taphozous longimanus*, and *Eonycteris spelaea* bats captured in Lai Chau, Son La, and Dong Thap provinces [2].

Plague first appeared in Vietnam in 1898 in Nha Trang due to the goods transportation from Hong Kong. With the efforts of medical experts, from 2003 to now, Vietnam has not recorded any cases of plague in humans or presence of *Y. pestis* DNA in rats and fleas. However, plague still exists in Wuhan, China, which borders many northern mountainous provinces of Vietnam, and the natural movement of rodents can carry pathogens [17, 18]. Therefore, studying the epidemiological surveillance of plague plays an important role in controlling plague resurgence. In our study, 2.13% collected rodents were positive with antibodies against F1 antigen of *Y. pestis*.

Among the studied pathogens, the results in Fig. 3 showed that the collected small animals have the highest positive rate for *Leptospira* spp. And *Y. pestis*. Therefore, in the following results, we focus on analyzing the distribution of these two pathogens according to the study area and sampling location.

3.5 The Prevalence of F1 Antibodies Against *Y. pestis* by Regions

The results in Fig. 4 indicated that F1 antibodies against *Y. pestis* prevalency rate was mainly in Lai Chau, with a rate of 3.59%, followed by Ha Giang at 2.79%. Subsequently, the areas of Lang Son account for 1.42%, Dien Bien for 1.27%, and Cao Bang for 1.2%, no positive cases detected in the Central and Southern regions. A case of plague was reported in June 2016 in Yunnan province, China, after an individual came into contact with a dead house mouse (*R. flavipectus*) [19].

Our study reveals that antibodies against F1 of *Y. pestis* are predominantly detected in the Northern region of Vietnam, expecially Ha Giang and Lai Chau provinces bordering China. The movement of rodents in the wild can transmit the pathogen. As a result, regular and continuous plague surveillance research is necessary to control the risk of disease re-emergence.

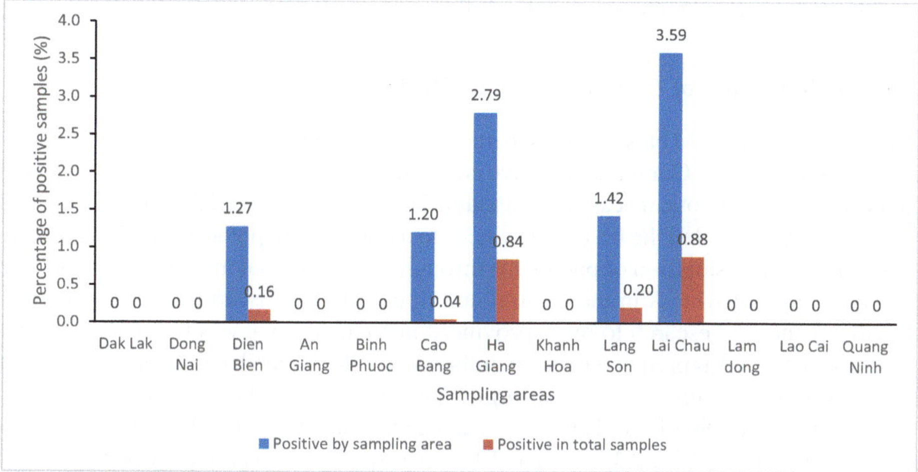

Fig. 4. Distribution of F1 antibodies against *Y. pestis* by regions.

3.6 Prevalence of Antibodies Anti F1 of *Y. Pestis* by Sampling Site

The results of the analysis of the prevalence of antibodies anti F1 of *Y. pestis*, based on the location of animal sampling in this study, are depicted in Fig. 5. The data indicated that antibodies against F1 of *Y. pestis* are predominantly present in two areas: artificial ecosystems account for 3.09% (30/971) and forests account for 2.98% (12/403). These rates are significantly higher compared to other areas such as bushes account for 0.71%

(2/281), areas around houses for 1.5% (5/334), and not determined group for 0.98% (4/408). All of animals collected in caves are bats which are not host of *Y. pestis* currently, that why there is no positive samples found in caves in this study.

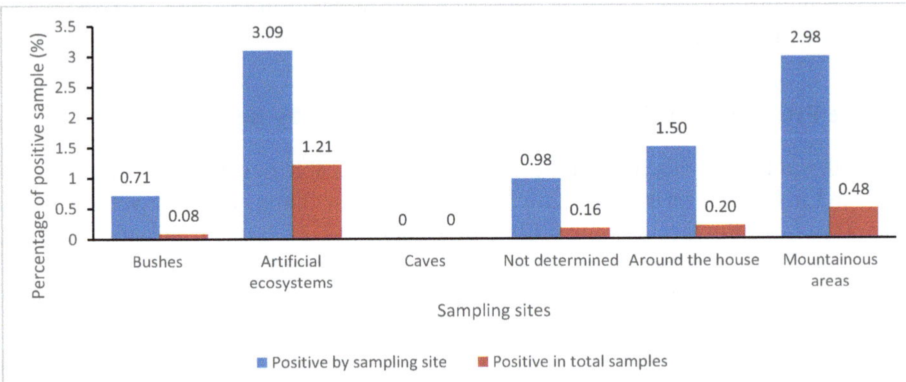

Fig. 5. Location of sampling site for *Y. pestis.*

The not determined group is the group of animals collected with the support of local people. However, information about the collection location was not precisely determined, possibly in the high mountains or on the hills.

3.7 The Prevalence of *Leptospira* spp. by Regions

As we know, rodents and bats are hosts to many pathogens including *Y. pestis, Leptospira, hantavirus* and others. Our main objective is to study the epidemiological surveillance of plague and other zoonotic diseases in Vietnam. The period 2019–2023 is the first stage of our study, so we selected the border provinces to answer the hypothesis of whether there is cross-border transmission of pathogens through natural animal movement or not? and expand to more provinces in the Central and Southern regions. In the coming period, we will expand the research to more provinces in Vietnam to assess the risk of disease transmission as well as predict the risk of outbreaks for the research pathogens. Analyzing the rate of positive samples with *Leptospira* per the total number of tested samples in Fig. 6 showed that Dien Bien, Ha Giang, Lai Chau, Lang Son have the highest positive rates, accounting for 2.13, 2.13,1.77, and 1.53% respectively. These provinces have the largest sample number, so the probability of a positive rate will be higher. However, analyzing the number of positive samples per the total number of samples collected in each province shows that Dong Nai, An Giang, Khanh Hoa are the provinces with the highest positive rates with *Leptospira,* accounting for 36.84% (14/38), 27.17% (25/92) and 20.83% (10/48), respectively.

Unlike the circulation of antibodies against the F1 antigen of *Y. pestis* in rodents which was only observed in the Northern region, DNA of *Leptospira* spp. was found into animal populations including rats and bats in both the Northern and Southern regions. In which, the rate of animals positive for *Leptospira* spp. in the South was higher than in the

North with P < 0.05. This result aligns with other studies. For instance, 18.3% of mice collected in Ho Chi Minh City tested positive for *Leptospira* by MAT, with 33% being *R. norvegicus*, 26.5% being *Bandicota indica*, 24.6% being *R. tanezumi*, 14.3% being *R. exulans*, and 7.1% being *R. argentiventer* [15]. Similarly, 22.7% of mice collected in Kien Giang (a province in Southern Vietnam) tested positive for *Leptospira* by MAT, with *R. norvegicus* accounting for 19.18% [19].

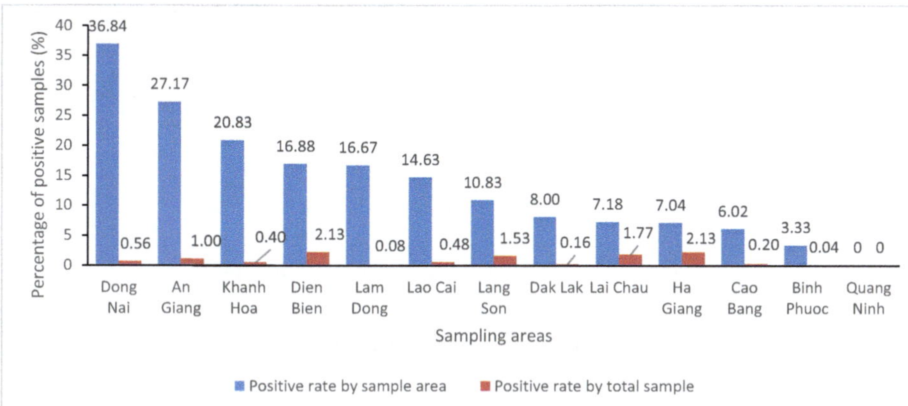

Fig. 6. Distribution of *Leptospira* spp. in the regions.

3.8 *Leptospira* Spp. Prevalence by Sampling Site

Similar to the *Y. pestis*, Leptospiral DNA were mostly found in rodents and bats collected in the artificial ecosystems (agrocenoses) and forest. The result in Fig. 7 indicated that *Leptospira* spp. Also tends to be found in similar locations, with artificial ecosystems accounting for the highest rate at 4.74%, followed by areas around houses at 1.73%. The rate of *Leptospira* spp. Detected in rodents in areas around people's houses is 8.65 times higher than the rate of antibodies against *Y. pestis* (as shown in Figs. 5 and 7), with a statistically significant difference (OR = 8.65, 95%CI: 3.9–18.6, p < 0.05). This suggests that people are at a high risk of *Leptospira* spp. Infection. The survey results on the serological positivity rate of *Leptospira* in serum samples collected from healthy individuals in Ha Giang and Cao Bang showed a positive rate of 17.5%. Additionally, interviews revealed that over 90% of people in the area were exposed to rodents around their houses and in the fields (in press).

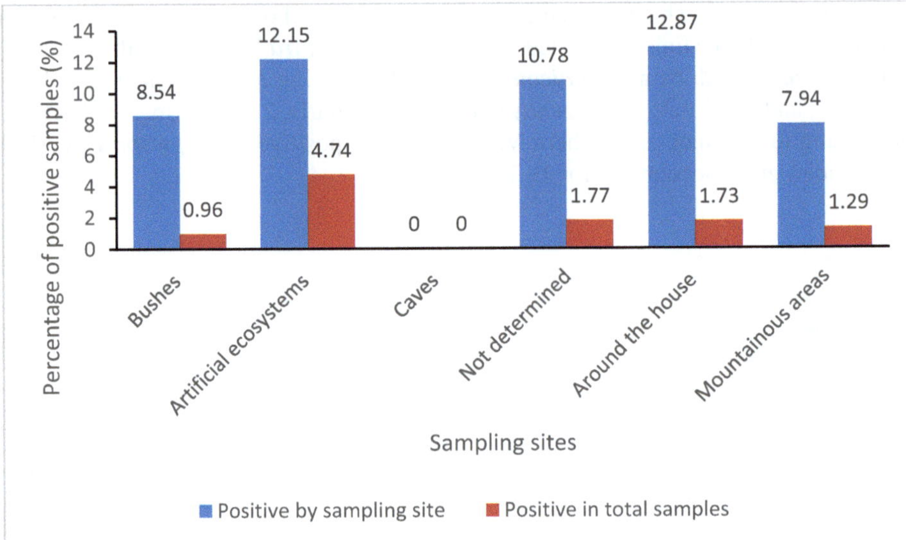

Fig. 7. Location of *Leptospira* spp. Sampling site.

The not determined group is the group of animals collected with the support of local people. However, information about the collection location was not precisely determined, possibly in the high mountains or on the hills.

3.9 The Prevalence of Pathogens in Rodents and Bats Collected in Lai Chau

Based on the analysis results, Lai Chau and Ha Giang are the provinces with the highest prevalence of hosts and pathogens. In this section, we will provide a detailed analysis of the prevalence of each type of host in Lai Chau and Ha Giang. Figure 8 shows that in Lai Chau, out of 32 surveyed species, 13 were positive for pathogens. Among these, *Rattus nitidus* had the highest prevalence (3.75%), which was 23.4 times higher than *Berylmys, Cynopterus sphinx, and Macroglossus sobrinus* with an odds ratio of 23.4, 95% confidence interval: 5.8–90.7, and p-value <0.05. Following this, *Rattus andamanensis* had a prevalence of 3.1%. Lai Chau also had the highest prevalence of *Leptospira* spp. (7.18% of the total 613 samples collected), which was statistically higher than the remaining species with p-value <0.05. In the research provinces, Lai Chau was the only province with bat samples for research, including 7 species: *Rhinolophus affinis, Rhinolophus sinicus, Rousettus amplexicaudatus, Rhinolophus pearsoni, Rhinolophus thomasi, Rhinolophus macrotis,* and *Rousettus leschenaultia.* However, only 4 samples of *Rousettus amplexicaudatus* were found positive for *Leptospira*, accounting for 18.18% of the total 22 samples collected. This result aligns with the study by Ekaterina V Radyuk et al., 2024 [2]. ELISA analysis of animal blood samples collected in Lai Chau to search for traces of agents including *Y. pestis*, *hantavirus* and *chikungunya virus* shows that no antigens of *hantavirus* and *chikungunya virus* were found. However, the circulation of antibodies against the F1 antigen of *Y. pestis* was detected, indicating that there had been exposure to *Y. pestis* in the rodent population in Lai Chau (Fig. 9). PCR results showed that DNA of pathogens including *Leptospira* spp. (in rodents and bats), *SARS_CoV_2* (in bats), *Y. pseudotuberculosis*, *Y. enterocolitica*, *B. burgdorgeri* (in ticks) were detected in Lai Chau. Among the studied agents, *Leptospira* accounted for the highest percentage of 7.18% (44/613), including 4 samples in bats and 40 samples in rodents. The remaining pathogens including *Y. pseudotuberculosis*, *Y. enterocolitica*, *B. burgdorgeri* accounted for low percentages of 0.16; 0.16; 1.63%, respectively. Additionally, out of the 18 tick samples collected in rodents that tested positive for *B. burgdorferi*, 10 were from Lai Chau. The present of IgM/IgG antibodies against *Borrelia miyamotoi* and *Borrelia burgdorferi sensu lato* in health people serum samples in Lai Chau in 2021 were reported with the ratio 25 and 2 out of 220 samples, respectively [2]. In this study, due to the limitation of real-time PCR kit, we only detect the DNA of *Borrelia burgdorferi* in ticks taking from the rodent samples which showed positive for antibody anti F1 *Y. pestis*, therefore the ratio was very low. If we tested in all of collected animal samples, the number of positive samples for *Borrelia burgdorferi* should be higher. But based on this data suggests that it may have the corelation or contacting between human and small mammals with *Borrelia* infection in Lai Chau.

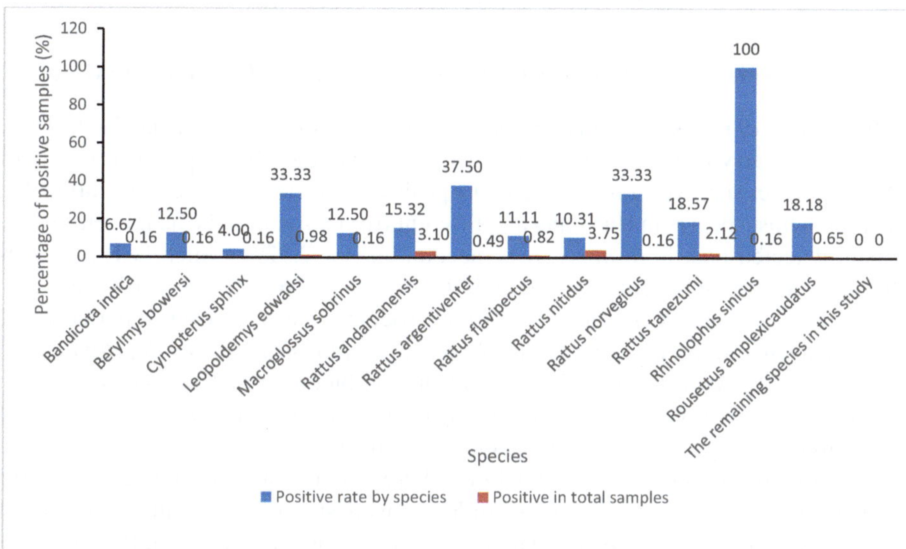

Fig. 8. Proportion of host species carrying the pathogens in Lai Chau by total samples.

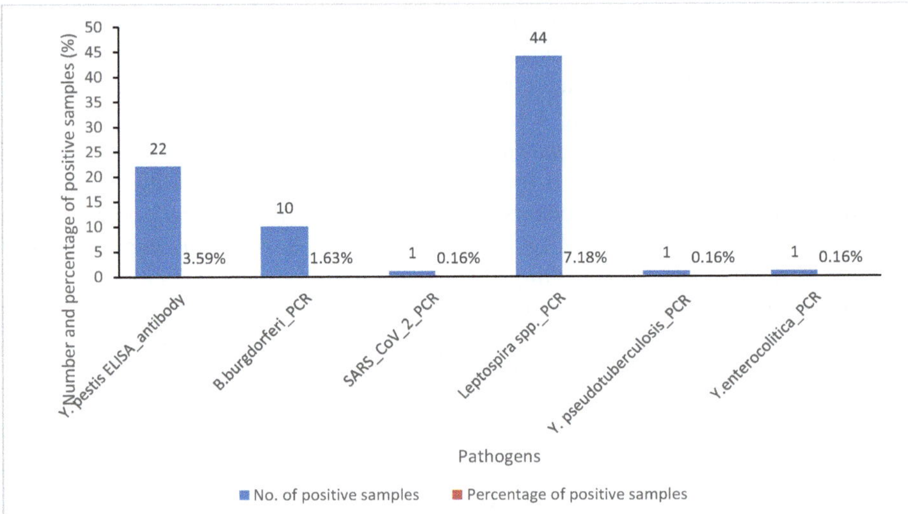

Fig. 9. Rate of tested pathogens in Lai Chau.

3.10 The Prevalence of Pathogens in Rodents and Bats Collected in Ha Giang

The distribution of host species in Ha Giang in Figs. 10 and 11 indicated that the diversity of host species carrying the pathogens is lower than that of Lai Chau. Specifically, out of a total of 32 species surveyed, only 9 species are positive for tested pathogens. Additionally, the common host species in Ha Giang are different from those in Lai Chau. In Ha Giang, the main detected host is *R. norvegicus*, accounting for 7.57%, which is 18.92 times higher than *R. tanezumi* (OR = 18.92, 95%CI: 8.1- 44.2, $p < 0.05$). Furthermore, *R. flavipectus* (1.2%) and *R. nitidus* (1.46%) are rodents that have been reported to be hosts of many infectious agents including *Rickettsia, Leptospira, Bartonella, Hantavirus, Borrelia,...* [3, 9]. Analysis of ELISA test results on blood animal samples found that 2.79% of them were positive for antibodies against the *Y. pestis* F1 antigen. Thus, there was exposure to *Y. pestis* in rodent specimens collected in Ha Giang and Lai Chau. These are two border provinces with China where plague was detected again in 2016 [19]. Thus, there may have been cross-border transmission through natural movement of rodents. Chikungunya virus and hantavirus antigens were detected in 0.13% of the rodent population collected in Ha Giang. Similarly, PCR testing results showed that DNA of *orthohantavirus* was detected in 0.53% of the rodent population. Thus, both PCR and ELISA showed that *orthohantavirus* causing the dangerous hemorrhagic fever with renal syndrome was circulating here. Among the studied agents, *Leptospira* spp. had the highest prevalence, accounting for 7.04% of the total 753 samples collected in Ha Giang by PCR, which is 54.15 times higher than the positive rate for *chikungunya virus* by ELISA. DNA of *Y. enterocolitica* detected in ticks only accounted for 0.13% with OR = 54.15, 95%CI: 17.3–162.3, $p < 0.05$. Additionally, of the 18 tick samples positive for *B. burgdorferi*, 6/18 were collected in Ha Giang. Co-infection with *Leptospira* and *Hantavirus* was observed in *R. norvegicus* mice collected in this province (data not shown).

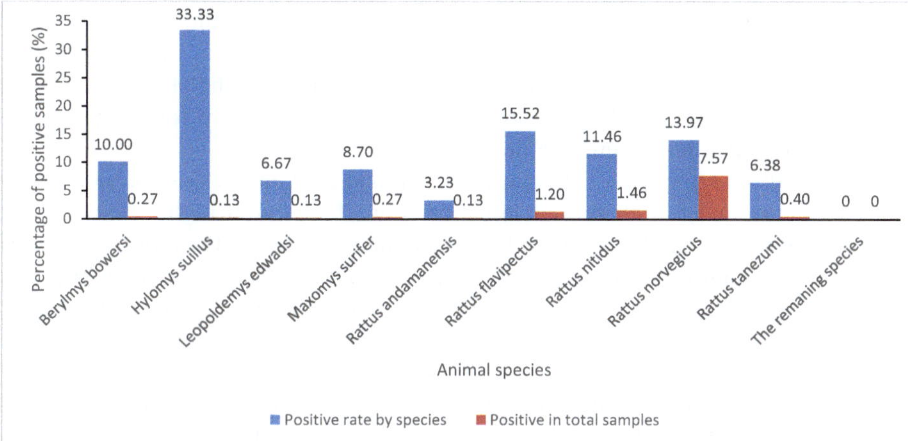

Fig. 10. Proportion of host species carrying the pathogens in Ha Giang by total samples.

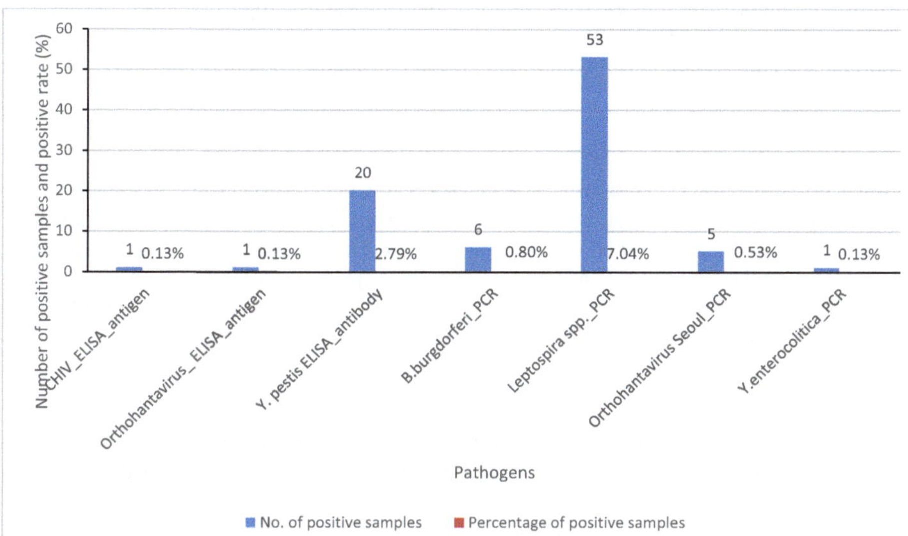

Fig. 11. Rate of tested pathogens in Ha Giang.

4 Conclusions

The analysis in this study has shown various assessments of the distribution of host species and disease-transmitting agents in certain areas of Vietnam. The results can be summarized as follows:

1. Host species surveyed are mainly concentrated in the northern region of Vietnam, with the highest prevalence in Dien Bien, Ha Giang, and Lai Chau. *Rattus* sp. is the dominant species, primarily found in artificial ecosystems, mountainous areas, and residential areas, posing a high risk of infection for people in these areas.

2. Among the surveyed pathogens, DNA of *Leptospira* spp. and antibodies against F1 of *Y. pestis* have the highest prevalence, mainly concentrated in Lai Chau and Ha Giang. Antibodies against *Y. pestis* are primarily found in artificial ecosystems and mountainous forests, while DNA of *Leptospira* spp. is mainly found in artificial ecosystems and around human settlements.
3. In the Northern regions of Vietnam, Lai Chau and Ha Giang exhibit diverse levels of circulation and a high risk of infection. Lai Chau shows higher diversity and prevalence rates than Ha Giang. In Lai Chau, the main host is the mouse *R. nitidus*, and the main pathogens are *Leptospira* spp., while in Ha Giang, the main host is *R. norvegicus*, and *Leptospira* spp. is the common pathogen. Co-infection of *Leptospira* spp. and *orthohantavirus* in rodents has been observed in Ha Giang.
4. Research shows that in the northern region of Vietnam there is a high risk of human infection with *Leptospira* spp., and the possibility of epizootic complications due to plague remains. The greatest danger is posed by rodents of the genus *Rattus*, and the highest risk of human infection is in areas of artificial ecosystems.

Acknowledgments. The authors thank the Joint Vietnam-Russia Tropical Science and Technology Research Center and FKUN Russian Anti-Plague Institute "Microbe" Rospotrebnadzor for supporting and encouraging this research. The research results are products of project M-1.1.1 in 2019–2023.

Author Contributions Le Thi Lan Anh, Kasyan ZH.A, Chumachkova E.A., Sharova I.N., Porshakova A.M., designed the study, conducted the experiments, analyzed the data, wrote and finalized the manuscript, and approved it for submission. Le Thi Van Anh, Le Van Quang analyzed and presented the data. Le Thi Lan Anh wrote the manuscript. Kasyan ZH.A, Chumachkova E.A., Sharova I.N., Porshakova A.M., Vo Viet Cuong, Trinh Van Toan, Le Van Quang, Le Thi Van Anh revised the manuscript. All authors read and approved the final manuscript.

Funding. This study was supported by Joint Vietnam–Russia Tropical Science and Technology Research Center and FKUN Russian Anti-Plague Institute "Microbe" Rospotrebnadzor under the project M-1.1.1 for the period 2019–2023.
Data Availability. The data of this study is available from the corresponding author.

Declarations.

Ethics Approval and Consent to Participate.. The study was approved by the Ethics Committee in Biomedical Research of the Vietnam - Russia Tropical Center (approval certificate No. 1046/CN-HĐĐĐ dated 04/5/2019 and No. 1261/CN-HĐĐĐ dated 05/5/2022).

Consent for publication. The authors agree to publish this article.

Competing Interests. The authors declare no competing interests.

Conflicts of Interest. The authors declare no the potential conflicts of interest concerning this article's study, authorship, and publication.

References

1. Petrusevski, A.B.: History of infectious diseases development in the old and the middle ages with the emphasis on the plague and leprosy. Vojnosanit. Pregl. **70**(7), 704–708 (2013)
2. Ekaterina, V.R., et al.: Leptospira infection in bats in Vietnam. Acta Trop. **257**, 107298 (2024)
3. Lan Anh, L.T., Alexander, E.B., Chau N.V.: Investigation of multiple infections with zoonotic pathogens of rodents in northern Vietnam. J. Vector Borne Dis. **58**(1), 47–53 (2021)
4. Stukolova, O., et al.: Seroprevalence of Borrelia, Rickettsia and Hantaviruses in North Vietnam. Int. J. Infect. Dis. **116**, S126 (2022)
5. Sergei, V.K.: Biodiversity of Vietnam: Bats of Vietnam. Available from. https://www.nhbs.com/biodiversity-of-vietnam-bats-of-vietnam-book. (Accessed: 2013)
6. Darrin, L., Truong Son, N.: An Identification Guide to the Rodents of Vietnam. Available from. https://www.amnh.org/content/download/139778/2272953/file/an-identification-guide-to-the-rodents-of-vietnam2.pdf. Accessed 2001
7. Toan, T.V., et al.: Detection of Rickettsia species in rodents collected in the northern provinces of Vietnam in 2020–2022. J. J. Infections and Diseases **1**, 162–167 (2024)
8. Koma, T., et al.: A survey of rodent-borne pathogens carried by wild Rattus spp. Northern Vietnam. Epidemiol Infect **141**(9), 1876–1884 (2013)
9. Toan, T.V., et al.: The situation of Rickettsia spp. infection on hosts and ectoparasites in the areas of some military units in the Northern region (from 2021–2023). J. Military Med. **5**(367) (2023)
10. Murphy, E.G., et al.: Detection of Seoul virus in wild brown rats (Rattus norvegicus) from pig farms in Northern England. Vet. Rec. **184**(17), 525 (2019)
11. Cueto, G.R., et al.: Prevalence of hantavirus infection in wild Rattus norvegicus and R. rattus populations of Buenos Aires City, Argentina. Trop Med. Int. Health **13**(1), 46–51 (2008)
12. Boey, K., Shiokawa, K., Rajeev, S.: Leptospira infection in rats: a literature review of global prevalence and distribution. PLoS Negl. Trop. Dis. **13**(8), e0007499 (2019)
13. He, W., et al.: Detection and genetic characterization of Seoul virus in liver tissue samples from rattus norvegicus and rattus tanezumi in urban areas of Southern China. Front Vet Sci **8**, 748232 (2021)
14. Kien, N.X., Chuyen, N.V.: Epidemiological characteristics, risk factors of scrub typhus in some key areas. Vietnam Med. J. **526**(2) (2023)
15. Loan, H.K., et al.: How important are rats as vectors of leptospirosis in the Mekong Delta of Vietnam? Vector Borne Zoonotic Dis. **15**(1), 56–64 (2015)
16. Dietrich, M., et al.: Leptospira and bats: story of an emerging friendship. PLoS Pathog. **11**(11), e1005176 (2015)
17. Shi, L., et al.: Reemergence of human plague in Yunnan, China in 2016. PLoS ONE **13**(6), e0198067 (2018)
18. Pham, H.V., et al.: Correlates of environmental factors and human plague: an ecological study in Vietnam. Int. J. Epidemiol. **38**(6), 1634–1641 (2009)
19. Shi, L., et al.: Reemergence of human plague in Yunnan, China in 2016. PLoS ONE **13**(6), e0198067 (2018). https://doi.org/10.1371/journal.pone.0198067

Identification and Sequence Analysis of Botulinum Neurotoxin Encoding Genes in Vietnamese Honey Samples

Hai Anh Chu[1], Thanh Thao Bui[1], Minh Vu[1], Tang Thi Nga[2], Nguyen Thuy Tram[2], Le Thi Trang[2], Minh Nguyet Tran[3], and Yen Pham[3(✉)]

[1] Faculty of Biology, Vietnam National University, University of Science, Hanoi, Vietnam
[2] Department of Bacteriology, National Institute of Hygiene and Epidemiology, Hanoi, Vietnam
[3] Key Laboratory of Enzyme and Protein Technology, Vietnam National University, University of Sciences, Hanoi, Vietnam
yenpb@vnu.edu.vn

Abstract. Background

Infant botulism cases are increasingly reported worldwide. Honey contaminated with *Clostridium botulinum* (*C. botulinum*) is the major type of food that causes this high-mortality disease. Notably, the soil in the close proximity to the apiaries has been identified as a main source of contamination.

Objective

This study aimed to explore enrichment-free DNA extraction and nucleic acid amplification methods for the identification of *C. botulinum* in honey and the associated soil samples.

Materials and methods

6 honey samples and the related soil samplies were collected from different apiaries in a northern midland province in Vietnam. The presence of *C. botulinum* was determined by the primer sets designed for the amplification of two botulinum neurotoxin encoding genes (*bont/A* and *bont/B*) by two nucleic acid-based techniques (polymerase chain reaction, PCR, and loop-mediated isothermal amplification, LAMP). The PCR products were sequenced, and the data was compared using the Nucleotide Basic Local Alignment Search Tool (BLAST).

Results

The results showed that all honey samples were positive for both BoNT encoding genes. Based on the sequence alignments, a single nucleotide polymorphism (SNP) was identified in the *bont/B* gene. Furthermore, the associated soil samples were also shown to be positive, indicating a possible contamination route.

Conclusion

This study provided the first evidence for the co-presence of *C. botulinum* in both honey and the apiary soil, presenting potential risks of botulism outbreaks in this province.

Keywords: *Clostridium botulinum* · Single nucleotide polymorphism · Honey · Apiary soil · LAMP · PCR

L. Vu et al. (Eds.): PBC 2025, 2026.
https://doi.org/10.1007/978-3-032-29579-8_6

1 Introduction

Botulism is a severe neurological condition widespread among different animal species, including fish, birds, cattle, and humans [1]. Botulism is caused by the most lethal toxin known, botulinum neurotoxin (BoNT), which is produced mainly by the anaerobic bacterium *Clostridium botulinum (C. botulinum)* [2] and classified into eight serotypes from A to H. In humans, botulism can be categorized into different clinical syndromes: infant, food-borne, wound, intestinal, iatrogenic, and inhalation botulism [3–5]. Depending on the geographical regions, the prevalence of each form is vastly different. For instance, according to the United States Center for Disease Control reports, most botulism cases are infant botulism (approximately 70%), followed by food-borne and wound forms [6]. In Italy, between 1986 and 2015, food-borne botulism was the most dominant, accounting for nearly 90% of the confirmed cases. It is likely that the later discovered form, *i.e.*, infant botulism, first recognized in 1976 [7], has been underreported, with only 26 countries confirming at least one case in the 1976–2006 period [8].

Clostridium botulinum has been found in honey from different countries with prevalences ranging from 4 to 22% [9]. Raw honey consumption was reported in 20% of the infant botulism cases in the United States [10]. The contamination sources and transmission routes of the bacterium in honey have been studied, indicating that bacterial spores in the apiary soil could be the major factor. According to Nevas *et al.* in a 2006 study, one third of the soil samples were contaminated, compared to 10–20% for the other possible contamination sources [11]. The number of spores in honey has been estimated to be 10^2–10^3 spores/kg, however, in cases of honey causing infant botulism, it could go up to 10^4 spores/kg [12]. The standard procedure for *Clostridium botulinum* determination must include an enrichment step due to the resistance of spores against the chemicals used in DNA extraction. Notably, certain substances in honey could inhibit bacterial growth and produce culture-negative results [13].

Direct extraction methods are ideal for bacterial detection; however, they often face significant inhibitory effects from high sugar contents [14]. In 2000, Notomi *et al.* developed the loop-mediated isothermal amplification (LAMP) using *Bst* polymerase, which is known for inhibitor resistance [15].

In this study, we examined Vietnamese honey samples and related apiary soil samples for the presence of *Clostridium botulinum* using both regular PCR and LAMP. Our results indicated the co-occurrence of the bacterium in both honey and the associated soil and supported the use of the LAMP technique for bacterial detection.

2 Materials and Methods

2.1 Sample Preparation

Honey and soil samples (designated H1-H6 and S1-S6, respectively) surrounding beekeeping areas were collected from 6 apiaries in Bac Giang province, Northern Vietnam. Fifty milliliters of honey from the manufacturer's containers were transferred to sterile containers labeled with the dates and the locations of the sampling sites. The samples were stored at room temperature, then transported to the laboratory for further experiments. The soil samples were aliquoted and stored at −20 °C immediately upon arrival at the laboratory (within 48 h from collection).

2.2 Template Preparation

The templates for amplification were obtained using four methods. With the GeneAll® Exgene™ Soil DNA minikit (South Korea) (E), 200 μL of sample were extracted according to the manufacturer's instructions (https://www.geneall.com/upload/Exgene%E2%84%A2%20Soil%20DNA%20mini%20(Ver%203.0). In the repeating freeze-thaw cycle (FT), 200 μL of sample was rapidly frozen with liquid nitrogen for 1 min, then transferred to a temperature-controlled bath at 100 °C for 1 min. This thermal shock process was repeated 5 times. The entire solution was centrifuged at a speed of 10,000 rpm for 5 min at room temperature to collect the supernatant. With the bead-beating method (B), 200 μL of honey was supplemented with 0.15 g of silica beads with a particle size of 0.1 mm and subjected to one cycle of thermal shock using liquid nitrogen and a temperature-controlled bath as described in FT method. The entire sample was then vortexed at the highest speed setting, approximately 4,500 rpm for 5 min, followed by centrifugation at 10,000 rpm for 5 min at room temperature to collect the supernatant. With the in-house extraction kit (S), a 200 μL volume of honey samples was added to the tube containing 25 μL of Proteinase K, 200 μL of ATL, and 10 μL of DDT and incubated at 56 °C for 1 h. The template was then extracted using the silica column method.

3 Molecular Methods

3.1 PCR Amplification

PCR reactions were performed in a volume of 10 μL, including 6 μL of 2x Master Mix (Thermo Fisher Scientific, United States), 10 μM each of AF3 and AB3, or 10 μM each of BF3 and BB3 primers. Template DNA was denatured by heating at 94 °C for 4 min, followed by 30 cycles of denaturation at 94 °C for 30 s, annealing at 56 °C for 50 s, and extension at 72 °C for 15 s, and a final extension at 72 °C for 5 min. The PCR products were analyzed by electrophoretic separation in 2% agarose gel.

3.2 LAMP Reaction

The 15 μL LAMP reaction mixtures contained 1x LAMP master mix (1x iBuffer, 8U GspSSD2.0 DNA polymerase, 3–5 mM $MgSO_4$, 0.4 mM dNTPs) (Optigene, United Kingdom), 0.2 μM AF3/AB3, 1.0 μM AFIP/ABIP, and 2.0 μM ALF/ALB for BoNT/A and 0.2 μM BF3/BB3, 1.0 μM BFIP/BBIP, and 0.8 μM FLF/FLB for BoNT/B. The LAMP reaction was run at a constant temperature of 63 °C for the *bont/A* gene and 65 °C for the *bont/B* gene. The amplification time for both genes was 30 min. The fluorescence signals were obtained during the amplification process using the Genie III instrument (Optigene, United Kingdom) (Table 1).

Table 1. *Clostridium botulinum* specific primers used in LAMP reactions [16].

Primer sets	Primer names	DNA sequences (5' to 3')	Length
BoNT/A	AF3	TCAATACATTAGATTTAGCCCA	250 bp
	AB3	CCCAAATGTTCTAAGTTCCT	
	AFIP	GTAGCAAATTTGCCTGCACCTAAAATTTTCATTTGGTTTTGAGGAGTCA	
	ABIP	AGCACATGAACTTATACATGCTGGATTTTGCTTACTTCTAACCCACTCA	
	ALF	GAGGATTTGTATCAACTTCAA	
	ALB	GCAATTAATCCAAATAGGGTTTT	
BoNT/B	BF3	GCCAGTTTTAAATGAAAATGAGAC	240 bp
	BB3	CTACTTTAATGCCATATAATCCATG	
	BFIP	CGCTTACATATTCTGGTTTTAATCATTTTGCATCAAGGGA	
	BBIP	TGTTCAAGAAAACAAAGGCGCAATTTTTAAGTTCATGCATTAATATCAAGGC	
	BLF	GCATTATACCCCCGAAGCCT	
	BLB	GTATATTTAATAGACGTGGATAT	

3.3 Sequence Analysis and Alignment

The PCR products were checked by electrophoresis on 2% agarose gel and then were sequenced at either Macrogen Co. Ltd. (South Korea) or using an in-house procedure that included ExoSAP-IT™ PCR Product Cleanup Reagent Purification Kit, BigDye® Terminator Sequencing Kit, and BigDye XTerminator® Purification. All of the sequences were aligned using the Nucleotide Basic Local Alignment Search Tool (BLAST), GenBank (http://www.ncbi.nlm.nih.gov).

4 Results

4.1 Detection of *Clostridium Botulinum* from Honey Samples

All of the honey samples were positive using the LAMP method; however, no PCR product could be detected on the agarose gel. In fact, honey samples contain many PCR inhibitors, including sugars (e.g., fructose and glucose, which account for at least 80% of honey), polyphenols, etc. [17, 18]. Thus, to isolate DNA suitable for PCR amplification, we used an in-house procedure employing a binding step onto silica columns. The results showed that 4/6 honey samples (H2-H5) (Fig. 1A) were positive for *C. botulinum* serotype A, and 5/6 samples (H2-H6) (Fig. 1B) were positive for *C. botulinum* serotype B by PCR method. In a previous study, mechanical extraction was used [19]; yet, the crude lysate still contained all of the inhibitors. In our method, a combination of chemical and enzymatic methods was used, which improved cell lysis efficiency. In addition, DTT facilitates DNA release from complexes by denaturing bound proteins. Furthermore, the DNA was concentrated in the elution step, providing more template molecules per reaction.

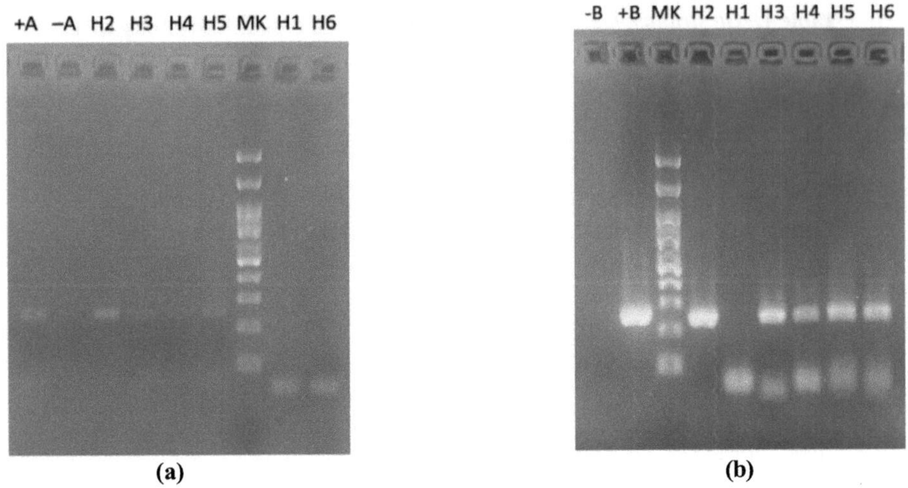

Fig. 1. A. PCR products using primer pairs AF3 and AB3 of 6 honey samples were electrophoresed on 2% agarose gel. (Lane MK: 1kb DNA ladder, lanes 1, 2: +A, −A: positive and negative control, H1-H6: honey samples). **B** PCR products using primer pairs BF3 and BB3 of 6 honey samples were electrophoresed on 2% agarose gel. (Lane MK: 1kb DNA ladder, lanes 1, 2: −B, +B positive and negative control, H1-H6: honey samples)

Gel electrophoresis results also indicated that *C. botulinum* in the H2-H5 samples carries both BoNT/A (~250 bp) and BoNT/B (~ 240 bp) encoding genes. In fact, there have been various studies reporting on *C. botulinum* strains producing more than one serotype of BoNT, for example, Af and Bf, in which the main toxin is denoted by an uppercase letter and the secondary toxin is denoted by a lowercase letter [20]. According to a report by Harris *et al.*, from 1978 to 2023, 173/4310 cases were identified with the presence of *C. botulinum*, of which 67/173 cases were identified as *C. botulinum* serotype A and 63/173 as *C. botulinum* serotype B [21]. Another study by Nevas *et al.* also showed that nearly half (9/20) of honey samples were positive for *C. botulinum* with both BoNT/A and BoNT/B encoding genes [13].

4.2 Multiple Sequence Alignment Analysis of BoNT Encoding Genes

Four out of six honey samples (H2-H5) that tested positive for the BoNT/A encoding gene had the same sequence and were similar to 88 strains listed on NCBI, including strain CB-27 (Accession Number MT199282) and strain NCTC 13319 (Accession Number CP046450.1). Notably, strain CB-27 has both BoNT/A and BoNT/B encoding genes, but it only produces BoNT/A, while BoNT/B is not expressed [22]. This result suggests that some *C. botulinum* strains may carry multiple toxin genes but only express one toxin while the other gene is silent (Fig. 2).

Fig. 2. A. Comparison of *bont/A* gene sequences of *C. botulinum* samples in Vietnam and worldwide, **B.** Comparison of *bont/B* gene sequences of *C. botulinum* in honey samples H2-H6

5/6 honey samples tested positive for the BoNT/B encoding gene, with 4/6 samples sharing the same sequence and similarity to 94 strains published on NCBI. The H3 honey sample has a different sequence from the other samples at 1 position (C < T); however, this mutation does not change the amino acid (ATC to ACC, both encoding methionine). This indicates that the mutation does not affect the function of the encoded protein.

4.3 Detection of *C. Botulinum* Bacteria from Soil Samples

Table 2. Amplified signal using 3 extraction methods by LAMP and PCR

Sample	LAMP result (Peak time)			PCR		
	GeneAll® Exgene™ Soil DNA minikit (E)	Repeating freeze-thaw cycle (FT)	Bead beating (B)	GeneAll® Exgene™ Soil DNA minikit (E)	Repeating freeze-thaw cycle (FT)	Bead beating (B)
S1	14:00	19:45	17:15	–	–	–
S2	13:15	24:45	–	+	–	–
S3	12:45	27:45	21:00	+	–	–
S4	14:45	16:30	13:45	+	–	–
S5	28:00	21:45	22:45	–	–	–

(continued)

Table 2. (*continued*)

Sample	LAMP result (Peak time)			PCR		
	GeneAll® Exgene™ Soil DNA minikit (E)	Repeating freeze-thaw cycle (FT)	Bead beating (B)	GeneAll® Exgene™ Soil DNA minikit (E)	Repeating freeze-thaw cycle (FT)	Bead beating (B)
S6	16:15	21:00	13:15	–	–	–

(+): DNA amplification
(−): No DNA amplification
S1, S2, S3, S4, S5, S6: Soil samples

The results showed that both amplification methods, LAMP and PCR, could detect *C. botulinum* serotype B bacteria in the studied soil samples. However, the PCR amplification method only detected 3/6 samples (S2-S4) extracted using the GeneAll® Exgene™ Soil DNA minikit, while the remaining 2 preparation methods, FT and B, did not form bands when electrophoresizing (Table 2). The commercial extraction kit specialized for soil samples was more effective than the other methods, FT and B. The LAMP method used in the study has better amplification efficiency than the conventional PCR method. This is because LAMP synthesizes a large amount of DNA, operating under isothermal conditions with the *Bst* DNA polymerase, which is less affected by inhibitors in the sample [15]; hence, it can lower the limit of detection. For example, Sharma *et al.* reported in the study of *Acinetobacter baumannii* detection that the detection limit of the LAMP assay for the primers ITS Ab1 and bla OXA-23 was 100 pg/μl and 1 pg/μl DNA, respectively [23], which were 10 and 100 times more sensitive than PCR [24].

5 Disscussion

Monofloral honey, especially lychee honey, has a unique flavor and is prioritized for local product development in Bac Giang, the largest lychee production region in Vietnam. It is a common practice to consume raw honey, and sterilization is avoided due to concerns regarding changes in honey quality. However, new sterilizing techniques have been studied to eliminate the bacteria present in honey without affecting its properties [25, 26]. It is important to note that, during the production and processing process, honey is very susceptible to contamination by microorganisms from the surrounding environments [27]. In particular, Clostridium spores are increasingly identified in honey, posing risks for infants and children with immune deficiencies [28].

We reported the identification of *C. botulinum* bacteria in honey and soil samples using different DNA template preparation procedures. It can be concluded that the LAMP method, due to the use of the *Bst* polymerase enzyme with strong strand separation activity, has the ability to amplify and is less affected by inhibitors than the PCR reaction. In addition, the use of chemical separation combined with enzyme lysis also helps to significantly reduce the inhibitory substances, such as sugars and polyphenols, in honey.

In this study, *C. botulinum* was identified in all samples using both direct LAMP and PCR amplification of extracted DNA. Preliminary sequence analysis indicated that the identified strain could carry an active *bont/A* gene and a silent *bont/B* gene. Furthermore, we found BoNT-encoding genes in the associated soil samples, indicating that spores could be transmitted between apiary soil and honey. Thus, sterilizing methods should be considered in the manufacturing process of honey to ensure safe consumption.

This study represents an initial investigation into the presence of *C. botulinum* in honey and soil samples in Vietnam. In future studies, we will expand both the number of samples and the geographical scope of sampling to improve the reliability and generalizability of the findings.

References

1. Rasetti-Escargueil, C., Lemichez, E., Popoff, M.R.: Human botulism in France, 1875–2016. Toxins **12**(5), 338 (2020). https://doi.org/10.3390/toxins12050338
2. Peck, M.W., et al.: Historical perspectives and guidelines for botulinum neurotoxin subtype nomenclature. Toxins **9**(1), 38 (2017). https://doi.org/10.3390/toxins9010038
3. Anniballi, F., et al.: Botulism in Italy, 1986 to 2015. Eurosurveillance **22**(24), 30550 (2017). https://doi.org/10.2807/1560-7917.ES.2017.22.24.30550
4. Chatham-Stephens, K., Fleck-Derderian, S., Johnson, S.D., Sobel, J., Rao, A.K., Meaney-Delman, D.: Clinical features of foodborne and wound botulism: A systematic review of the literature, 1932–2015. Clin. Infectious Diseases **66**(Suppl_1), S11–S16 (2018). https://doi.org/10.1093/cid/cix811
5. Rao, A.K., Sobel, J., Chatham-Stephens, K., Luquez, C.: Clinical guidelines for diagnosis and treatment of botulism, 2021. MMWR. Recommendations and Reports, **70**(2), 1–30 (2021). https://doi.org/10.15585/mmwr.rr7002a1
6. Bennett, M.R., Gurnani, P., Hill, P.J., Alexander, C., Rawson, F.J.: Iron-catalysed radical polymerisation by living bacteria. Angew. Chem. Int. Ed. **59**(12), 4750–4755 (2020). https://doi.org/10.1002/anie.201915084
7. Pickett, A.: Commentary: a cell line for detection of botulinum neurotoxin type B. Front. Pharmacol. **9**, 1056 (2018). https://doi.org/10.3389/fphar.2018.01056
8. Fox, C.K., Keet, C.A., Strober, J.B.: Recent advances in infant botulism. Pediatr. Neurol. **32**(3), 149–154 (2005). https://doi.org/10.1016/j.pediatrneurol.2004.10.001
9. Grenda, T., Grabczak, M., Kwiatek, K., Bober, A.: Prevalence of *C. botulinum* and *C. perfringens* spores in food products available on Polish market. J. Veterinary Res. **61**(3), 287–291 (2017). https://doi.org/10.1515/jvetres-2017-0038
10. Van Horn, N.L., Street, M.: Infantile botulism. Unpublished work (n.d.)
11. Nevas, M., Lindström, M., Hörman, A., Keto-Timonen, R., Korkeala, H.: Contamination routes of *Clostridium botulinum* in the honey production environment. Environ. Microbiol. **8**(6), 1085–1094 (2006). https://doi.org/10.1111/j.1462-2920.2006.001000.x
12. Lindström, M., Korkeala, H.: Laboratory diagnostics of botulism. Clin. Microbiol. Rev. **19**(2), 298–314 (2006). https://doi.org/10.1128/CMR.19.2.298-314.2006
13. Nevas, M., Hielm, S., Lindström, M., Horn, H., Koivulehto, K., Korkeala, H.: High prevalence of *Clostridium botulinum* types A and B in honey samples detected by polymerase chain reaction. Int. J. Food Microbiol. **72**(1–2), 45–52 (2002). https://doi.org/10.1016/S0168-1605(01)00615-8
14. Rossen, L., Nørskov, P., Holmstrøm, K., Rasmussen, O.F.: Inhibition of PCR by components of food samples, microbial diagnostic assays and DNA-extraction solutions. Int. J. Food Microbiol. **17**(1), 37–45 (1992). https://doi.org/10.1016/0168-1605(92)90017-W
15. Notomi, T., et al.: Loop-mediated isothermal amplification of DNA. Nucleic Acids Res. **28**(12), e63 (2000). https://doi.org/10.1093/nar/28.12.e63

16. Sakuma, T., Kurosaki, Y., Fujinami, Y., Takizawa, T., Yasuda, J.: Rapid and simple detection of *Clostridium botulinum* types A and B by loop-mediated isothermal amplification. J. Appl. Microbiol. **106**(4), 1252–1259 (2009). https://doi.org/10.1111/j.1365-2672.2008.04084.x
17. Lalhmangaihi, R., Ghatak, S., Laha, R., Gurusubramanian, G., Kumar, N.S.: Protocol for optimal quality and quantity pollen DNA isolation from honey samples. J. Biomol. Tech. **25**(4), 92–95 (2014). https://doi.org/10.7171/jbt.14-2504-001
18. Dodd, S., et al.: Detection of sugar syrup adulteration in UK honey using DNA barcoding. Food Control **167**, 110772 (2025). https://doi.org/10.1016/j.foodcont.2024.110772
19. Nguyet, T.T.M., Thao, B.T.T., Duy, N.J., Nga, T.T., Tram, N.T., Yen, P.B.: Establishment of a rapid direct DNA preparation procedure for the detection of *Clostridium botulinum* serotype B from honey. Vietnam J. Food Control **5**(3), 533–541 (2022)
20. Hutson, R.A., Zhou, Y., Collins, M.D., Johnson, E.A., Hatheway, C.L., Sugiyama, H.: Genetic characterization of *Clostridium botulinum* type A containing silent type B neurotoxin gene sequences. J. Biol. Chem. **271**(18), 10786–10792 (1996)
21. Harris, R.A., Dabritz, H.A.: Infant botulism: In search of *Clostridium botulinum* spores. Curr. Microbiol. **81**(10), 306 (2024). https://doi.org/10.1007/s00284-024-03828-0
22. Jeon, J.H., et al.: Genetic Characterization of *Clostridium botulinum* Isolated from the First Case of Infant Botulism in Korea. Ann. Lab. Med. **41**(5), 489–492 (2021). https://doi.org/10.3343/alm.2021.41.5.489
23. Sharma, A., Gaind, R.: Development of loop-mediated isothermal amplification assay for detection of clinically significant members of *Acinetobacter calcoaceticus-baumannii* complex and associated carbapenem resistance. Front. Mol. Biosci. **8**, 659256 (2021). https://doi.org/10.3389/fmolb.2021.659256
24. Yamamoto, N., et al.: Clinical specimen-direct LAMP: a useful tool for the surveillance of blaOXA-23-positive Carbapenem-resistant Acinetobacter baumannii. PLoS ONE **10**(7), e0133204 (2015). https://doi.org/10.1371/journal.pone.0133204
25. Bera, A., Almeida-Muradian, L.B., Sabato, S.F.: Effect of gamma radiation on honey quality control Radiat. Phys. Chem. **78**(7–8), 583–584 (2009). https://doi.org/10.1016/j.radphyschem.2009.03.013
26. Khatun, M.A., Razzak, M., Hossain, M.A., Hossain, A., Islam, M., Shahjalal, M., Khan, R., Huque, R.: Gamma radiation processing of honey of mustard, black seed and lychee flower: measurement of antioxidant, antimicrobial, and Fourier transform infrared (FT-IR) spectra. SSRN Electron. J. (2022)
27. Vázquez-Quiñones, C.R., Moreno-Terrazas, R., Natividad-Bonifacio, I., Quiñones-Ramírez, E.I., Vázquez-Salinas, C.: Microbiological assessment of honey in México. Rev. Argent. Microbiol. **50**(1), 75–80 (2018). https://doi.org/10.1016/j.ram.2017.04.005
28. Snowdon, J.A., Cliver, D.O.: Microorganisms in honey. Int. J. Food Microbiol. **31**(1–3), 1–26 (1996). https://doi.org/10.1016/0168-1605(96)00970-1

Sequence Investigation and Activity Characterization of the Major Sulfur-Oxidizing Enzymes of a Purple Photosynthetic Bacteria Strain Isolated in Vietnam

Phuc T. Do[1,2](✉), Yen Pham[1,2], Bach-Lien Le-Dam[2], Phuong- Anh Ha[2], Hai T. Pham[1], Phuong M. Nguyen[3], and Ngoc-Anh Trinh-Thi[1,2]

[1] Faculty of Biology, VNU University of Science, Hanoi, Vietnam
dothiphuc@hus.edu.vn

[2] Key Laboratory of Enzyme and Protein Technology (KLEPT), VNU University of Science, Hanoi, Vietnam

[3] Faculty of Environmental Sciences, VNU University of Science, Hanoi, Vietnam

Abstract. Hydrogen sulfide (H_2S) has been considered as dangerous as carbon dioxide in causing environmental pollution. When present in high concentrations, H_2S is extremely poisonous, resulting in serious damage to multiple organs, even leading to death. Various H_2S bioremediations, which utilize microorganisms and are being the friendliest method to the environment, have become widely studied. Photosynthetic purple bacteria that belong to the sulfur-oxidizing bacteria (SOB) group have been found to be a promising solution for H_2S pollution, attributed to different enzymes involved in various pathways that metabolize sulfur compounds. In this study, we investigated the DNA sequences, predicted the conserved motifs in the putative protein sequences, and characterized the activity of the major sulfur-oxidizing enzymes of *Rhodobacter sediminis* strain PAM37 isolated in Vietnam. The target enzyme-encoding genes included sulfide:quinone oxidoreductase (SQR), and flavocytochrome c sulfide dehydrogenase (FCC). Based on the sequencing analyses, the amplified fragments had similarities between 95% and 98% with the published data in GenBank, and then were used to find conserved motifs. Further fractionation of the cultured bacteria for the sulfide oxidase assay revealed that sulfate formation activity was found mostly in the membrane and the periplasmic fractions, where the studied enzymes are located. The activity was inhibited up to 50% by KCN, a flavoprotein inhibitor, indicating the possible presence of SQR and FCC enzymes. These preliminary results provided important information for future applications of these strains for H_2S removal in polluted areas.

Keywords: *Rhodobacter sediminis* · Hydrogen sulfide · Sulfide:quinone oxidoreductase · Flavocytochrome c sulfide dehydrogenase · KCN

L. Vu et al. (Eds.): PBC 2025, 2026.
https://doi.org/10.1007/978-3-032-29579-8_7

1 Background

Naturally, H_2S gas can be produced by different organisms or through geothermal activity below the earth surface. Besides, H_2S can be produced by human activities (anthropogenic), including agricultural and industrial operations. These activities are shown to contribute a significant quantity of H_2S gas to the environment. However, anthropogenic gas is more difficult to document than naturally occurring H_2S due to a lack of monitoring and reporting, despite its direct impact on human health [1].

In human cells, at low concentrations (0.01 μM to 1 μM), sulfide is an electron donor for complex II of the electron transport chain in the mitochondria, thereby stimulating ATP production. However, when the concentration is tripled, sulfide become toxic by binding to cytochrome c and inhibiting their activity in complex IV of the electron transport chain, preventing ATP production, causing cells to be deprived of oxygen, leading to dangerous consequences like respiratory failure [2]. The extent of H_2S impact on human health has been examined based on its concentration. Generally, H_2S can be smelled as unpleasant "rotten egg" odor even in low concentrations (0.0005 ppm to 0.03 ppm). And at the concentrations beyond 100 ppm, the effects become more severe, especially, from 500 ppm onwards, which cause death gradually or suddenly depending on the dose and period of exposure [3]. Currently, there is no effective treatment for H_2S poisoning, so it is crucial to reduce and remove H_2S from the residences.

Various methods have been implemented for the treatment of H_2S, but in general, they can be divided into three main categories: physical, chemical, and biological. Even though being developed later, the biological strategies have obtained remarkable achievements. The major advantages of biological-based methods are environmentally friendly and flexible usages. Within this category, microbial technology is possibly the most promising method, with the example of sulfur-oxidizing bacteria (SOB), which are capable of converting toxic H_2S gas into non-toxic forms. These bacteria can be divided into two groups, photoautotrophs (PET) and chemolithotrophs (CET). Cyanobacteria and purple bacteria are the two main members of the PET group, which typically exist in an anaerobic environment.

There have been many studies on the applications of photosynthetic bacteria to treat H_2S in wastewater. Using a combination of primarily two strains of sulfur and non-sulfur bacteria, Ferrera et al. were able to remove 95 - 97.9% of wastewater with an input sulfide level of 26–112 mgS/L [4]. Photosynthetic purple bacteria *Ectothiorhodospira magna* achieved a sulfide removal rate of 99.99% after 72 h of testing at pH 8.0, 35 °C [5].

The enzyme pathways involved in the oxidation of sulfur compounds SOB possess various enzymatic systems including five major different types: sulfide oxidation enzymes (S^{2-}); elemental sulfur oxidation enzymes (S^0); sulfite oxidation enzymes (${SO_3}^{2-}$); thiosulfate oxidation enzymes (${S_2O_3}^{2-}$); tetrathionate oxidation enzymes (${S_4O_6}^{2-}$) [17]. Compared to the other oxidation products, elemental sulfur (S^0) is the most ideal compound to be converted from H_2S; hence, the sulfide oxidizing enzyme system, which catalyzes the conversion of sulfide to elemental sulfur, has a strong potential in removing H_2S. This enzyme system consists of two main enzymes: sulfide:quinone oxidoreductase (SQR) and flavocytochrome c-sulfide dehydrogenase (FCC). SQR is a protein from the flavoprotein disulfide reductase (FDR) superfamily, has been discovered as a single polypeptide chain with a molecular mass of approximately 50–55 kDa and is generally related to the coenzyme FAD in biochemical studies [31]. FCC belongs to the class of oxidoreductase enzymes and uses the sulfur group as an electron donor. FCC is composed of a flavoprotein (fccB), which is for FAD binding, and a diheme cytochrome (fccA). Based on the peptide chain composition, FCC is identified as a soluble enzyme, mostly found in the periplasm [6], with the flavoprotein subunit having a molecular mass of approximately 40 kDa and the fccA at approximately 21 kDa [7].

There have not been much noticeable studies on SQR and FCC enzymes yet because of some difficulties encountered in the purification process. It is notably hard to achieve SQR and FCC alone in stable and active forms due to their complexity in structure [8, 9].

In Vietnam, several studies related to SOB group have been carried out, mostly with the main goal of being to treat wastewater contaminated by H_2S. An in depth research carried out by Do Thi Lien et al. [10] described 35 strains of photosynthetic purple bacteria isolated from mud and water samples in aquaculture water bodies around Vietnam, and investigated on the identification of isolates, the *sqr* gene characteristics, and cloning. Three isolates of the highest recorded in growth and sulfide removal activity were then applied in the testing and production of bioproducts to treat polluted water sources with positive results when the toxin content was significantly reduced [10]. Other photosynthetic purple bacteria products, such as vasoexcitor material (VEM), are capable of oxidizing sulfides in the bottom sediment of aquaculture ponds [11].

Nevertheless, there have been few molecular biology investigations of SOB in Vietnam to date, particularly with regard to SQR and FCC enzymes. In order to better understand the sulfur oxidizing activity of the two purple strains that were isolated from wastewater samples in Vietnam, we conducted this investigation with the goal of identifying the genes encoding key enzymes in sulfur metabolic pathways.

2 Materials and Methods

2.1 Materials

The strain *Rhodobacter sediminis* PAM 37 was isolated and stored at GREENLAB (Laboratory of Bioelectrochemical Technologies for Green Growth), University of Science, Vietnam National University (HUS-VNU), Hanoi, Vietnam [12].

2.2 Methods

2.2.1 Primer Design

Due to the lack of research data on *R. sediminis* and the high similarity in 16S rRNA encoding region between *R. sediminis* and *R. capsulatus* strains [13], the primers were designed based on the published sequences of *R. capsulatus* species. All sequences used for primer design were obtained from KEGG and NCBI database. The designed primers were aligned with conservative sequences outside the opening reading frame (ORF).

2.2.2 Total DNA Extraction

The bacterial cells were harvested from the petri disk into a 1.5 ml tube containing 200 μl CTAB buffer. The sample was mixed with 20 μl of lysozyme (30 mg/ml) and incubated at 37 °C for 1 h. Subsequently, 60 μl of 10% SDS and 10 μl of Proteinase K were added and immediately incubated at 55 °C for 2 h. After incubation, 300 μl of CI solution (Chloroform:Isoamyl with a ratio of 24:1) was added into the mixture, the tube was then centrifuged at 12000 rpm, 4 °C for 10 min. The supernatant in the top layer was collected into a fresh tube with 1 equal volume of cold Isopropanol, and left on ice for at least 1 h. The mixture was centrifuged at 12000 rpm, 4 °C for 10 min, then the white pellet was kept and washed twice with 1 ml of 70% cold Ethanol. The sample was left to dry at room temperature afterward, then resuspended in TE buffer, and stored at -20 °C for future use. The extracted DNA was checked on agarose gel under the UV light. This method was also used to examine PCR products.

2.2.3 PCR for *sqr* and *fccb* Gene Amplification

Components in a PCR mixture contained 0.416 μM for each primer, 2x DreamTaq PCR Master Mix (Thermo Scientific, USA). Thermal cycle for amplification of sqr gene included 95°C for 3 min; 35 cycles of 95°C for 45 s, 60.2°C for 30 s, 72°C for 90 s; 72°C for 8 min. Thermal cycle for amplification of fccb gene included 95°C for 3 min; 35 cycles of 95°C for 45 s, 61°C for 30 s, 72°C for 90 s; 72°C for 8 min. After the gel electrophoresis examination, the PCR products were sequenced by 1st BASE (Malaysia) to compare with available sequences in NCBI Genbank.

2.2.4 In Silico Analysis on Gene and Putative Protein Sequences

The PCR amplicons which were sent for sequencing by 1st BASE (Malaysia were analyzed using Bioedit software. Similar DNA sequences were searched on NCBI Genbank via the BLASTn tool while sequences of homologous amino acids were searched using the BLASTp tool. Alignment results of deduced amino acid sequences from other species were performed using the MUSCLE tool (EMBL-EBI, European Molecular Biology Laboratory, UK). Species sequences with higher than 95% identity were chosen for phylogenetic analysis. The phylogenetic tree constructions of sqr and fccb sequences used Molecular Evolutionary Genetics Analysis (MEGA) software version 11, based on the Maximum Likelihood method, General Time Reversible (GTR) model with bootstrap values of 500.

2.2.5 Cellular Fractionation of Purple Photosynthetic Bacteria

Cell pellets were harvested from liquid cultures when the optical density value at 600 nm reached 1 and then followed the procedure described by Klotz (1992) to collect three different cellular fractions: the periplasmic, cytoplasmic, and the membrane fractions. The periplasmic and the cytoplasmic fluids were concentrated using ethanol precipitation method. The Bradford assay was carried out to determine the protein concentration of all the samples, including three extracted fractions and the total protein fluid. To examine the presence of these enzymes in their localization, the fractions were subjected to SDS-PAGE and visualized by Coomassie Blue staining.

2.2.6 Sulfide Oxidase Activity Assay

The enzyme activity was determined through the measuring of the enzymatic reaction product in the form of sulfate (Hirano et al. 1996) method. The reaction mixture contained 0.45 ml of 0.1 M sodium acetate (pH 5.6) and 0.1 ml sample contained the fraction, then 0.05 ml freshly prepared sodium sulfide solution (0.06 g Na_2S, 0.16 g NaOH, 0.02 g $Na_2EDTA.2H_2O$, 2 ml glycerol, and 40 ml distilled water). The mixture was incubated at 30°C for 30 min and stopped with the addition of 0.15 ml NaOH 1.0 M. After the addition of 0.005 M barium chloride solution and 60 s of constant vortex, the barium sulfate precipitate would form in the mixture and was proportional to the sulfate concentration in the sample. The turbidity was measured by the spectrophotometer at 450 nm to evaluate the concentration of sulfate ions formed during sulfide oxidase assay. One unit of sulfide oxidase activity was defined as an amount of the enzyme required to produce 1 μmol sulfate per hour/ml (U/ml).

An experiment with an inhibitor was performed simultaneously, in which the protein fraction was mixed with 0.02 M KCN for 5 min before performing the same procedure as previously described. The pH activity profile of SQR and FCC were determined between pH 4.5 and 9.5 through the adjustment of sodium acetate buffer at 30 °C. The optimum pH value obtained from this assay was used in the temperature activity profile. The effect of temperature on SQR activity was measured at 6 temperature points between 20 °C and 90 °C.

3 Results and Discussion

3.1 Primer Design and Gene Amplification

Table 1. Primer sequences used in this study for *R. sediminis* PAM37

Gene	Strain	Amplicon size	Primer name	Primer sequence (5' - 3')
sqr	PAM37	1374 bp	sqr-37F	CAAGAAGGCCGAGCGGATGATG
			sqr-37R	ACATCCTCGCCTTCCGCTTCG
fccb	PAM37	1216 bp	fcc-37F	ATCTAYCGCCCGATCCATCCC
			fcc-37R	CCGATCTGGGYGATGAAAATGTCG

With specific designed primers (in Table 1), the amplified products on gel electrophoresis of *sqr* and *fccb* genes of strain PAM37 were approximately 1400 bp and 1200 bp, respectively, which matched the expected sizes without nonspecific bands (Fig. 1). These results indicated the initial success in the amplification of the two target genes. For further confirmaton, the PCR products were sequenced by 1st BASE. The obtained nucleotide sequences were deposited in the NCBI GenBank database under accession numbers PX677470 for the *sqr* gene and PX677471 for the *fccb* gene.

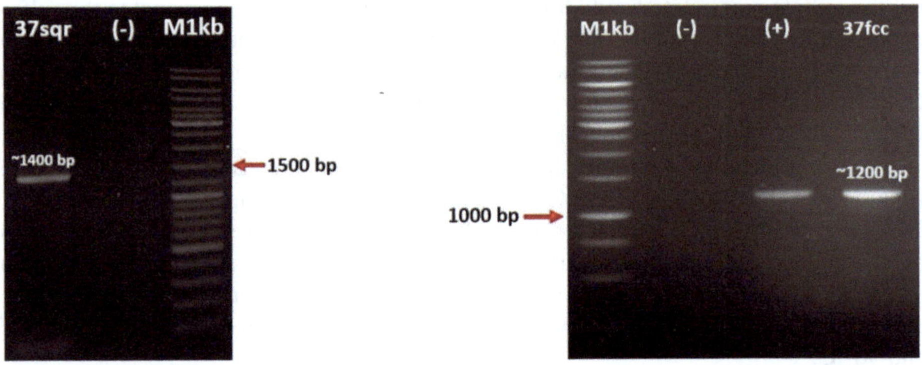

Fig. 1. Gel electrophoresis results of PCR products of genes *sqr* (left) and *fccb* (right) of *R. sediminis* PAM37

Using NCBI BLAST tool, the nucleotide sequences with high similarity to the coding sequences of *sqr* and *fccb* in *R. sediminis* strain PAM37 were shown in Fig. 2. Based on the results, it can be seen that the query sequence had a fairly high similarity with the *sqr* and *fccb* sequences in other strains of the same species with the percentage of identity over 90%. The results in Fig. 2 showed that the *sqr* and *fccb* sequences of this strain belonged to its respective species, especially, the sequence of *R. sediminis* PAM37 and other *R. capsulatus* strains were highly related. To further confirm, the deduced amino acid sequences from DNA sequences were simutalneously analyzed to find important domains of the enzymes.

PAM37 - *fccb* gene

Description	Scientific Name	Max Score	Total Score	Query Cover	E value	Per. Ident	Acc. Len	Accession
Rhodobacter capsulatus SB 1003, complete genome	Rhodobacter ca...	2046	2046	100%	0.0	98.91%	3738958	CP001312.1
Rhodobacter capsulatus strain A12 chromosome, complete genome	Rhodobacter ca...	1463	1463	100%	0.0	89.84%	3616128	CP061202.1
Rhodobacter capsulatus strain 37b4 chromosome, complete genome	Rhodobacter ca...	1452	1452	100%	0.0	89.66%	3912233	CP119563.1
R.capsulatus genes rnfA - rnfF, fdxC, fdxN, Orf14 and Orf 10	Rhodobacter ca...	760	1420	91%	0.0	93.30%	12480	X72888.1
R.capsulatus rnfA, rnfB, rnfC, rnfD, rnfG, rnfE, rnfH genes	Rhodobacter ca...	239	239	14%	1e-57	96.18%	5700	Y11913.1

PAM37 - *sqr* gene

Description	Scientific Name	Max Score	Total Score	Query Cover	E value	Per. Ident	Acc. Len	Accession
Rhodobacter capsulatus SB 1003, complete genome	Rhodobacter ca...	2027	2027	99%	0.0	97.31%	3738958	CP001312.1
Rhodobacter capsulatus strain 37b4 chromosome, complete genome	Rhodobacter ca...	1796	1796	99%	0.0	93.84%	3912233	CP119563.1
Rhodobacter capsulatus strain A12 chromosome, complete genome	Rhodobacter ca...	1750	1750	99%	0.0	93.14%	3616128	CP061202.1
Rhodobacter sp. LPB0142, complete genome	Rhodobacter xa...	590	590	93%	1e-163	76.49%	3462926	CP017781.1
Ciceribacter thiooxidans strain F43B chromosome, complete genome	Ciceribacter thio...	296	296	48%	4e-75	75.85%	3661327	CP059896.1
Rhodopseudomonas palustris strain ATCC 17000 chromosome, complete genome	Rhodopseudom...	283	349	77%	5e-71	73.10%	5305909	CP140155.1
Martelella sp. AD-3, complete genome	Martelella sp. AD-3	273	273	53%	4e-68	74.35%	4554702	CP014275.1
Rhodopseudomonas palustris HaA2, complete genome	Rhodopseudom...	267	339	77%	2e-66	72.73%	5331656	CP000250.1
Rhodopseudomonas palustris strain PS3 chromosome, complete genome	Rhodopseudom...	258	322	53%	2e-63	74.95%	5269926	CP019966.1
Martelella lutilitoris strain ZRK001 chromosome, complete genome	Martelella lutilitoris	256	256	53%	6e-63	73.86%	4374534	CP066786.1
Rhodopseudomonas palustris strain P1 chromosome, complete genome	Rhodopseudom...	252	316	53%	9e-62	74.77%	5318559	CP088006.1
MAG: uncultured proteobacterium isolate MFD03044 bin.1.21 genome assembly, chromosome: 1	uncultured Pseu...	248	248	57%	1e-60	73.18%	3999070	OY969816.1
Rhodopseudomonas palustris strain KTSSR54 chromosome, complete genome	Rhodopseudom...	241	305	53%	3e-58	74.40%	5330578	CP076678.1
Rhodopseudomonas palustris strain VNW64 chromosome, complete genome	Rhodopseudom...	241	299	53%	3e-58	74.40%	5462690	CP076679.1
Rhodopseudomonas palustris strain PP803 chromosome, complete genome	Rhodopseudom...	241	305	53%	3e-58	74.40%	5316238	CP076677.1
Stenotrophomonas acidaminiphila strain T25-65 chromosome, complete genome	Stenotrophomon...	229	229	56%	8e-55	72.77%	3915662	CP043570.1
Rhodopseudomonas palustris strain YSC3 chromosome, complete genome	Rhodopseudom...	223	293	53%	4e-53	73.84%	5371816	CP019967.1
Rhodopseudomonas palustris strain TLS06 chromosome, complete genome	Rhodopseudom...	223	293	53%	4e-53	73.84%	5194203	CP076676.1
Rhodopseudomonas palustris strain GJ-22 chromosome, complete genome	Rhodopseudom...	212	212	57%	1e-49	72.67%	5042906	CP041387.1

Fig. 2. Sequence comparison using BLAST tool (NCBI) of genes *sqr* (lower) and *fccb* (upper) of isolated *R. sediminis* strain PAM37

3.2 Protein Sequence Analyses of *R. sediminis* Strain PAM37

The deduced protein sequence of SQR was calculated to be 376 amino acids in length. The phylogenetic tree showed that the SQR of *R. sediminis* strain was closely related to the SQR of *R. capsulatus* and could belong to the SQR type VI (Fig. 3).

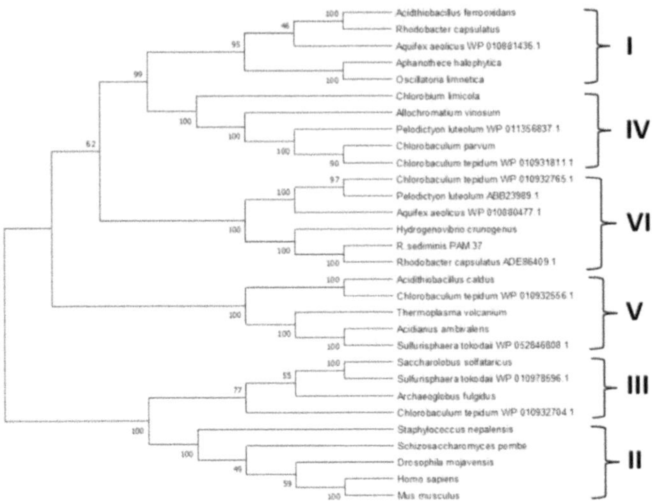

Fig. 3. Phylogenetic tree of six SQR types based on amino acid sequences. The comparative analysis consists of 30 amino acid sequences of SQR groups of different species.

Although there were differences in some amino acid position, the SQR sequence of *R. sediminis* PAM37 with the three FAD-binding motifs of type VI SQR were conserved, and was highly similar to *R. capsulatus*. This indicated that *R. sediminis* and *R. capsulatus* were more closely related to each other than the other species. In SQR FAD-binding motifs, the results in Fig. 4 were shown the stability of FAD-binding domain which were conserved in all 3 motifs and other residues. Accordingly, the deduced amino acid FCC sequence was also conserved in *R. sediminis* PAM 37, except for the lack of Cysteine residues at the execution site function, possibly due to mutation or interference of FCCB from other pathways.

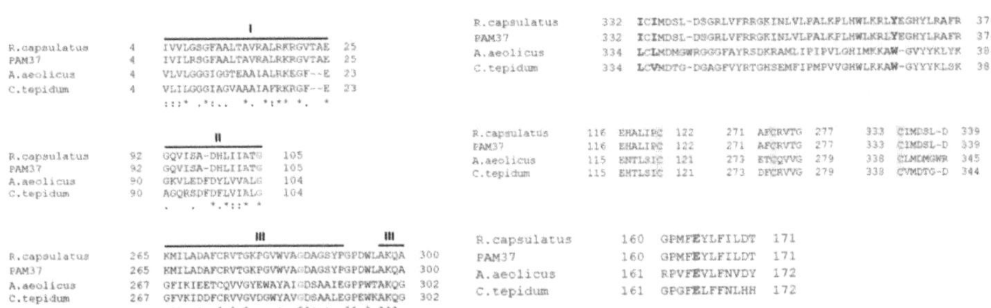

Fig. 4. Three conserved FAD-binding motifs in type VI SQRs of *R. sediminis* strain PAM37. Residues participating in quinone binding of SQR VI. Cysteine residues involved in sulfide oxidation in type VI SQR. The glutamic acid residue was conserved in SQR type VI.

Some other residues are conserved in type VI SQR such as G105, which is the end position of the β-fold structure, A103 is predicted to interact with the phosphate residue attached to the ribityl chain of FAD, and G285 is related to the Cα chain containing phosphate group of FAD. Based on the SQR type VI sequences identified in *A. aeolicus* and *C. tepidum,* I332, I334 and Y367 possibly played roles in quinone binding as marked in red bold letters (numbered according to amino acid order of *R. sediminis* PAM37).

The three Cys residues involved in sulfide oxidation were C122, C272 and C333, which C122 is responsible for covalent binding with the FAD cofactor, and C333 plays a role in the semi-oxidation reaction by binding and oxidizing sulfur. While the exclusive C272 in VI participates in the catalytic process, ensuring that the sulfide oxidation process occurs normally like other SQRs. Residue E165 marked in bold purple is conserved in *R. sediminis* near the N5 position of FAD and was found to cause a pK shift of sulfite bound to FAD.

The deduced FCCB protein sequence had 340 amino acids, starting from methionine (M) and ended in alanine (A). Motif I had 6 conserved amino acid residues, motif II had 5 conserved amino acids, and motif III had 9 conserved amino acids (Fig. 5).

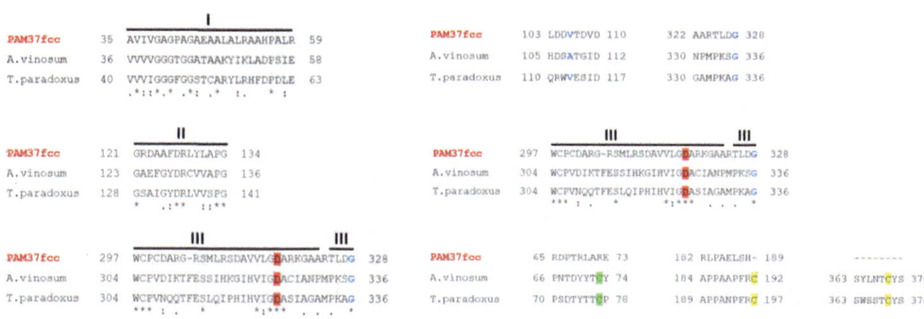

Fig. 5. Three FAD-binding motifs are conserved in FCC of *R. sediminis* PAM37. Glutamic acid residue was conserved in FCCBs. Several amino acid residues stabilized FAD in FCCB. Proprietary aspartate base available only from FCCs. Cysteine residues were conserved in FCCB.

Position E64 of *R. sediminis* was marked with bold orange letters that plays a role in creating a hydrogen bond with the O3 atom of FAD. It was found that position E63 of *R. sediminis* was similar to the E65 residue of *A. vinosum,* but in *T. paradoxus* it was the N residue. Some conserved residues in *R. sediminis* play a role in keeping the configuration of FAD stable, such as: V106, G328 (marked in bold green letters). The G328 residue was conserved in all three FCCs, but the V106 residue in *A. vinosum* was replaced by the Alanine residue. The number of absolute conserved amino acid residues in FCCB was not much, which indicated that FCCBs from different species did not have high similarity.

In particular, in the FCC of *R. sediminis,* there was an aspartate at position D317 (corresponding to D324 in *Allochromatium vinosum*) that was believed to have the function of interacting with the ribityl chain of FAD, which appeared in all glutathione reductase family flavoproteins but not appear in SQR. In theory, FCCB have a Cysteine residues forming a covalent bond with FAD and two other Cysteine residues forming a redox couple in the active site dynamic. However, when aligning the deduced amino acid sequence of *R. sediminis* with the known FCC sequence, Cysteine residues did not appear at the execution site function. This was caused due to the function of the FCCB by another mechanism, or due to mutation and failure to perform function sulfur oxidation capacity.

3.3 Evidence for the Presence of *sqr* and *fcc* in *R. sediminis* Strain PAM37

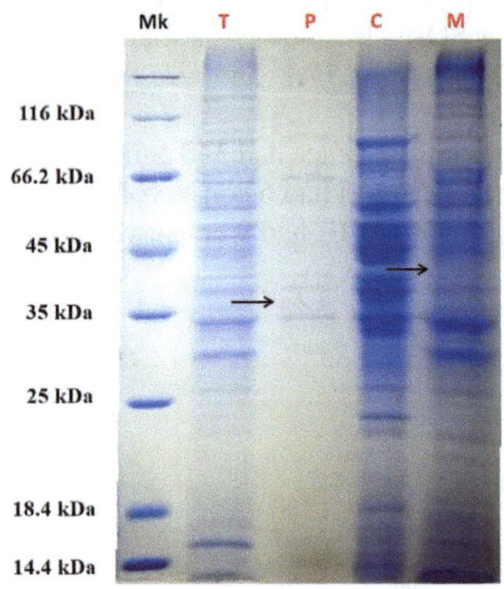

Fig. 6. SDS-PAGE electrophoresis of SQR and FCC proteins in different cellular fractions of *R. sediminis* PAM 37. Mk: Unstained protein marker; T: Total cell lysate; P: Periplasmic fraction; C: Cytoplasmic fraction; M: Membrane fraction; Arrow: possible position of targeted proteins (FCC in P and SQR in M).

In Fig. 6, the number of protein bands found in the total cell lysate sample was the highest, followed by the cytoplasmic(C), membrane (M), and lastly, periplasmic (P) fractions. The band patterns of M fraction resembled those of T fraction, possibly indicating the less effective sonication lysis. It has been reported that in bacterial cells, for example, *E. coli* K12, the proteins in the cytoplasmic fraction account for more than 66%, and the membrane and periplasmic fractions for 26% and 8%, respectively. This is consistent with what we found in this research.

In the periplasmic fraction, there was a protein band at the position of approximately 40 kDa. According to theoretical calculations based on the FCC gene sequences, FCC sizes were 37 kDa. Similarly, in the membrane fractions, there were bands of approximately 48 kDa in size, compared to the calculated 41 kDa. There were differences between the calculated MW and the actual positions of protein bands because of many reasons, including the protein structure, post-translation modifications, aggregation, degradation, etc. For example, the SQR of *R. capsulatus* should be 47 kDa based on the number of amino acids; however, when recombined with E. coli and analyzed by SDS-PAGE electrophoresis, the result was 55 kDa. Another problem was the possibility of different proteins with close/similar positions, making it hard to find the targeted proteins. Thus, even though SDS-PAGE electrophoresis was used to initially investigate the presence of FCC and SQR in protein fractions, further experiments were needed to confirm the presence of enzyme SQR and FCC in their respective compartments.

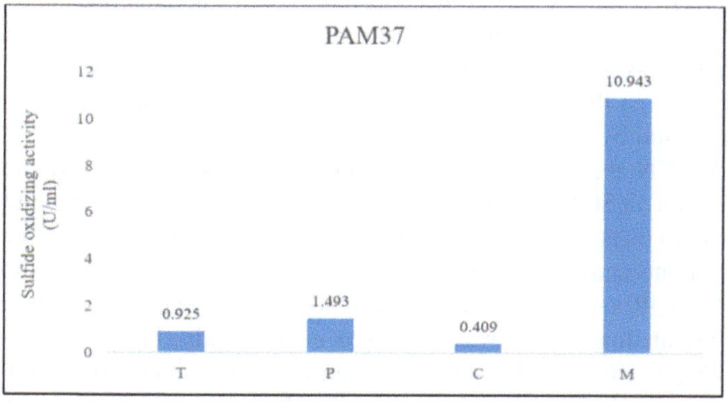

Fig. 7. Enzyme activities of different fractions from *R. sediminis* strain PAM37. T: Total cell lysate; P: Periplasmic fraction; C: Cytoplasmic fraction; M: Membrane fraction.

Overall, the membrane fractions had the highest activity with 11.118 U/ml, and the lowest records were found in cytoplasmic fractions, with an activity of 0.409 U/ml (Fig. 7). There were also enzyme activities recorded in the periplasmic fraction at 1.493 U/ml, which was significantly low compared to other fractions. There was also low activity recorded in the total lysates sample despite its high concentration and number of proteins, which could imply that our enzymes account for a small part of the total protein of the bacteria. The activity in cytoplasmic fractions was low, which meant each fractions were separated from each other. The membrane fractions had much higher enzyme activity compared to the periplasmic fraction, which implied that the SQR enzyme has stronger activity than the FCC enzyme. To further confirm the presence of the target enzymes, the inhibitory test using KCN was carried out (Fig. 8).

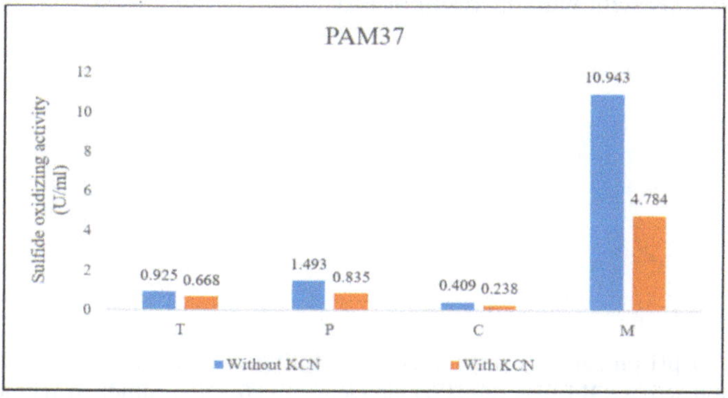

Fig. 8. Sulfide oxidation activities of *R. sediminis* strain PAM 37 in the presence of KCN. T: Total protein sample, P: Periplasmic fraction, C: Cytoplasmic fraction, M: Membrane group

In both periplasmic and membrane fractions, the activity was decreased by more than half compared to the normal ones. The membrane fraction activity went from 10.943 to 4.784 u/ml, equal to 56% of the membrane fraction was inhibited, and the periplasmic fraction went from 1.493 to 0.835 u/ml, equal to 44% of the activity had gone. Meanwhile, the activity in both the total lysates sample and the cytoplasmic fraction decreased by only 27%, much less than the other two samples. These results indicated that there is a possibility of FCC in the periplasmic and SQR in the membrane fraction, and both of them were inhibited by KCN.

Other than KCN, there are several other inhibitors for SQR, mostly quinone-analogs. The research done on SQR from *Oscillatoria limnetic* compared the inhibitor sensitivity of SQR activity in its isolated form with the membrane-bound SQR [4]. Aurachin C and NQNO, as quinone-analogs, were used in this experiment. The results show that both analogs inhibit SQR and SQR membrane-bound, which means they remained sensitive toward quinone-analogs. KCN was also used in this study; however, they only did it with isolated SQR. KCN needed a much higher concentration (12000 nM) to reach the same result compared to its counterparts (< 72 nM). So researchers in the future may explore the inhibition ability of quinone-analogs on membrane fraction to paint a more complete picture.

With FCC, certain substances have been investigated to find more inhibitors; yet there is not much promising news, therefore KCN remains the most common FCC inhibitor [14].

Furthermore, in a range of pH 4.5 to 9.0, the enzyme activity in the periplasmic fraction was present in all tested pH values with the lowest activity at pH 4.5 (0.04 U/ml), suggesting that the enzyme activity in the periplasmic fraction could be inhibited completely at pH lower than 4.5. The highest recorded was found at the pH of 7.5 at 4.90 U/ml. With the presumption that this fraction activity came from the enzyme FCC, the optimum pH that we found was lower than the previously found optimum pH of FCC, which was 8.6. From Fig. 9B, it seemed that SQR in the membrane fractions of PAM 37 was still active at a broader range of pH, maybe even outside of our chosen one, and the highest activity was recorded at a pH of 6.5 with the result of 16.21 U/ml. The optimum pH for purified SQR enzyme active is 7.0 in previous study [15], which was higher than our result and consistent with the result of FCC.

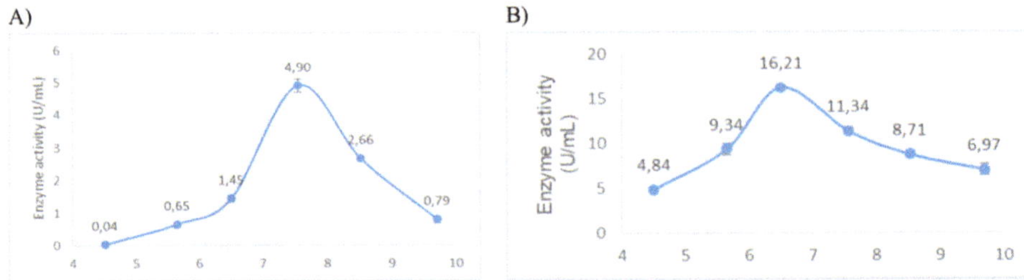

Fig. 9. Effect of pH on sulfide oxidation activity. **A** Effect of pH on the periplasmic fraction sulfide oxidation activity. **B** Effect of pH on the membrane fraction sulfide oxidation activity

The results in Fig. 10 showed the highest enzyme activity of the periplasmic fraction was 7.20 U/ml and was achieved at 60 °C. Meanwhile, the highest activity of the membrane fraction was 15.03 U/ml at 30 – 40 °C, the same range for PAM 37's best growth condition.

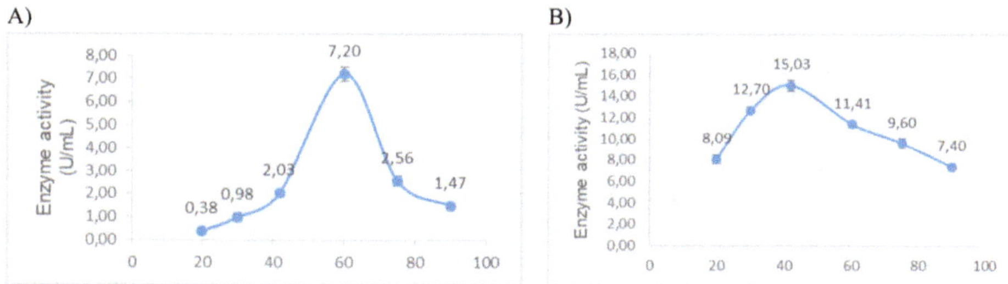

Fig. 10. Effect of temperature on sulfide oxidation activity. **A** Effect of temperature on the periplasmic fraction sulfide oxidation activity. **B** Effect of temperature on the membrane fraction sulfide oxidation activity

4 Conclusions

In tropical countries, especially Vietnam, the average temperature varies depending on geographical configuration. From this research, the enzyme SQR of *R. sediminis* PAM 37 can work effectively anywhere, but the same statement could not be assumed on the FCC enzyme. With the possible mechanism of sulfide oxidizing capability of *R. sediminis* PAM37, this bacterial strain could be further optimized regarding working conditions for applications of H2S bioremediation.

References

1. Environmental toxicology of hydrogen sulfide: Malone Rubright S.L., L.L. Pearce. J. Peterson Nitric Oxide **71**, 1–13 (2017)
2. Szabo, C., C. Ransy, K. Módis, M. Andriamihaja, B. Murghes, C. Coletta, G. Olah, K. Yanagi, F. Bouillaud: Regulation of mitochondrial bioenergetic function by hydrogen sulfide. Part I. Biochemical and physiological mechanisms. Br J Pharmacol, 171 (8), pp. 2099–2122 (2014)
3. Osha: "Hydrogen Sulfide: Health Hazards."(2014)
4. Ferrera, I., et al.: High-diversity biofilm for the oxidation of sulfide-containing effluents. Appl. Microbiol. Biotechnol. **64**, 726–734 (2004)
5. Zhao, X., et al.: Enhancement of COD, ammonia, phosphate and sulfide simultaneous removal by the anaerobic photosynthetic bacterium of *Ectothiorhodospira magna* in batch and sequencing batch culture. Water Sci. Technol. **78**(9), 1852–1860 (2018)
6. Dolata, M.M., Van Beeumen, J.J., Ambler, R.P., Meyer, T.E., Cusanovich, M.A.: Nucleotide sequence of the heme subunit of flavocytochrome c from the purple phototrophic bacterium, *Chromatium vinosum.* A 2.6-kilobase pair DNA fragment contains two multiheme cytochromes, a flavoprotein, and a homolog of human ankyrin. J. Biol. Chem, **268** (19), pp. 14426–14431 (1993)
7. Osipov, E.M., et al.: Structure of the flavocytochrome c sulfide dehydrogenase associated with the copperbinding protein CopC from the haloalkaliphilic sulfur-oxidizing bacterium *Thioalkalivibrio paradoxus*ARh 1. Acta Crystallogr D Struct. Biol. **74**(Pt 7), 632–642 (2018)

8. Duzs, Á., Miklovics, N., Paragi, G., Rákhely, G., Tóth, A.: Insights into the catalytic mechanism of type VI sulfide:quinone oxidoreductases. Biochim. Biophys Acta Bioenerg **1862**(2), 148337 (2021)
9. Kusai, A., Yamanaka, T.: Cytochrome c (553, *Chlorobium thiosulfatophilum*) is a sulphide-cytochrome c reductase. FEBS Lett. **34**(2), 235–237 (1973)
10. Lien D.T.: Research on the application of purple phototrophic bacteria for sulfide removal in polluted water sources, Doctoral Thesis (in Vietnamese) (2016)
11. Lien, D.T., Phuong, N.T.D., Thuy, N.T.B., Uyen, D.T.T., Khang, D.D.: Effects of bioproduct of purple photosynth bacteria on environmental quality of pond under intensive tilapia farming. J. Sci. Dev., **12**(3), pp. 379–383 (in Vietnamese) (2014)
12. Minh, D.B., Hai P.T.: Isolation and selection of purple nonsulfur bacteria for nutrient rich biomass production from wastes. VNU J. Sci., **37**(2) (2021)
13. Subhash, Y., Lee, S.S.: *Rhodobacter sediminis* sp. nov., isolated from lagoon sediments. Int. J. Syst. Evol. Microbiol, **66**(8):2965–2970 (2016)
14. Visser J.M., G.a.H. De Jong, L.A. Robertson, J.G. Kuenen: A novel membrane-bound flavocytochrome c sulfide dehydrogenase from the colourless sulfur bacterium *Thiobacillus* sp. W5. Arch Microbiol, **167** (5), pp. 295–301 (1997)
15. Wakai, S., Tsujita, M., Kikumoto, M., Manchur, M.A., Kanao, T., Kamimura, K.: Purification and characterization of sulfide:quinone oxidoreductase from an acidophilic iron-oxidizing bacterium, *Acidithiobacillus ferrooxidans*. Biosci. Biotechnol. Biochem. **71**(11), 2735–2742 (2007)

Making Healthy Foods from Plant By-Products with Sun Drying: Bamboo Leaf Tea & Dried Shoot—A Case Study

Nguyen Thi Nhu Tam[1], Hoang Hieu Nga[2], Pham Duc Trinh[3], Hoang Ngoc Nhung[4], and Tran Chi Thanh[5](✉)

[1] International School, Vietnam National University, Ho Chi Minh, Vietnam
[2] Vietnam National University, University of Law, Ho Chi Minh, Vietnam
[3] College of Foreign Economic Relations, Ho Chi Minh City, Vietnam
[4] HUTECH University, Ho Chi Minh, Vietnam
[5] Sac Moc Tinh Co., Ltd., Ho Chi Minh, Vietnam
sacmoctinh@gmail.com

Abstract. Sun-drying is a traditional preservation technique commonly used in rural Vietnam. However, direct exposure to sunlight can degrade sensory qualities and pose food safety risks. This study evaluates key challenges in traditional sun-drying and proposes simple, practical improvements. Methods included field observations, and sample analysis conducted at NaKa Garden Eco-Agriculture Cooperative in Moc Chau, Son La. Two bamboo-based products—bamboo leaf tea and dried bamboo shoots—were compared using traditional and improved drying methods involving a 2.5% acid or alkaline salt treatment. Results indicated that improved products exhibited superior color, flavor, and aroma. Notably, white dried bamboo shoots had significantly higher protein, starch, and fat content than yellow shoots. Bamboo leaf tea met microbiological and heavy metal safety standards (TCVN 7975:2008). The semi-enclosed sun-drying technique with pH treatment effectively reduced microbial contamination while preserving nutritional quality, all without machinery. In conclusion, this improved method is feasible, cost-effective, and scalable for rural cooperatives, enhancing product quality for both domestic consumption and export markets.

1 Introduction

Sun-drying food, especially post-harvest agricultural produce, is an ancient preservation technique used for thousands of years due to its low cost and high efficiency. It is particularly suitable for small-scale agricultural areas because of its affordability, ease of implementation, and the availability of abundant solar energy. However, directly sun-drying food under sunlight can negatively affect the color, aroma, and nutritional value of the product [1].

In Vietnam, many agricultural products such as rice, corn, pepper, coffee, and fruits are dried directly on yards, tarps, or roadsides—especially in the Central and Central

L. Vu et al. (Eds.): PBC 2025, 2026.
https://doi.org/10.1007/978-3-032-29579-8_8

Highlands regions [2]. However, this method has many limitations: it depends on weather conditions, lacks temperature and humidity control, and is susceptible to contamination by microorganisms and impurities, which can compromise food quality and safety [3, 4].

In the context of climate change, increasing market demand, and strict food safety standards for export, researching, improving, and proposing solutions for traditional sun-drying methods is necessary. Specifically in Vietnam, Article 10 of the Food Safety Law (No. 55/2010/QH12) stipulates that food must be produced, processed, preserved, and transported under hygienic and safe conditions. This applies to all processing facilities, including households and cooperatives, requiring elevated, covered platforms with measures to prevent insects, dust, and environmental contamination [5]. Circular 17/2018/TT-BNNPTNT further specifies that drying areas must be clean, elevated, and not directly on the ground to avoid cross-contamination. Protective measures such as nets or shelters must be used when drying outdoors [6]. In addition, Vietnamese Standard TCVN 5603:2008 (Codex—HACCP) strongly recommends preventive microbiological controls in traditional food processing, particularly during drying. Without proper safeguards, products dried under direct sunlight face high risks of microbial contamination from pathogens like E. coli, Salmonella, or molds [7].

Potential solutions include improved drying techniques (e.g., solar houses, elevated racks, semi-enclosed drying), controlled use of solar energy, and guidance for farmers in applying good practices in processing, drying, and storage [8].

2 Research Objectives

1. Describe the main challenges related to the quality, safety, and efficiency of the sun-drying process.
2. Evaluate technical solutions to improve the quality and safety of traditional sun-drying methods.

3 Methodology

The pH treatment method developed by Sac Moc Tinh Co., Ltd. Effectively removes anti-nutritional compounds in plant by-products while preserving valuable nutrients. This method enables the transformation of discarded plant materials into safe and nutritious food products [9].

This study used a descriptive approach and measured technical parameters in the sun-drying process for two specific agricultural products: bamboo leaf tea and white dried bamboo shoots, conducted at the NaKa Garden Eco-Agricultural Cooperative in Moc Chau, Son La.

- Traditional sun-drying protocol: Plant materials are directly put on tray or spare land after being rinsed for sun drying. The total time of drying is vary depending on the weather, hence the plant materials are defined "dried" by farmers' feeling of drying things by hand. In our study, we requested to dry at least 2 days with an average of 10 h of sun drying.

- pH treatment protocol was mentioned in our previous publication at the reference number 9 in the reference list and the alkaline/acidic salts formula is kept confidentially for the future business purpose.
- Product sampling: Dried products (bamboo leaf tea and white dried bamboo shoots) were collected with 3 samples for laboratory test and 1 sample for the confirmatory test to compare results between traditional and improved pH-treated drying methods from Sac Moc Tinh Co., Ltd. All samples were analyzed for microbiological (Coliform, mold with plate count method); moisture, and impurity indicators per FAO/WHO (2014) standards; Tannin (AOAC 955.35, 2019); and Heavy metals (AOAC 999.11 and 971.21).

Data Analysis

- Quantitative: SPSS software was used to process descriptive statistics (means, frequencies) and compare traditional versus improved groups using t-tests or ANOVA.
- Qualitative: Interview content was analyzed using thematic coding to identify key challenges and technical improvement needs.

4 Results and Discussion

(1) Comparison of Product Characteristics Between Two Sun-Drying Methods

Table 1 summarizes differences in sensory characteristics between traditional processing and traditional processing with raw material treatment for two bamboo-based products: bamboo leaf tea and dried bamboo shoots. For bamboo leaf tea, the improved method enhanced color (green vs. brown), taste (sweet vs. bitter), and aroma (distinct bamboo leaf scent vs. burnt, sun-exposed smell). For dried bamboo shoots, the improved method produced lighter colored products (ivory-brown), thinner slices that are easier to use, with natural sweetness and the aroma of fresh bamboo—features not achieved by traditional methods. These findings confirm the positive effects of raw material treatment in improving the sensory quality of bamboo-based products.

(2) Evaluation of Bamboo Leaf Tea Quality (TCVN 7975)

Table 2 presents physical and microbiological parameters of bamboo leaf tea (*Bambusa vulgaris*) compared to the limits set by TCVN 7975:2008. Results show that the product meets safety standards: total aflatoxin and *Salmonella* spp were not detected; Coliform and mold/yeast levels were significantly below permissible thresholds. The tea had low moisture (3.37%) ensuring preservation, and contained a moderate tannin content (0.60 g/100g)—a compound beneficial for health. Although the total ash content (15.9%) exceeded the standard based on tea leaves from *Camellia sinensis*, this may be attributed to the natural properties of bamboo leaf which contains high silica content, often as a major component of the inorganic fraction, along with other oxides such as CaO, MgO, K_2O, and P_2O_5. These mineral oxides do not decompose during burning and thus contribute substantially to the total ash yield [10].

Table 1. Product characteristics between two sun-drying methods

Product—Attributes	Traditional Sun drying	Traditional Sun drying with pH treatment
Bamboo leaf tea Color Flavor Aroma	Brownish-yellow Bitter Burnt, sun-scorched smell	Green Sweet Distinct bamboo leaf aroma
Dried bamboo shoots Color Texture Flavor Aroma	Brownish-yellow Whole dried bamboo shoots Mildly sweet and slightly salty; absorbs flavors during cooking No aroma of fresh bamboo shoot	Ivory-brown Thinly sliced dried bamboo shoots Naturally sweet; retains the distinctive taste of fresh bamboo Distinct aroma of fresh bamboo shoot

Heavy metals (Pb, Cd, As) were found at safe, low levels, and mercury (Hg) was not detected. Overall, the product met food safety standards and shows potential as a functional food from bamboo.

Table 2. Bamboo leaf tea – lab test and quality assurance test results

Parameter	Lab Test		Quality Assurance (pH treatment)	Standard TCVN 7975: 2008
	No pH treatment	pH treatment		
Humidity (%)	3.55 ± 0.81	3.45 ± 0.72	3.37	10.0
Total ash (%)	14.20 ± 2.10	15.0 ± 1.50	15.90	8.0
Fiber (%)	17.50 ± 0.90	17.50 ± 0.80	18.00	NA
Tannin (g/100g)	0.72 ± 0.05	0.57 ± 0.04	0.60	NA
Pb (mg/kg)	NA	NA	0.116	2.0
Hg (mg/kg)	NA	NA	Not detected	0.05
Cd (mg/kg)	NA	NA	0.025	1.0
As (mg/kg)	NA	NA	0.043	1.0
Total Aflatoxin (ug/kg)	NA	NA	Not detected	No presence
Salmonella spp (Positive/25g)	Not detected	Not detected	Not detected	No presence
Coliforms (CFU/g)	3.9 x10^6 ± 1,1 x 10^6	< 10	< 10	1 x 10^3
Yeasts and molds (CFU/g)	7.8 x 10^5 ± 0.5 x 10^5	1,5 x 10^3 ± 0.7 x 10^3	1.1 x 10^3	1.0 x 10^4

* *NA = Not Available at our laboratory*

This simple processing method does not require machinery and utilizes direct solar energy for drying. At the same time, it complies with current quality standards for herbal products on the market, including indicators such as total ash content, fiber content, absence of heavy metals, absence of aflatoxins, and microbiological safety parameters. Specifically, Salmonella spp. Was not detected, coliform levels were below the threshold of 10 CFU/g, and yeast and mold levels were within acceptable limits. Notably, the tannin content in our product is lower than that reported in Arowolo's method ($0.6 < 0.77$) [11].

(3) **Nutritional Comparison Between Yellow and White Dried Bamboo Shoots**

Table 3 compares the basic chemical composition between yellow and white dried bamboo shoots in 3 samples. The results show that white bamboo shoots are significantly more nutritious than yellow ones in key indicators, particularly in protein (20.5 g/100g vs. 1.0 g/100g) and starch (34.5 g/100g vs. 2.0 g/100g). White bamboo shoots also contain higher fat and fiber, making them more energy-rich and digestion-friendly. Ash content in both types is similar, reflecting relatively stable mineral levels. These differences may be due to differences in raw materials, processing methods, or bamboo species, indicating the potential to improve nutritional value through improved processing technology.

Table 3. Nutritional comparison between yellow and white dried bamboo shoots ($n = 3$)

Parameter (g/ 100 g)	Yellow-dried Bamboo shoots	White-dried Bamboo shoots	p-value (Welch t-test)*
Total ash	7.27 ± 0.19	7.88 ± 0.25	0.03
Protein	1.20 ± 0.15	20.00 ± 0.50	< 0.001
Starch	2.00 ± 0.15	33.60 ± 0.87	< 0.001
Fat	Not detected	3.82 ± 0.95	NA
Fiber	26.56 ± 0.21	29.56 ± 0.06	< 0.001
Cyanuric compounds	Not detected	Not detected	NA

* *Welch t-test is used for small sample sizes*
* *NA = Not Available*

To improve the quality of current yellow dried bamboo shoots, a 2.5% acid salt solution was applied to fresh *Dendrocalamus velutinus* shoots. This was compared with a similar method without the acid treatment. Preliminary results show the white bamboo shoot method produced superior quality across all evaluated indicators. While both types had similar ash (8.05g vs. 7.13g) and fiber content (29.6g vs. 26.8g), white shoots had significantly higher protein (20.5g vs. 1.0g), starch (34.5g vs. 2.0g), and contained 3.82g fat, whereas yellow shoots had none. A major factor in the nutritional loss of yellow shoots is the traditional boiling time of up to 2 h, whereas white shoots only required 10 min of boiling. Notably, this method does not use machines and utilizes direct solar drying, helping retain nutrients during processing.

5 Conclusion

This study demonstrates that improving traditional sun-drying by using acid or alkaline salt treatment significantly enhances the sensory quality, nutritional value, and microbial safety of agricultural products. White dried bamboo shoots and bamboo leaf tea processed with the improved method meet food safety standards while retaining beneficial nutrients. The method is simple, cost-effective, and suitable for households and cooperatives in Vietnam.

Acknowledgment. The research team would like to thank NaKa Garden Eco-Agricultural Cooperative for supporting the research activities and providing relevant data.

References

1. Ankit, S., Abhishek, A., Amritanshu, S., Atul, S.: Chapter 9 - Solar applications for drying of agricultural and marine products. In *Solar Energy Advancements in Agriculture and Food Production Systems*, ScienceDirect, pp. 313–350 (2022). Available: https://www.sciencedirect.com/book/9780323898669/solar-energy-advancements-in-agriculture-and-food-production-systems?via=ihub=#book-description
2. Nguyen, T.L., Pham, V.K., Tran, M.D.: Drying practices and post-harvest losses of agricultural products in vietnam: a case study in central highlands. Vietnam J. Agric. Sci. Technol. **24**(3), 45–53 (2021)
3. Food and A. Organization, *Small-scale Postharvest Handling Practices: A Manual for Horticultural Crops.* Rome: FAO and UC Davis, (2014)
4. Müller, J., Heindl, A.: Drying of medicinal plants and herbs. Stewart Postharvest Rev. **2**(6), 1–6 (2006)
5. National Assembly of Vietnam.: *Law on Food Safety No. 55/2010/QH12.* Hanoi, Vietnam. (2010). Available: https://vanban.chinhphu.vn/default.aspx?pageid=27160&docid=93513
6. Ministry of Agriculture and Rural Development.: *Circular No. 17/2018/TT-BNNPTNT on food safety management for agricultural, forestry, and fishery products.* Hanoi, Vietnam. (2018). Available: https://thuvienphapluat.vn/van-ban/thuong-mai/Thong-tu-17-2018-TT-BNNPTNT-quan-ly-an-toan-thuc-pham-nong-lam-thuy-san-402744.aspx
7. Ministry of Science and Technology.: *TCVN 5603:2008 – Code of practice for general principles of food hygiene (HACCP – Codex).* Hanoi, Vietnam. (2008). Available: https://vanban.chinhphu.vn/default.aspx?pageid=27160&docid=151027
8. Esper, A., Mühlbauer, W.: Solar drying—an effective means of food preservation. Renew. Energy **15**(1–4), 95–100 (1998)
9. Nguyen, D.H., Hoang, N.N., Vuong, T.L.A., Nguyen, V.H., Tran, C.T.: Technology for stabilizing the quality of Chlorophyll and nutrition in the production of bamboo leaf tea. Artic. Sci. Technol. Dev. Vietnam. **10**(4), 12–16 (2024)
10. Lucas Henrique Pereira Silva, Jacqueline Roberta Tamashiro, Fábio Friol Guedes de Paiva, Luis Fernando dos Santos, Silvio Rainho Teixeira, Angela Kinoshita, Patricia Alexandra Antunes, Bamboo leaf ash for use as mineral addition with Portland cement. J. Build. Eng, **42**, 102769, (2021) ISSN 2352-7102, https://doi.org/10.1016/j.jobe.2021.102769
11. Arowolo, A., Taiwo, A., Adepoju, M.: Quality evaluation of tea produced from Bamboo, Plantain and Zobo Leaves. **6**(5), (2021)

Preliminary Assessment of Oligo-Chitosan's Impact on the Microbial Composition Within Soybean Rhizosphere Using Full-Length 16S rRNA Sequencing

Hoang Dai Phuc Tran[1,4], Nhat-Thong Le[2,4], Tran Vinh Bao Ngoc[3], Doan Minh Trung Duong[2,4], Nguyen Phuong Thao[1,2,4], Hoang Phan[5,6], and Minh-Thong Le[1,4](✉)

[1] School of Biotechnology, International University, Ho Chi Minh City, Vietnam
lmthong@hcmiu.edu.vn
[2] Research Center for Infectious Diseases, International University, Ho Chi Minh, Vietnam
[3] Applied Biomedical Science, School of Biotechnology, West of England University, International University Campus, Ho Chi Minh, Vietnam
[4] Vietnam National University, Ho Chi Minh, Vietnam
[5] School of Biotechnology, Tan Tao University, Long An, Vietnam
[6] School of Science, Engineering and Technology, RMIT University, Ho Chi Minh, Vietnam

Abstract. The soil microbiome plays a vital role in plant health by influencing nutrient availability, disease resistance, and overall plant growth. Chitosan derivatives, including oligo-chitosan and chitooligosaccharides (COS) have been recognized for their various roles as biocontrol agents, noted for their non-toxic nature, biodegradability, and biocompatibility. Nonetheless, the specific effects of COS on soil microbial composition remain unclear. This study explores microbial community shifts in the soybean rhizosphere in the presence and absence of COS treatment. Treatments without (control) and with COS (200 ppm) were studied using full-length 16S rRNA sequencing. Across six samples, we identified 182 species covering 124 genera and 14 phyla. Although microbial diversity did not noticeably vary across treatment groups, COS treatment show trend of reducing sample variance and selectively enriched *Chujaibacter soli* and *Mizugakiibacter sediminis*, while decreasing *Castellaniella defragrans*, *Luteitalea pratensis*, *Rhodanobacter caeni*, and Rhodothermaceae bacterium RA. Collectively, these preliminary findings may suggest that COS may beneficially modulate rhizosphere microbial composition. Further research is needed to elucidate microbial dynamics throughout soybean development and assess COS's potential in sustainable agriculture.

Keywords: 16S rRNA · Nanopore sequencing · Oligo-chitosan · Rhizosphere microbial community · Soybean

L. Vu et al. (Eds.): PBC 2025, 2026.
https://doi.org/10.1007/978-3-032-29579-8_9

1 Introduction

Chitin and chitosan are essential biopolymers derived from crustacean shells, such as those of shrimp, squid, and crab, by removing minerals and proteins, thereby utilizing resources that were previously considered waste [1]. These substances not only improve plant nutrition [2] but also possess the capability to penetrate the cell membranes of microorganisms, including fungi, bacteria, and viruses, thereby offering protection to plants [3]. This aligns with the growing demand for sustainable agriculture practices, interest in chitosan is rising due to its ability to (1) promote plant growth and defense [4], (2) provide protection against pathogens [5], (3) offer environmental benefits due to its biological origin, and (4) biodegrade easily in soil [1]. The recent development of chitooligosaccharide (COS), a derivative of chitosan and chitin, demonstrates not only enhanced physicochemical properties but also reveals a range of additional biological activities [3]. It is assumed that COS retains the chitosan functions but has a smaller atomic mass, allowing it to dissolve quickly in both acidic and neutral environments, thereby increasing its ability to be absorbed by plants [6].

The rhizosphere is defined as the plant-root interface with three zones based on their relative proximity and influence from the plant root. Rhizosphere microbial communities play a crucial role in nutrient cycling, regulating gas exchange, inducing microaggregation, and modifying the soil's biochemical environment [7, 8]. Additionally, various microbial functions significantly contribute to the sustainability of soil health by providing essential nutrients, including carbon and nitrogen. This interaction significantly impacted the species composition, abundance, and distribution of microorganisms in the soil [9]. As a result, the diversity and richness of soil biology are regarded as fundamental aspects of soil health, serving as a vital biological indicator for assessing soil fertility.

On the other hand, microorganisms rely on nutrients derived from plants for their survival, either by utilizing existing compounds or by transforming them into essential elements necessary for their existence. The introduction of specific nutrients into soil during cultivation can lead to alterations in microbial composition and the dynamics of interactions between microbes and plants [10, 11]. Furthermore, studies indicate that treatments incorporating chitosan promote microbial diversity, leading to enhanced soil fertility [2]. The overall trend suggests a beneficial influence on soil health. Moreover, chitosan has demonstrated the ability to enhance resistance to saline stress and bolster immunity against fungal and bacterial pathogens [5, 12]. COS, derived from organic waste, presents a promising alternative to chemical fertilizers; however, there is a significant lack of research on the regulatory mechanisms of COS as a potential agricultural prebiotic in the interactions between crops and root-associated microorganisms.

In this study, we investigated the effects of COS derived from shrimp co-products on the rhizosphere microbial communities of soybeans. The aim was to obtain foundational insights of the microbial community through microbial analysis using full-length 16S rRNA amplification in conjunction with long-read Oxford Nanopore Sequencing. The outcome was anticipated to yield preliminary insights into the differentiation of taxonomic distinctions at genus and species levels, thereby improving our comprehension of the influence of COS in practical applications.

2 Materials and Methods

2.1 Planting Design

The experiment design was illustrated in Fig. 1. Commercial soybean MTD-176 seeds were incubated in the laboratory until germination occurred. The germinated seed will subsequently be transferred to greenhouse conditions for planting in TRIBAT soil (Vietnam). On day 14, treatments were applied to two experimental groups: a control group receiving only water (n = 3, A1-A3), and one receiving 200 ppm COS (Vietnam Food - VNF, Vietnam) (n = 3, B1-B3). With 200 ppm of COS concentration is equivalent to 200 mg/L.

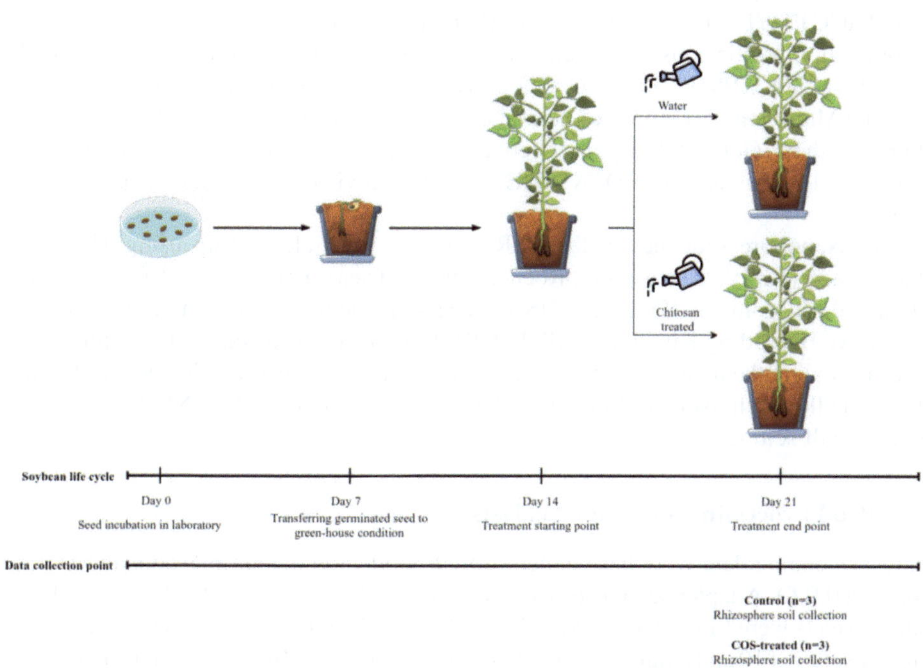

Fig. 1. Experimental design in this study

Environmental conditions were monitored throughout the experiment, including temperature (35–40 °C), humidity (55–60%), and watering frequency, to ensure optimal growth. Rhizosphere soil samples were collected from each pot on day 21 and preserved in DESS buffer [13], then stored at − 80 °C until DNA extraction.

2.2 Total DNA Extraction from Rhizosphere Soil

A 150 mg (wet weight) soil sample was subjected to DNA extraction. After removing the DESS buffer via centrifugation, DNA was extracted using the UniGene Extraction Kit (Salagene, Vietnam). Briefly, the soil was washed with PBS, homogenized in lysis buffer with 0.5 mm glass beads, and then incubated with Proteinase K and RNase A. After brief centrifugation to remove soil debris, the supernatant was collected and purified through a silica column following the manufacturer's instructions. DNA quality was assessed via gel electrophoresis and quantified using SimpliNano™ Spectrophotometer (Biochrom, UK). The extracted DNA was stored at −20 °C for subsequent experiments.

2.3 Amplification of the Full-Length 16S rRNA Gene and Nanopore Sequencing

The amplification of full-length 16S rRNA gene was carried out in a 20-µl reaction mixture containing 30 ng of genomic DNA, 0.4 µM of universal 16S rRNA bacterial primers 27F (5'-AGAGTTTGATCCTGGCTCAG-3') and 1392R (5'-GGTTACCTTGTTACGACTT-3'), and 2X PCR mastermix (Salagene, Vietnam). The amplification profile consisted of an initial denaturation at 95 °C for 5 min, followed by 35 cycles of 30s at 95 °C, 30s at 56 °C, 2 min at 72 °C and a 5-min final extension at 72 °C by using MultiGene™ OptiMax Thermal Cycler (Labnet, NJ, USA). PCR products were stained with 6X GelRed (Salagene, Vietnam) and analyzed by electrophoresis on a 1.5% agarose gel in Tris-acetate-EDTA (TAE) buffer, then visualized under ultraviolet (UV) light.

For Nanopore sequencing, the PCR products were cleaned up using 1V KAPA Pure Beads (KAPA Biosystems, Roche, South Africa) and quantified using QuantiFluor ST Fluorometer (Promega, USA). The sequencing library was prepared using the Native Barcoding Kit 24 V14 (SQK-NBD114.24) (Oxford Nanopore Technologies, UK), following the manufacturer's instructions. Sequencing was conducted on a Flongle (R10.4.1) flow cell using a MinION Mk1B device operated by MinKNOW (v24.11.10) with default settings.

2.4 Read Processing and Data Analysis

Raw sequencing data were base-called in SUP mode and demultiplexed using Dorado (version 0.9.6). Adapter sequences were trimmed with Porechop (v0.2.4) [14], and low-quality reads were filtered using NanoFilt (v2.8.0) [15], retaining reads with a minimum quality score of 10 and lengths ranging from 800 to 1600 bp. High-quality Fastq files were passed to taxonomy assignment using Emu [16]. All downstream statistical analyses were performed and visualized using R (v4.4.1). For taxonomy counting, raw counts were obtained followed by filtering to remove taxa in fewer than two replicates per treatment ensuring a more reliable representation.

The total number of reads per sample was compositional transformed prior to analysis to ensure comparability of microbiome abundances. In this study, compositional transformation prefers to converting raw counts into relative abundance as microbiome data are naturally compositional [17]. Alpha diversity indices, including Richness, Evenness, Shannon, and Simpson, were computed. Results are presented as means ± standard deviation (SD). For beta diversity, species abundance was used to calculate Bray-Curtis dissimilarity distances. These results were then used as input for Principal Coordinate Analysis (PCoA). The control and treated samples were performed using Student's t-test, with statistical significance set at $p < 0.05$.

3 Results

Sequencing data was generated for six samples, categorized into groups Control (A) and COS-treated (B). The total read depth varied across samples, ranging from 4,118 reads (Sample B2) to 14,580 reads (Sample A3). Despite differences in total sequencing yield, alignment to the reference was consistently highly efficient. Across all samples, over 99.6% of reads were successfully mapped, with the number of unmapped reads remaining negligible (ranging from 6 to 25 reads per sample). Table 1 details the read counts and mapping statistics for each sample.

Table 1. Read count per sample before and after filtering with respective mean quality

Sample	No. Reads	Mapped reads	Unmapped reads
A1	4504	4490	10
A2	5649	5636	11
A3	14580	14552	25
B1	9145	9126	17
B2	4118	4109	6
B3	6306	6288	16

3.1 Overview of Bacterial Diversity Among Samples in the Study

Across six rhizosphere soil samples, we identified 182 bacterial species with ≥ 1% relative abundance belonging to 124 genera and 14 phyla (Fig. 2). Both treatments were shown to share 110 species. In comparison, 60 species and 12 species were observed to be unique in control and COS-treated samples (Fig. 3A). The influence of COS on bacterial diversity was summarized in Table 2, Fig. 3B. Our data indicated a richness of 99.33 ± 59.37 (mean ± SD) in the control group and 87 ± 28.51 in the COS-treated group. Shannon indices were shown at 2.826 ± 0.579 and 2.937 ± 0.352 in the control and COS-treated groups, respectively. In addition, no significant difference was observed via

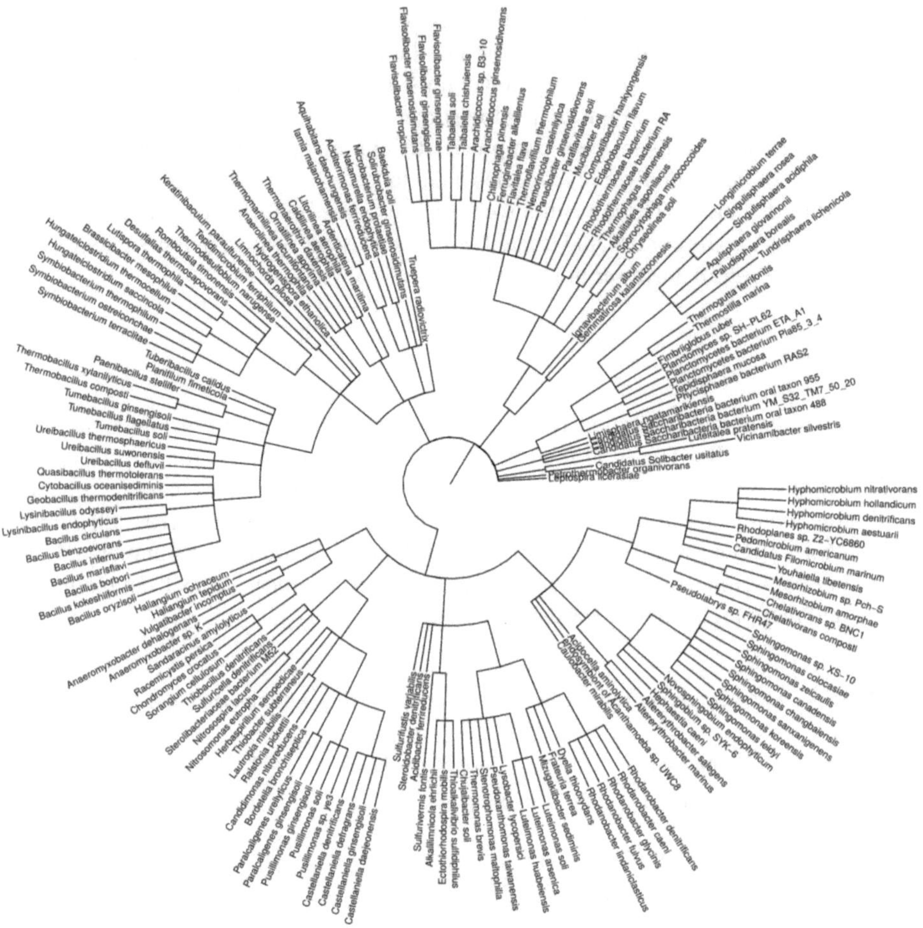

Fig. 2. Phylogenetic tree of total identified species through full-length 16S rRNA.

alpha diversity (Fig. 3B), we observed a shaping effect of COS on the soil microbiome when sample variances were reduced (Table 2).

The relative abundance of the most dominant bacteria was shown in Fig. 3C. Among the 20 dominant bacteria, *Chujaibacter soli* was observed to have the highest abundance, followed by *Castellaniella defragrans* and *Rhodanobacter denitrificans.* Consistent with previous observations, when COS was shown to affect the rhizosphere microbial community (Table 5), our data suggested two distinct patterns for control and COS-treated samples (Fig. 3).

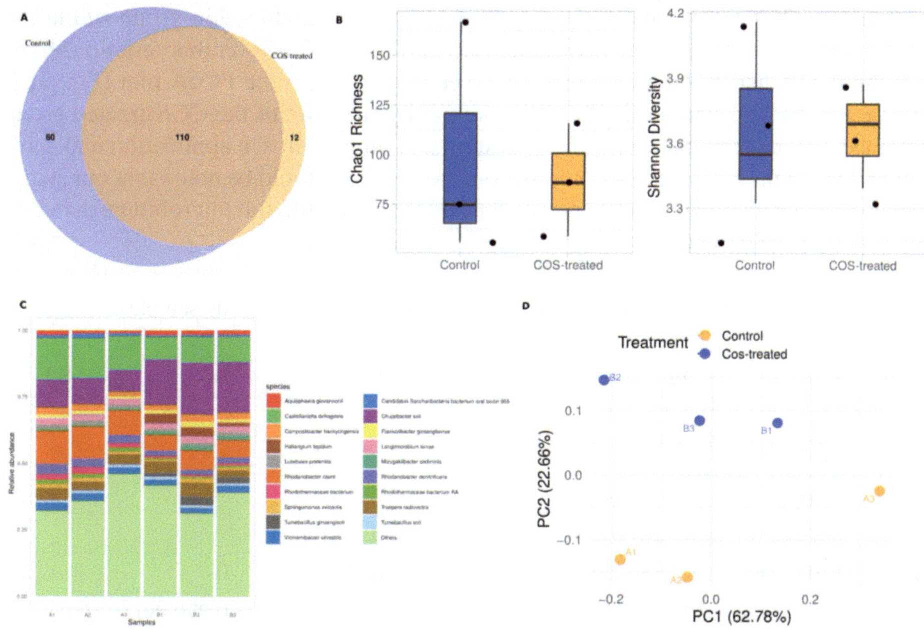

Fig. 3. General metagenomic statistics of studied samples. (A) Venn diagram of identified species from control and COS samples. (B) Samples richness and Shannon diversity index among control and COS samples. (C) Overall relative abundance of top 20 most abundant species. (D) Principle Coordinate Analysis (PCoA) of studied samples

Table 2. Alpha diversity indices

Treatment group	Richness	Evenness	Shannon	Simpson
Control	99.33 ± 59.37	0.647 ± 0.202	2.826 ± 0.579	0.942 ± 0.014
COS-treated	87 ± 28.51	0.647 ± 0.202	2.937 ± 0.352	0.938 ± 0.007

Beta diversity analysis was conducted to evaluate the structural differentiation of microbial communities between the COS-treated and Control groups. Principal Coordinate Analysis (PCoA) based on Bray-Curtis dissimilarity revealed that the first two principal coordinates explained a total of 85.44% of the compositional variation (Fig. 3D). The separation between COS and control groups is explained by PC2 (22.66%), whereas PC1 (62.78%) captures within-group variability, suggesting that factors other than treatment account for most of the variance.

Pairwise-Curtis dissimilarity matrix (Table 3) showed high within-group similarity. While there was some variation in distance (A3-A1; A3-A2) however this variation aligns with the broad dispersion of samples observed along PC1 in the PCoA plot (Fig. 3D). Despite this background variability, the community structure of the COS-treated group remained distinct from the Control, suggesting that oligo-chitosan application may have some effect on the microbiome composition. However, it should be noted that our statistics suggest that the treatment has not shown any impact on the soil microbiome diversity ($R^2 = 0.22$, p-value = 0.4).

Table 3. Bray-Curtis dissimilarity distance matrix across study samples

	A1	A2	A3	B1	B2	B3
A1	0					
A2	0.1943351	0				
A3	0.5446278	0.4314120	0			
B1	0.4003551	0.3261497	0.3089791	0		
B2	0.3073949	0.3659089	0.5851811	0.4030740	0	
B3	0.2921542	0.2765051	0.4130441	0.2204766	0.2707802	0

3.2 Species Composition Analysis

Taxonomic analysis revealed that Proteobacteria was the most prevalent group, accounting for 59.20 ± 2.47% in control samples and 58.73 ± 1.54% in COS-treated samples (Table 4). Bacteroidetes was the second most common, with 11.768 ± 1.529 in control samples and 10.113 ± 1.372 in COS-treated samples. The other nine phyla each had an abundance of more than 1% (Table 4).

Table 4. Differential relative abundance analysis at phylum levels among treatment groups

Phylum	Control		COS-treated		p-value
	Mean	SD	Mean	SD	
Acidobacteria	4.594	0.381	3.491	0.399	0.02591367989
Actinobacteria	2.883	0.078	2.165	0.45	0.1057794699
Bacteroidetes	11.768	1.529	10.113	1.372	0.2359133314
Candidatus Saccharibacteria	1.513	0.283	1.021	0.121	0.07840690026
Chloroflexi	3.68	0.859	3.758	1.149	0.9298021283
Deferribacteres	0.142	0.134	0	0	0.2072937705
Deinococcus-Thermus	3.852	0.729	3.917	1.246	0.94294271
Firmicutes	7.016	3.476	10.196	1.935	0.2566413285

(continued)

Table 4. (*continued*)

Phylum	Control		COS-treated		p-value
	Mean	SD	Mean	SD	
Gemmatimonadetes	2.234	0.137	2.285	0.268	0.7880608168
Ignavibacteriae	0.641	0.183	0.213	0.211	0.05806679998
Planctomycetes	2.444	0.367	3.892	0.998	0.1156636566
Proteobacteria	59.202	2.47	58.73	1.542	0.7950593238
Spirochaetes	0	0	0.061	0.106	0.4226497308
Verrucomicrobia	0.031	0.054	0.159	0.276	0.5068055889

A notable change ($p < 0.05$) in the Acidobacteria phylum was observed, where COS treatment resulted in this phylum becoming less abundant. While no large differences ($p < 0.05$) were noted in other phyla, minor changes of at least 1% were observed post-COS treatment in Bacteroidetes (a decrease of 1.655%), Planctomycetes (a increase of 1.448%), and Firmicutes (a increase of 3.18%). Although phylum-level abundances remained unchanged overall, the use of long-read Nanopore sequencing enabled high-resolution taxonomic classification down to the species level.

Eighteen species across sixteen genera exhibited possible changes in relative abundance ($p < 0.05$) following COS treatment (Fig. 4, Table 5). Acidobacteria previously exhibited a significant change in phylum, attributed to a notable reduction in *Luteitalea pratensis* (0.449%, Table 5). However, the other two Acidobacteria members, *Candidatus Solibacter usitatus* and *Vicinamibacter silvestris*, exhibited nonsignificant changes. Data indicated a significant enrichment of *Chujaibacter soli* (8.928%) and the genus *Chujaibacter* (Fig. 4A, B, Table 5), while the abundance of the related phylum Proteobacteria showed no significant change. Additionally, several species were identified to significantly contribute at least 0.2%, including *Mizugakiibacter sediminis, Sorangium cellulosum, Pseudolabrys sp. FHR47, Thermomonas brevis, Rhodoplanes sp. Z2-YC6860* (phylum Proteobacteria), *Planctomyces sp. SH-PL62* (phylum Planctomycetes), *Hungateiclostridium saccincola* (phylum Firmicutes), and *Sporocytophaga myxococcoides* (phylum Bacteroidetes) (Fig. 4A, C). A notable change was seen in *Hungateiclostridium saccincola*, but there was no significant difference among the other *Hungateiclostridium* genus member, such as *Hungateiclostridium clariflavum* and *Hungateiclostridium thermocellum* ($p > 0.05$, Fig. 4B).

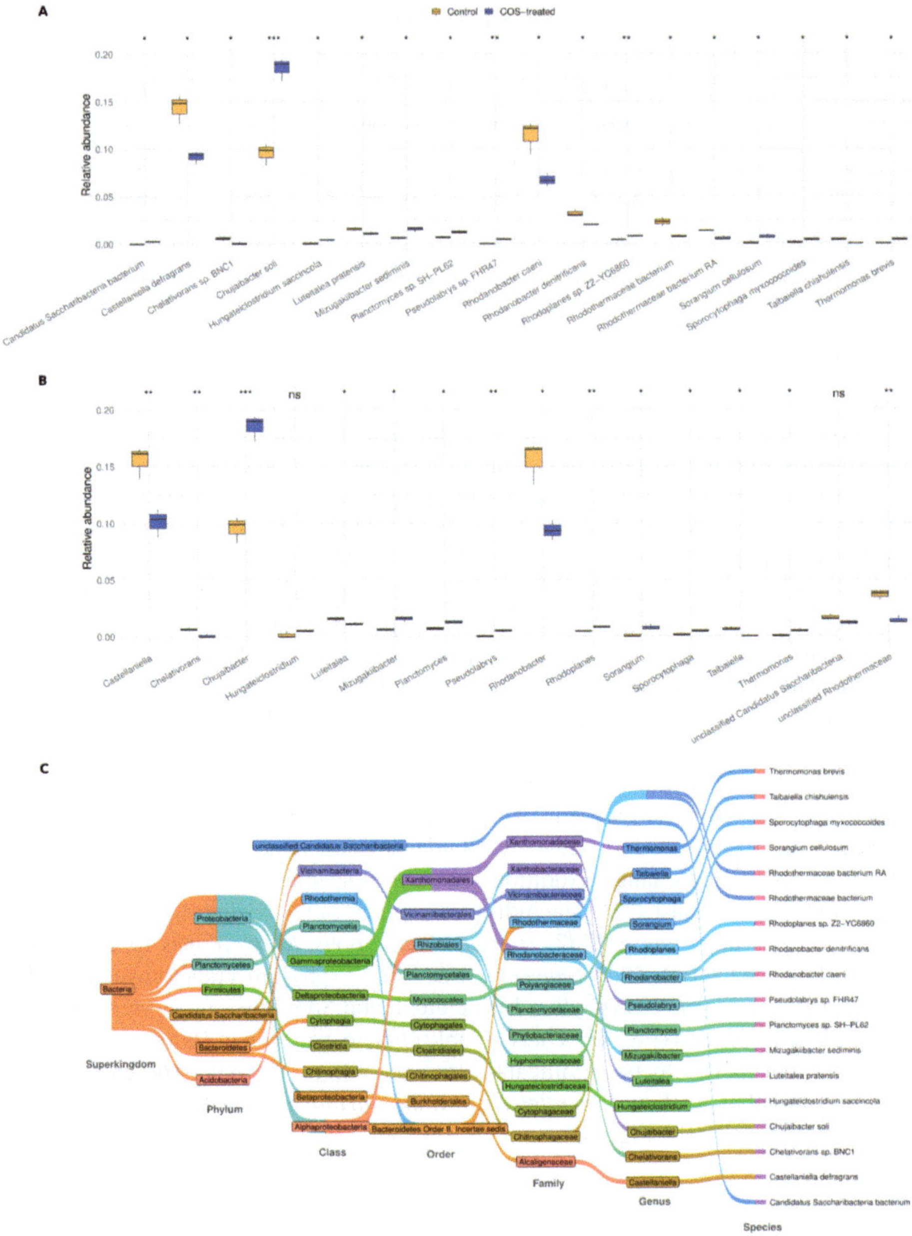

Fig. 4. Differential microbial composition analysis among control and COS-treated groups. (A) Species exhibited significant relative abundance after COS treatment, and (B) their belonging genus (C) Sankey plot illustrating the lineages of species with significant relative abundance differences. 'ns': $p > 0.05$, '*': $p \leq 0.05$, '**': $p \leq 0.01$, '***': $p \leq 0.001$, '****': $p \leq 0.0001$

Table 5. Candidate species exhibit possible changes in relative abundance across treatment groups.

Species	Control		COS-treated		p-value
	Mean	SD	Mean	SD	
Candidatus Saccharibacteria bacterium	0.048	0.083	0.265	0.087	0.03595782725
Castellaniella defragrans	14.358	1.531	9.167	0.684	0.01554247823
Chelativorans sp. BNC1	0.591	0.197	0.078	0.136	0.02540813102
Chujaibacter soli	9.57	1.179	18.498	1.219	0.0008060080362
Hungateiclostridium saccincola	0.084	0.145	0.432	0.039	0.04465957868
Luteitalea pratensis	1.557	0.177	1.108	0.11	0.02773610831
Mizugakiibacter sediminis	0.609	0.112	1.561	0.266	0.01432500261
Planctomyces sp. SH-PL62	0.693	0.135	1.253	0.164	0.01110579248
Pseudolabrys sp. FHR47	0.041	0.071	0.477	0.098	0.004570175384
Rhodanobacter caeni	11.384	1.73	6.696	0.738	0.02816303306
Rhodanobacter denitrificans	3.183	0.384	2.002	0.078	0.02930814088
Rhodoplanes sp. Z2-YC6860	0.442	0.052	0.816	0.071	0.002594986987
Rhodothermaceae bacterium	2.284	0.477	0.795	0.115	0.02706178342
Rhodothermaceae bacterium RA	1.382	0.086	0.577	0.215	0.01331232837
Sorangium cellulosum	0.075	0.129	0.741	0.227	0.01924937837
Sporocytophaga myxococcoides	0.117	0.132	0.448	0.133	0.03791955274
Taibaiella chishuiensis	0.444	0.107	0	0	0.01892134122
Thermomonas brevis	0.044	0.075	0.446	0.122	0.01284801826

Conversely, our analysis revealed that eight species, including *Taibaiella chishuiensis, Rhodothermaceae bacterium RA, Rhodothermaceae bacterium* (phylum Bacteroidetes), *Luteitalea pratensis* (phylum Acidobacteria), *Chelativorans sp. BNC1, Rhodanobacter caeni, Castellaniella defragrans, Rhodanobacter denitrificans* (phylum Proteobacteria) exhibited a significant decline in relative abundance following COS treatment. The species with the greatest alteration, *Castellaniella defragrans*, had a 5.191%

decrease, followed by *Rhodanobacter caeni*, which had a 4.688% decrease (Fig. 4A, Table 5). Following COS treatment, alterations in other species ranged from 0.444% to 1.489%.

Although Acidobacteria was the sole group identified with notable changes, a detailed examination at the species level indicated that the supplement seemed to favor only specific bacterial species in the rhizosphere that are responsive to COS. These results suggest that a higher taxonomic resolution could be effectively applied in future studies.

4 Discussion

Chitin and its derivative, chitosan, are biodegradable and non-toxic substances that have garnered increasing attention for their potential to enhance soil and substrate quality while promoting plant growth [18]. Chitosan undergoes decomposition through specific enzymatic processes, resulting in the formation of COS, which are low-molecular-weight oligosaccharide products. This enzymatic processing enhances COS's water solubility, bioactivity, and bioavailability, making it more readily absorbed and utilized by organisms [19]. In this study, we employed full-length 16S rRNA to conduct metagenomic analysis and investigate the impact of COS on the microbial population in the soybean rhizosphere.

In total, 110 common species were identified (Fig. 3A), with a mean richness of 99.33 $\pm$ 59.37 for the control group and 87 $\pm$ 28.51 for the COS-treated group (Table 2). The Shannon index was recorded at 2.826 $\pm$ 0.579 for the control group and 2.937 $\pm$ 0.352 for the COS-treated group. Within the controlled greenhouse environment, three small ecosystems were established as replicates for each treatment group. The analysis conducted using the Shannon index revealed comparable diversity across the simulated ecosystems. Nevertheless, minor discrepancies were noted for COS-treated samples (0.352) in contrast to control samples (0.579). This suggests a potentially more stable microbial community in the COS-treated group, likely influenced by the structural shaping effects of COS, aligning with earlier findings on tomato [20]. To further interpret these community shifts, Principal Coordinate Analysis (PCoA) was employed to visualize the dissimilarity between untreated and COS-treated soil (Fig. 3D). While distinct patterns emerged, the first principal coordinate (PCoA1) accounted for a substantial 62.78% of the variance, which appears largely unrelated to the treatment effect. We attribute this variation to the inherent spatial heterogeneity of the soil environment and stochastic factors in microbial assembly, which are often more pronounced in small-scale pilot studies ($n = 3$). Consequently, validating these spatial dynamics with larger sample sizes will be a critical objective for future investigations.

Subsequently, a differential analysis of the microbiome is conducted at the phylum level for the groups being examined. As a result, Acidobacteria displayed significant changes. Acidobacteria have been recognized as essential decomposers involved in the metabolic processes of nitrogen, carbohydrates, and cellulose [21, 22]. However, the data presented here reveal a decline in Acidobacteria, indicating that COS could favor changes in other decomposed bacterial phyla. In addition, although a single notable change was detected at the phylum level (*Acidobacteria,* Table 2), the abundance analysis at the lower taxonomic levels within these phyla revealed a contrasting scenario. Consequently, our preliminary findings highlighted the need for a transition to a higher taxonomic resolution to elucidate significant changes ($p < 0.05$).

Prior studies have shown that chitin promotes the growth of chitin-decomposing microorganisms from the phylum Proteobacteria, Bacteroidetes, and Firmicutes [23]. The COS has shown enhanced characteristics [19] and acts as a carbon source for specific microorganisms [24]. This study identified *Chujaibacter soli*, a member of Proteobacteria, as the species with the most significant alteration, exhibiting an increase of 8.928% (Fig. 4A, Table 5). Furthermore, four additional members of the Proteobacteria phylum, namely *Mizugakiibacter sediminis*, *Sorangium cellulosum*, *Pseudolabrys sp. FHR47, Thermomonas brevis*, and *Rhodoplanes sp. Z2-YC6860*, has demonstrated an enrichment of at least 0.2% following COS treatment. Additionally, the phylum of Firmicutes showed an increase of 3.180%. A notable increase in Firmicutes, specifically *Hungateiclostridium saccincola*, has been documented (Fig. 4A, Table 5). One member of the Bacteroidetes, *Sporocytophaga myxococcoides*, has been shown to have a significantly increased abundance, while the overall abundance of Bacteroidetes has declined (Table 4, 5).

This species-level analysis provides a nuanced understanding of COS's selective influence on rhizosphere microbial communities. The increase in certain members of Proteobacteria aligns with a decrease in *Chelativorans sp. BNC1*, *Rhodanobacter caeni*, *Castellaniella defragrans*, and *Rhodanobacter denitrificans*, all of which belong to the Proteobacteria phylum. Furthermore, *Taibaiella chishuiensis*, *Rhodothermaceae bacterium RA*, and other *Rhodothermaceae* bacteria within the Bacteroidetes phylum, along with *Luteitalea pratensis* from the Acidobacteria phylum, were noted to decline, despite an overall increase in abundance. Although current research on the impact of COS on the microbiome community may be preliminary, our findings provide significant insights that enhance the existing understanding of COS supplementation in agriculture. Additionally, the implementation of Oxford Nanopore Technology in this research enables a more detailed species-level insight of COS effect on soil microbiome which is not achievable if using the short-read approach. This is due to the short-read sequence-only section of the 16S rRNA gene for identification (e.g., V3-V4). While incorporating additional regions may improve classification accuracy, this comes at the expense of increased sequencing costs. Whereas long-read sequencing, with its ability to capture the entire full-length 16S rRNA gene, provides higher resolution and more detailed information for understanding COS-species interactions. It is worth noting that the use of ONT enabled full-length 16S rRNA sequencing, providing higher taxonomic resolution than short-read methods. While ONT is still evolving in terms of cost and accuracy, recent improvements in its flow cell and chemistry have achieved up to ~ 99% accuracy [25], making it increasingly viable for microbiome studies. This study also contributes more information to COS related research on different plant host systems. Combined with the consistently expanding databases (e.g., BacDive, GenBank, SILVA), better interpretation of COS-species characteristics and associated functions on soil health could be achieved. The broader coverage of long-read technology holds great potential for metagenomic fields, particularly in biostimulants' effect on microbiome research, which could support sustainable agriculture practice for example circulating fertilizer and green farming methods.

5 Conclusion

In summary, this study demonstrated a high-resolution approach providing initial information about the effect of COS on the soybean rhizosphere microbial community at the species level. Furthermore, this study explored the effect of COS on the rhizosphere microbiome and may provide deeper insights into species-specific responses when supplemented with COS. As bacteria constitute the majority of the soil microbiome, these findings may offer more information on COS-related microbial interventions aimed at enhancing soil health and supporting sustainable agricultural practices.

6 Limitations

This work acknowledges certain limitations, primarily its nature as a preliminary assessment designed to address a gap in existing data. Specifically, the sample size (n = 3 per group) restricts the statistical power required to confirm minor shifts, particularly among rare taxa. Nevertheless, the utilization of full-length 16S sequencing provides high-resolution taxonomic insights that establish a robust baseline, justifying the need for further large-scale investigation.In addition, as each treatment consists of only three replicates, a larger sample size is needed to observe the shaping effects of COS on the microbiome accurately and whether these changes may significantly affect the soybean biologically. COS was applied once at a concentration of 200 ppm, and soil samples collected during the mid-phase of the soybean life cycle only reflect the short-term responses of COS treatment. Further studies with extended COS application and sampling across different growth stages are required to clarify potential long-term effects. Consequently, conclusions drawn from this study regarding long-term effects require further experiments involving prolonged COS application across various stages of the soybean life cycle. Secondly, we evaluated soybean phenotypic variables, including root length, shoot length, and dry weight during COS treatment. However, the data were not statistically significant (data not shown). While the growth-promoting effects of COS have been demonstrated in previous studies [26–29], the potential agronomic impact of COS treatment should not be overlooked.

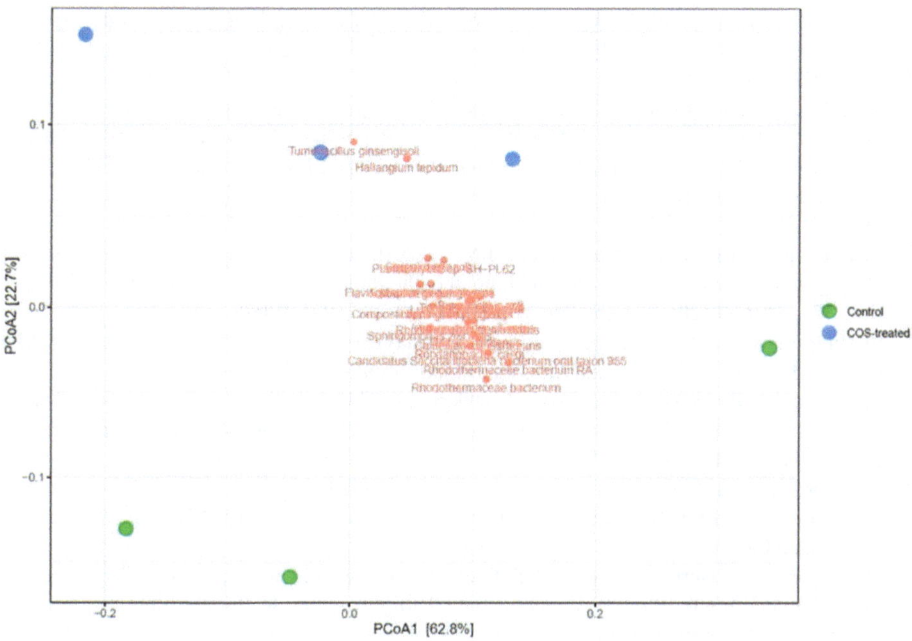

Fig. S1. Biplot graph of top 20 contributing taxa with species name label in red

Acknowledgement. This research is funded by Vietnam National University HoChiMinh City (VNU-HCM) under grant number A2023–28-02.

References

1. El Knidri, H., Belaabed, R., Addaou, A., Laajeb, A., Lahsini, A.: Extraction, chemical modification and characterization of chitin and chitosan. Int. J. Biol. Macromol. **120**, 1181–1189 (2018). https://doi.org/10.1016/j.ijbiomac.2018.08.139
2. Malerba, M., Cerana, R.: Chitosan Effects on Plant Systems. Int. J. Mol. Sci. **17**, 996 (2016). https://doi.org/10.3390/ijms17070996
3. Guan, Z., Feng, Q.: Chitosan and Chitooligosaccharide: The Promising Non-Plant-Derived Prebiotics with Multiple Biological Activities. Int. J. Mol. Sci. **23**, 6761 (2022). https://doi.org/10.3390/ijms23126761
4. Ni, X., Wang, J., Yue, Y., Cheng, W., Wang, D., Han, G.: Enhanced Antibacterial Performance and Cytocompatibility of Silver Nanoparticles Stabilized by Cellulose Nanocrystal Grafted with Chito-Oligosaccharides. Materials **11**, 1339 (2018). https://doi.org/10.3390/ma11081339
5. Riseh, R.S., Hassanisaadi, M., Vatankhah, M., Babaki, S.A., Barka, E.A.: Chitosan as a potential natural compound to manage plant diseases. Int. J. Biol. Macromol. **220**, 998–1009 (2022). https://doi.org/10.1016/j.ijbiomac.2022.08.109
6. El-Sawy, N.M., Abd El-Rehim, H.A., Elbarbary, A.M., Hegazy, E.-S.A.: Radiation-induced degradation of chitosan for possible use as a growth promoter in agricultural purposes. Carbohyd. Polym. **79**, 555–562 (2010). https://doi.org/10.1016/j.carbpol.2009.09.002
7. Mikha, M.M., Rice, C.W.: Tillage and manure effects on soil and aggregate-associated carbon and nitrogen. Soil Sci. Soc. Am. J. **68**, 809–816 (2004)

8. White, P.M., Rice, C.W.: Tillage Effects on Microbial and Carbon Dynamics during Plant Residue Decomposition. Soil Sci. Soc. Am. J. **73**, 138–145 (2009). https://doi.org/10.2136/sssaj2007.0384
9. de la Fuente, C.C., et al.: An extended root phenotype: the rhizosphere, its formation and impacts on plant fitness. Plant J. **103**, 951–964 (2020). https://doi.org/10.1111/tpj.14781
10. Kuzyakov, Y., Blagodatskaya, E.: Microbial hotspots and hot moments in soil: Concept & review. Soil Biol. Biochem. **83**, 184–199 (2015). https://doi.org/10.1016/j.soilbio.2015.01.025
11. Mendes, R., Garbeva, P., Raaijmakers, J.M.: The rhizosphere microbiome: significance of plant beneficial, plant pathogenic, and human pathogenic microorganisms. FEMS Microbiol. Rev. **37**, 634–663 (2013). https://doi.org/10.1111/1574-6976.12028
12. El Hadrami, A., Adam, L.R., El Hadrami, I., Daayf, F.: Chitosan in Plant Protection. Mar. Drugs **8**, 968–987 (2010). https://doi.org/10.3390/md8040968
13. Pavlovska M, Prekrasna I, Parnikoza I, Dykyi E (2021) Soil Sample Preservation Strategy Affects the Microbial Community Structure. Microbes Environ 36:ME20134. https://doi.org/10.1264/jsme2.ME20134
14. Wick, R.R., Judd, L.M., Gorrie, C.L., Holt, K.E.: Completing bacterial genome assemblies with multiplex MinION sequencing. Microb Genom **3**, e000132 (2017). https://doi.org/10.1099/mgen.0.000132
15. De Coster, W., D'Hert, S., Schultz, D.T., Cruts, M., Van Broeckhoven, C.: NanoPack: visualizing and processing long-read sequencing data. Bioinformatics **34**, 2666–2669 (2018). https://doi.org/10.1093/bioinformatics/bty149
16. Curry, K.D., et al.: Emu: species-level microbial community profiling of full-length 16S rRNA Oxford Nanopore sequencing data. Nat. Methods **19**, 845–853 (2022). https://doi.org/10.1038/s41592-022-01520-4
17. Gloor, G.B., Macklaim, J.M., Pawlowsky-Glahn, V., Egozcue, J.J.: Microbiome datasets are compositional: and this is not optional. Front. Microbiol. **8**,(2017). https://doi.org/10.3389/fmicb.2017.02224
18. Shahrajabian, M.H., Chaski, C., Polyzos, N., Tzortzakis, N., Petropoulos, S.A.: Sustainable agriculture systems in vegetable production using chitin and chitosan as plant biostimulants. Biomolecules **11**, 819 (2021). https://doi.org/10.3390/biom11060819
19. Xia, W., Liu, P., Zhang, J., Chen, J.: Biological activities of chitosan and chitooligosaccharides. Food Hydrocolloids **25**, 170–179 (2011). https://doi.org/10.1016/j.foodhyd.2010.03.003
20. Liu, Z., Li, X., Yu, J., Dai, X., Sun, J., Lu, L.: Chitooligosaccharides promoted growth of tomato seedlings by shaping the structure and metabolism of rhizosphere microbial communities. Food Biosci. **63**, 105703 (2025). https://doi.org/10.1016/j.fbio.2024.105703
21. de Chaves, M.G., Silva, G.G.Z., Rossetto, R., Edwards, R.A., Tsai, S.M., Navarrete, A.A.: Acidobacteria subgroups and their metabolic potential for carbon degradation in sugarcane soil amended with vinasse and nitrogen fertilizers. Front. Microbiol. **10**,(2019). https://doi.org/10.3389/fmicb.2019.01680
22. Kalam, S., et al.: Recent understanding of soil acidobacteria and their ecological significance: a critical review. Front. Microbiol. **11**, 580024 (2020). https://doi.org/10.3389/fmicb.2020.580024
23. Donderski, W., Brzezinska, M.S.: The utilization of N-acetyloglucosamine and chitin as sources of carbon and nitrogen by planktonic and benthic bacteria in Lake Jeziorak. Pol. J. Environ. Stud. **12**, 685–692 (2003)
24. Mehboob, I., Naveed, M., Zahir, Z.A.: Rhizobial association with non-legumes: mechanisms and applications. Crit. Rev. Plant Sci. **28**, 432–456 (2009). https://doi.org/10.1080/07352680903187753

25. Zhang, T., et al.: The newest Oxford Nanopore R10.4.1 full-length 16S rRNA sequencing enables the accurate resolution of species-level microbial community profiling. Appl. Environ. Microbiol. **89**, e00605-e623 (2023). https://doi.org/10.1128/aem.00605-23
26. Rojas-Pirela, M., Carillo, P., Lárez-Velásquez, C., Romanazzi, G.: Effects of chitosan on plant growth under stress conditions: similarities with plant growth promoting bacteria. Front. Plant Sci. **15**, (2024). https://doi.org/10.3389/fpls.2024.1423949
27. Sangwan, S., Sharma, P., Wati, L., Mehta, S.: Chapter 4 - Effect of chitosan nanoparticles on growth and physiology of crop plants. In: Husen, A. (ed.) Engineered Nanomaterials for Sustainable Agricultural Production, pp. 99–123. Academic Press, Soil Improvement and Stress Management (2023)
28. Suwanchaikasem, P., Idnurm, A., Selby-Pham, J., Walker, R., Boughton, B.A.: The impacts of chitosan on plant root systems and its potential to be used for controlling fungal diseases in agriculture. J. Plant Growth Regul. **43**, 3424–3445 (2024). https://doi.org/10.1007/s00344-024-11356-1
29. Wang, M., et al.: Effects of chitosan oligosaccharides on the yield components and production quality of different wheat cultivars (Triticum aestivum L.) in Northwest China. Field Crop Res **172**, 11–20 (2015). https://doi.org/10.1016/j.fcr.2014.12.007

A 3-Gene Signature for Early Diagnosis, Prognosis and Response Prediction in Non-small Cell Lung Cancer

Duc Phuc Ngo[1], Ngoc Dien Nguyen[1], Thanh Dat Nguyen[2,3], and Minh Nam Nguyen[3,4](✉)

[1] Faculty of Pharmacy, University of Health Sciences, Vietnam National University Ho Chi Minh City, Ho Chi Minh City, Vietnam
nndien.d2020@uhsvnu.edu.vn

[2] Department of Artificial Intelligence, Chang Gung University, Taoyuan City, Taiwan
M1461025@cgu.edu.tw

[3] Research Center for Genetics and Reproductive Health, University of Health Sciences, Vietnam National University Ho Chi Minh City, Ho Chi Minh City, Vietnam
nmnam@uhsvnu.edu.vn

[4] Department of Biomedical Engineering, Faculty of Medicine, University of Health Sciences, Vietnam National University Ho Chi Minh City, Ho Chi Minh City, Vietnam

Abstract. Lung cancer is the second most common type of cancer globally in terms of newly diagnosed cancer-related cases. It is primarily classified into small cell lung cancer (SCLC) and non-small cell lung cancer (NSCLC), with NSCLC accounting for approximately 85% of cases. Despite its prevalence, research for searching biomarkers that can be simultaneously applied in early diagnosis, prognosis, and treatment response in NSCLC remains limited. This study aims to identify biomarkers serving these multiple purposes in NSCLC. Data were collected from the Gene Expression Omnibus, ArrayExpress and the Genomics of Drug Sensitivity in Cancer databases. Separate gene signatures were identified for early diagnosis and prognosis, and a common gene signature was selected through Venn diagram. The shared gene signature was utilized to build models for diagnosis, prognosis, and treatment response prediction, which were validated using independent datasets. In the diagnostic model, a 3-gene signature (*SLC2A1, ADARB1*, and *COL10A1)* demonstrated excellent discrimination ability, with high area under the curve values (AUC $\geq$ 0.900) across all testing sets. The 3-gene signature effectively stratified patients into high-risk and low-risk groups and was confirmed as an independent prognostic factor in both univariate and multivariate Cox regression analyses. In treatment response prediction, the high-risk groups were highly associated with 3 and 8 potential therapeutic drugs in the GSE31210 and GSE50081 datasets, respectively. In conclusion, the 3-gene signature is a robust biomarker for early diagnosis with high sensitivity and specificity, serves as a strong prognostic predictor, and provides valuable insights into therapeutic options for high-risk patients.

Keywords: Non-small cell lung cancer · Biomarkers · Early diagnosis · Prognosis · Treatment response

L. Vu et al. (Eds.): PBC 2025, 2026.
https://doi.org/10.1007/978-3-032-29579-8_10

1 Introduction

According to GLOBOCAN (2020), lung cancer was the second most common cancer worldwide in the incidence of cancer-related new cases [1]. Lung cancer is generally divided into two categories: small cell lung cancer (SCLC) and non-small cell lung cancer (NSCLC) [2]. Diagnosis and treatment of lung cancer remain significant challenges as the disease is usually not symptomatic in its early stages [3]. Meanwhile, screening and early diagnosis are important for the effective treatment of NSCLC, as they significantly impact on treatment outcomes [4, 5]. Furthermore, early intervention not only improves clinical outcomes but also decreases the financial burden for patients [6]. Despite the advancement of medical technology, approximately 57% of patients are diagnosed after the metastasis of the cancer [7]. Besides that, the rate of false positive results can cause unnecessary invasive procedures, emphasizing the need for the integration of biomarkers with clinical findings to enhance diagnostic accuracy [7, 8].

Many studies have been conducted to develop gene signatures for the diagnosis and prognosis of lung cancer [9–11]. However, most of these studies have mainly focused on either predicting survival outcomes or diagnosing the disease, rather than addressing both simultaneously. For example, Pan et al. identified miR-33a-5p and miR-128-3p as tumor-suppressive microRNAs that were significantly downregulated in both lung cancer tissues and peripheral blood of patients, including those at early TNM stages [12]. Ma et al. identified a prognostic 15-mRNA signature that effectively classified patients into two risk groups. Expression of both the genes and their related proteins was lower in stage I than in stage II. Additionally, integrating the 15-gene model with tumor stage improved precision, compared to using stage only [9]. In another study, an 8-gene signature showed a promising ability in both prognosis and survival-time-based chemotherapy response prediction. The study also found that *EGFR*-mutated patients had poor prognosis in high-risk group than in low-risk group while this trend was not recorded in *KRAS*-mutated patients [10]. Currently, Food and Drug Administration (FDA) has approved biomarkers for targeted therapies based on patients' clinical traits like *EGFR* mutation while biomarkers for immunotherapies are more restrictive [13]. However, combining these biomarkers is essential to accurately determine the potential efficacy of a given therapeutic intervention. Regarding predicting treatment responses, the Genomics of Drug Sensitivity in Cancer (GDSC) database provides useful information on drug sensitivity across various cancer cell lines. GDSC can facilitate the identification of therapeutic biomarkers and link the effects of compounds to genomic features to discover new potential therapeutic compounds for treatments [14]. Moreover, besides finding biomarkers for new compounds, the genomic feature-based screenings are also useful for drug repurposing strategies, increasingly recognized for their cost-effectiveness and proven safety profiles [15]. Therefore, the integration of cell lines' genomic features and drug response capabilities can open a novel approach to drug repurposing in clinical practice.

As lung cancer remains one of the leading causes of cancer-related mortality, there is a critical need for reliable biomarkers that can be utilized throughout various stages of disease management. Despite improvements in medical research, the simultaneous application of biomarkers for diagnosis, prognosis, and treatment response prediction is currently not common in clinical practice. Therefore, this study aims to develop a gene signature that can be concurrently applied for early diagnosis, prognosis, and response prediction in patients with non-small cell lung cancer. The gene signature will help to advance the improvement of patient outcomes as well as reduce the global burden of lung cancer on healthcare systems.

2 Materials and Methods

2.1 Data Collection and Preprocessing

Clinical and gene expression data were retrieved from the Gene Expression Omnibus (GEO) database (https://www.ncbi.nlm.nih.gov/geo/). Clinical data included basic information on health conditions (e.g., cancer or healthy), gender, and age. Regarding information for specific purposes, tumor stage (stage I and II) and survival/disease-free time were required for early diagnosis and prognosis, respectively. Both HG-U133_Plus_2 and HG-U133A platforms were utilized in the study due to consistent probesets.

Raw gene expression data were preprocessed using robust multiarray averaging (RMA) and normalized using z-score methods. The GSE19188 dataset ($n = 156$) was utilized for both the screening of potential diagnostic genes and the training of the diagnostic model. For model testing, three additional datasets GSE19804 ($n = 120$), GSE18842 ($n = 91$), and GSE33532 ($n = 100$) were utilized.

For the prognostic model, the GSE37745 dataset ($n = 196$) was used for the screening of prognostic genes and model training. Independent datasets, including GSE14814 ($n = 133$), GSE29013 ($n = 55$), GSE30219 ($n = 293$), GSE31210 ($n = 226$) and GSE50081 ($n = 181$), were used as testing sets for the prognostic model.

For drug sensitivity prediction analysis, the oncoPredict package in R was used for in vivo response prediction in cancer patients [16]. Raw gene expression data of 47 NSCLC cell lines were retrieved from the ArrayExpress database (https://www.ebi.ac.uk/arrayexpress/) under accession of E-MTAB-783, preprocessed by RMA and normalized by z-score methods. To collect drug sensitivity information (IC50) of corresponding cell lines, GDSC1 data were downloaded from the Genomics of Drug Sensitivity in Cancer database (GDSC) (https://www.cancerrxgene.org/downloads/bulk_download), followed by a conversion step of ln(IC50) into IC50 values.

2.2 Gene Screening for Early Diagnosis and Prognosis

The GSE19188 dataset was utilized to screen for genes with potential diagnostic value in early-stage lung cancer. For this analysis, only patients diagnosed at stages I and II were included. Gene expression profiles were combined with clinical data to identify differentially expressed genes (DEGs) between these groups (normal versus cancer), following established microarray analysis procedure [17]. Each gene's diagnostic performance was evaluated by calculating the area under the curve (AUC) based on the maximum Youden index method. Genes that met the criteria of $|\log_2 FC| \geq 1$ and AUC ≥ 0.900 were selected for further analysis.

For assessing prognosis, the GSE37745 dataset was analyzed using the supervised principal components (SPCs) method [18, 19]. The analysis was conducted using default parameters from the oncoPredict package in R, and univariate regression coefficients were calculated for each gene. Those genes with regression coefficients exceeding a predefined absolute threshold of $\theta = 2$ were selected. Subsequently, patients were divided into high-risk and low-risk groups according to their computed risk scores.

2.3 Model Development

Venn diagram analysis (https://bioinfogp.cnb.csic.es/tools/venny/) was used to identify overlapping genes between the diagnostic and prognostic gene sets. To determine the most potential genes for the diagnostic model, Bayesian model averaging (BMA) method was applied. A logistic regression algorithm was utilized to construct the diagnostic model having the highest posterior probability among BMA-determined models. These genes were also used to construct the prognostic model using the SPCs method [18, 19].

2.4 Model Testing

The diagnostic model's performance was assessed through ROC curve analysis [12]. For evaluating prognostic model, gene expression and survival data from the test sets were used. Risk scores were calculated for each patient, and patients were classified into high-risk and low-risk groups using the Youden index. Those with risk scores above the group-stratifying threshold were assigned to the high-risk group and vice versa. Kaplan-Meier survival curves were performed following group stratification. Univariate and multivariate Cox proportional hazard regression analyses were also conducted to identify other independent survival-related factors.

2.5 Drug Sensitivity Prediction Analysis

A total of 12 drugs (vinorelbine, docetaxel, gemcitabine-1393, pemetrexed, paclitaxel, cisplatin-1005, crizotinib, alectinib, erlotinib, gefitinib, afatinib-1032, dabrafenib) were selected in drug response analysis, including 11 of which were FDA-approved for NSCLC treatment, with cisplatin currently undergoing clinical trials. DRUG_IDs of 1393, 1005 and 1032 were added because these drugs had multiple experimental IDs in GDSC1 data. Drug response predictions were carried out using the GSE31210 and GSE50081 datasets. By integrating IC50 values and gene expression levels from cell lines, the estimated IC50 values of the selected drugs were calculated for each patient across the two risk groups.

2.6 Statistical Analysis

All statistical analyses were performed using R software version 4.3.2. Log-rank test was used to compare survival outcomes between two risk groups. Differences in drug response were assessed using the Wilcoxon test. In all analyses, a p-value of less than 0.05 was considered statistically significant.

3 Results

3.1 Screening of Diagnostic and Prognostic Genes

The clinical characteristics of the datasets used for diagnostic and prognostic analyses are summarized in Supplementary Tables S1 and S2, respectively. In the screening of diagnostic genes, the high-throughput AUC-based screening identified 2401 probesets with AUC values ≥0.90. In addition, differential gene expression analysis between normal and cancerous tissues revealed 531 probesets with $|\log_2 FC| \geq 1$ (Fig. 1A). For the prognostic gene screening, a total of 168 probesets were selected, which demonstrated significant patient stratification into two risk groups from Kaplan-Meier analysis ($p < 0.0001$, Fig. 1B). To further evaluate the prognostic capability of these probesets and clinical parameters, both univariate and multivariate Cox regression analyses were conducted. The results indicated a significant association between these probesets and survival outcomes in both univariate (HR = 2.110, 95% CI: 1.504–2.960, $p =$ 1.52e-05) and multivariate analyses (HR = 2.001, 95% CI: 1.406–2.846, $p =$ 1.15e-04; Supplementary Table S3).

Subsequently, a Venn diagram was used to identify common probesets between the diagnostic and prognostic screenings (Fig. 1C). The analysis identified eight probesets corresponding to six genes: *SLC2A1*, *ADARB1*, *COL10A1*, *CCNB1*, *CDCA7*, and *CGNL1*. Bayesian model averaging (BMA) method was then performed to evaluate the potential of these genes for developing the diagnostic model (Table 1). The BMA analysis revealed that a model comprising *SLC2A1*, *ADARB1*, and *COL10A1* had the highest posterior probability of 0.262, leading to the selection of these three genes for model training in both diagnosis and prognosis.

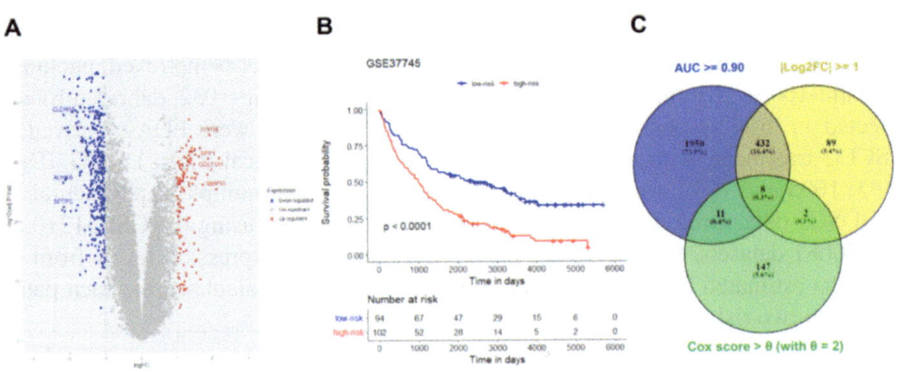

Fig. 1. Screening biomarkers. **A.** volcano plot for diagnostic screening in the training set GSE19188 (n = 156). **B.** Kaplan-Meier curve analysis of 168 prognostic probesets in GSE37745 dataset (n = 196), *p*-value calculated by the Log-rank test. **C.** Venn diagram of common probesets.

Table 1. Evaluation of potential models by BMA method

	Model 1	Model 2	Model 3	Model 4	Model 5
Intercept	1.228	3.532	0.759	3.249	2.450
SLC2A1	2.743	.	3.714	.	.
ADARB1	−2.679	−3.916	−2.940	−2.909	−3.112
COL10A1	1.368	1.960	.	1.837	1.763
CDCA7	.	.	.	.	1.096
CGNL1	.	.	.	.	.
CCNB1	.	.	.	1.542	.
nVar	3	2	2	3	3
BIC	−721.407	−720.436	−720.422	−719.051	−718.768
post prob	0.262	0.161	0.160	0.081	0.070

nVar: number of variables; BIC: Bayesian Information Criteria; post prob.: posterior probability

3.2 Diagnostic Model

The model's diagnostic ability was assessed through ROC curve analysis with an AUC of 0.980 in the training GSE19188 dataset, indicating significant discrimination between normal and cancerous conditions (Fig. 2A). Calibration of the model was evaluated using the bootstrap method with 100 repetitions, which resulted in a low mean absolute error of 0.010, showing the established model was well-calibrated (Fig. 2B).

Three independent datasets were employed to validate the diagnostic performance. In ROC curve analysis, AUC values were calculated for each dataset, showing the robust diagnostic performance of the 3-gene signature with AUC values of 1.000 in both GSE18842 and GSE33532, and 0.985 in GSE19804 (Fig. 2C).

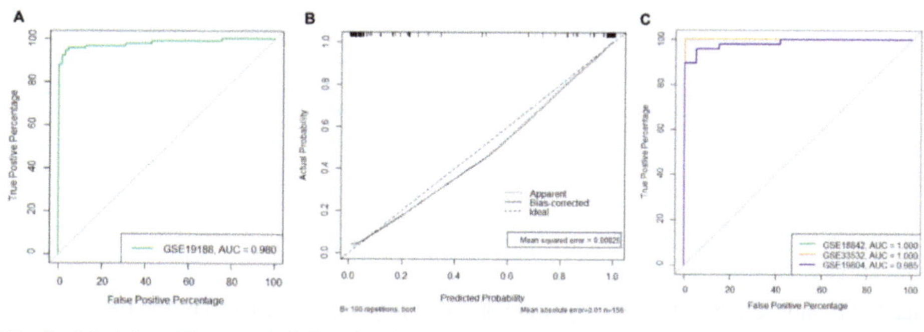

Fig. 2. Model performance of 3-gene signature on training and testing sets. **A**. ROC curve of 3-gene signature on training set GSE19188 (n = 156). **B**. calibration curve of 3-gene signature on training set GSE19188 (n = 156). **C**. ROC curve of 3-gene signature on the independent datasets (GSE18842, n = 91; GSE33532, n = 100; GSE19804, n = 120).

3.3 Prognostic Model

Patients were stratified into high-risk and low-risk groups according to their risk scores. Kaplan-Meier analysis showed the significant stratification capability of the 3-gene signature ($p = 0.00012$, Fig. 3A). Moreover, both univariate and multivariate Cox regression analyses indicated that the 3-gene signature could serve as an independent prognostic factor when adjusted with other clinical parameters. The results revealed that 3-gene signature had survival association in both univariate (HR = 1.892; 95% CI: 1.359–2.633, $p = 0.0002$) and multivariate analyses (HR = 1.892; 95% CI: 1.3354–2.681, $p = 0.0003$). Additionally, stage and age were also significant prognostic factors associated with survival status (Table 2). To validate the prognostic model, five independent datasets (GSE14814, GSE29013, GSE30219, GSE31210, and GSE50081) were used as validation sets (Fig. 3B, C, D, E and F). Kaplan-Meier analyses revealed that the 3-gene signature significantly stratified patients into risk groups in four out of five datasets, with the exception of GSE14814 ($p = 0.069$; Fig. 3G). Patients in the low-risk group consistently exhibited better survival probabilities than those in the high-risk group ($p < 0.01$). Univariate and multivariate Cox regression analyses further confirmed the 3-gene signature as an independent prognostic factor in these validation datasets. Notably, in the GSE14814 dataset, although the signature did not show a significant association with survival in univariate analysis (HR = 0.616; 95% CI: 0.364–1.043; $p = 0.072$), multivariate analysis revealed that patients in the high-risk group had worse prognoses (HR = 0.477; 95% CI: 0.277–0.821; $p = 0.008$) (Supplementary Table S4). The results for other datasets similarly indicated that the high-risk group had poorer outcomes (Supplementary Table S5, S6, S7, S8 and S9).

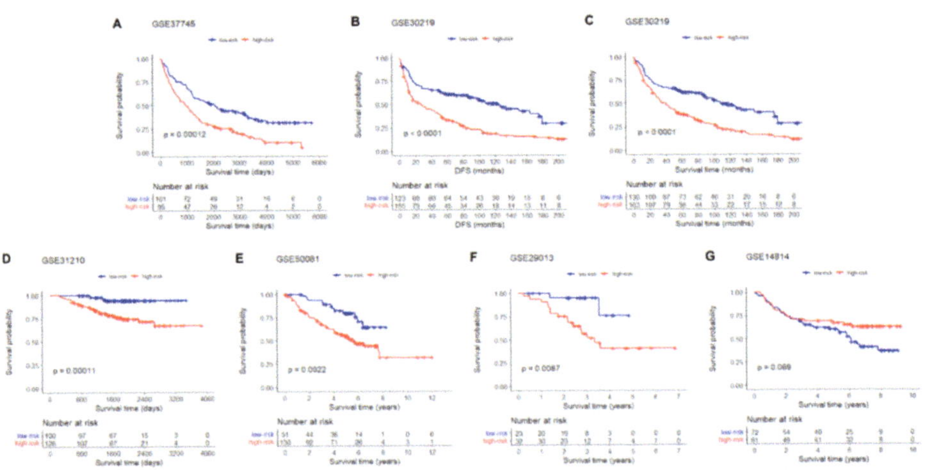

Fig. 3. Prognostic value of 3-gene signature on datasets, p-value calculated by the Log-rank test. **A**. Kaplan-Meier curve analysis on the training set GSE37745. **B, C**. Kaplan-Meier curve analysis on the test set GSE30219 by DFS (disease-free survival) and overall survival, respectively. **D–G** Kaplan-Meier curve analysis on the four testing sets: GSE31210, GSE50081, GSE29013 and GSE14814, respectively.

Table 2. Univariate and multivariate Cox proportional hazard regression analyses in the training set GSE37745, based on 3-gene signature

Variable		Univariate		Multivariate	
		HR (95% CI)	*p*	HR (95% CI)	*p*
Age	≥ 65 vs. < 65	1.438 (1.033–2.001)	0.032	1.407 (1.007–1.964)	0.0452
Gender	Female versus male	0.912 (0.657–1.268)	0.585	1.100 (0.778–1.556)	0.5906
Stage	III-IV versus I-II	1.707 (1.122–2.597)	0.013	1.772 (1.162–2.703)	0.0079
Risk score	High versus low	1.892 (1.359–2.633)	0.0002	1.892 (1.335–2.681)	0.0003

Stage is classified according to TNM classification

3.4 Drug Response Analysis in Risk Groups

Drug response prediction was performed on two risk groups using the GSE31210 (high-risk, n = 126; low-risk, n = 100) and GSE50081 (high-risk, n = 130; low-risk, n = 51) datasets. In GSE31210 dataset, high-risk patients exhibited more sensitive to three drugs including afatinib, crizotinib and gefitinib (Fig. 4A, B and C), while the low-risk group showed sensitivity to four drugs (Supplementary Fig. S1). In GSE50081 dataset, high-risk patients exhibited sensitivity to eight drugs including afatinib, alectinib, crizotinib, docetaxel, gefitinib, gemcitabine, paclitaxel and pemetrexed (Fig. 4D, E and F, Supplementary Fig. S2A, B, C, D and E), whereas the low-risk group was sensitive to only two drugs (Supplementary Fig. S2F, G).

Drug response prediction was also performed on the *EGFR*-positive and non-mutated groups in the GSE31210 dataset. In the *EGFR*-positive group, high-risk patients were predicted to be responsive to afatinib and gefitinib, while low-risk patients showed sensitivity to cisplatin, dabrafenib, erlotinib and vinorelbine (Fig. 5). In non-mutated group, high-risk individuals were more sensitive to three drugs, in contrast to only two drugs predicted to benefit the low-risk group (Supplementary Fig. S3).

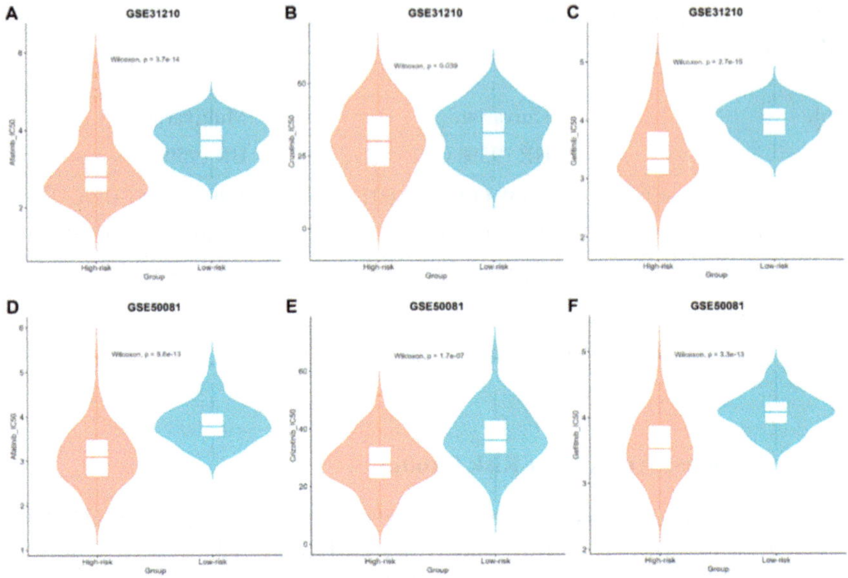

Fig. 4. Response prediction of two risk groups to 3 out of 12 drugs, *p*-value calculated by the Wilcoxon rank-sum test. **A–C**. Response prediction of two risk groups in GSE31210 dataset (high-risk, n = 126; low-risk, n = 100). **D–F**. Response prediction of two risk groups in GSE50081 dataset (high-risk, n = 130; low-risk, n = 51).

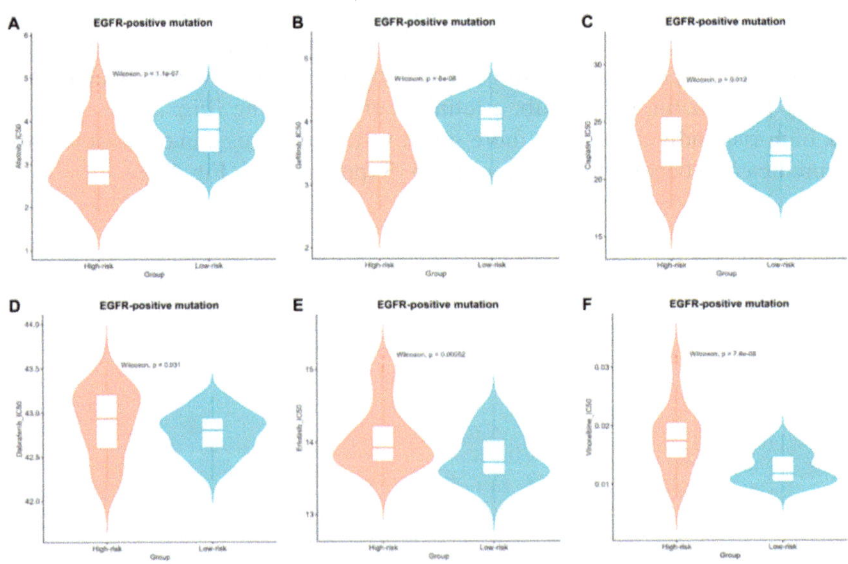

Fig. 5. Response prediction of two risk groups with *EGFR*-positive in GSE31210 dataset (high-risk, n = 48; low-risk, n = 79), *p*-value calculated by the Wilcoxon rank-sum test.

4 Discussion

In this study, we developed a 3-gene signature that not only improved early diagnosis and prognosis but also therapy response prediction for non-small cell lung cancer. To perform gene screening for diagnosis, we utilized a multi-step strategy based on Log_2FC analysis from DEGs and AUC-based high-throughput screening. In terms of prognostic gene screening, many earlier studies have typically followed a stepwise selection approach, starting with Log_2FC filtering and univariate Cox regression. However, we propose that such strategies may inadvertently overlook certain survival-associated genes, particularly those without obvious expression differences [9, 10]. To address this, we adopted SPCs method as both a screening and modeling approach for prognosis. This technique computed a Cox score for each gene, with optimized screening thresholds, allowing to capture candidate genes that traditional high-score-driven methods might bypass. The optimal model was afterward identified using the Bayesian Model Averaging (BMA) method, based on common genes derived from the previous result of Venn diagram analysis. This approach not only reduced the number of candidate genes but also identified the three most optimal genes for diagnostic model development. In contrast, other studies have used methods like protein-protein interaction networks and enrichment analysis to identify diagnostic candidates [11]. While previous methods have effectively highlighted biological interactions, our strategy prioritized the selection of genes with strong potential to build a robust diagnostic model.

In developing the diagnostic model, we performed both discrimination and calibration tests to assess thoroughly the model performance. While discrimination tests reveal the model's ability to distinguish between groups, calibration tests, such as mean absolute error, estimate the precision of the predicted results [20]. Our findings showed that the diagnostic model exhibited excellent discrimination and high precision, demonstrated by a high AUC value and low mean absolute error.

The prognostic model further validated the 3-gene signature's utility, with Kaplan-Meier analysis demonstrating significantly lower survival probabilities in the high-risk group compared to the low-risk group. These results are consistent with previous studies, validating the prognostic significance of our gene signature [9, 10]. Whereas these studies classified patients based on median index value of risk scores, we applied the Youden method to identify optimal thresholds to enable more precise patient stratification.

Initial diagnosis is important for defining prognosis and treatment outcomes in clinical practice [4]. ROC analyses confirmed the strong diagnostic potential of the 3-gene signature, which displayed excellent discrimination between normal and cancerous tissues. However, the relatively small sample sizes in the test datasets highlight the need for further validation in larger cohorts. Upon reviewing the independent genes of the signature, previous studies reported their significance in NSCLC. For instance, *ADARB1* has been shown to have low expression in adenocarcinoma (ADC) and squamous cell carcinoma (SCC) [21, 22], while *SLC2A1* is overexpressed in these cancer types [23]. In the same way, *COL10A1* was highly expressed in NSCLC, although this has only been reported in females [24]. These findings supported the robustness of our selected gene signature. Moreover, comparative analysis with other diagnostic biomarkers in NSCLC underscores the superiority of our 3-gene signature. For example, *TTYH3* reported as a potential diagnostic biomarker, achieved an AUC of more than 0.800 in different test sets, including an AUC of 0.829 in the GSE19188 dataset [25]. In contrast, our 3-gene signature demonstrated a higher AUC of 0.980 in the same dataset. Similarly, *UCK2*, which is another potential diagnostic and prognostic marker, showed an AUC of 0.965 in the GSE33532 dataset, while our signature achieved a perfect AUC of 1.000 [26]. Although non-invasive biomarkers like *LUNX* and *SPARC* have been explored in NSCLC, their diagnostic values remain limited [24, 27]. Further clinical analyses are necessary to determine if our 3-gene signature outperforms these biomarkers.

Several studies have pointed to the involvement of the genes in our panel in different aspects of NSCLC progression, especially in the context of metastasis. Overexpression of *SLC2A1* was involved with poor prognosis, and inhibiting its expression has been shown to reduce cell proliferation and metastasis in ADC cell lines [23, 28]. Conversely, downregulation of *ADARB1* had been associated with poor prognosis, particularly in metastatic ADC cell lines, while *COL10A1* overexpression had been related with poor outcomes and increased metastatic potential [21, 29]. The dysregulation of these genes could influence multiple downstream signaling pathways. The interaction between *COL10A1* and *DDR2* could facilitate the focal adhesion kinase signaling pathway, thereby remodeling extracellular matrix and promoting the metastasis of lung adenocarcinoma [29]. Additionally, *COL10A1* upregulation also caused the increase in cancer-associated fibroblasts (CAFs), Epithelial-Mesenchymal Transition (EMT) was afterward induced due to the simultaneous presence of *COL10A1* and CAFs. EMT, as a result, could trigger tumor metastasis [30]. Regarding the encoded protein of *SLC2A1*, a study showed that GLUT1 involves in lung cancer progression processes such as anti-apoptosis, cell proliferation, invasion and metastasis. GLUT1 interacts with p-EGFR and protects EGFR protein from degradation via ubiquitin-mediated proteolysis, thereby activating EGFR signaling and its downstream pathways such as AKT, ERK [31]. ADAR2 enzyme, encoded by *ADARB1*, converts miR-3167 into ed-miR-3167 via A-to-I RNA editing, activating Hippo signaling and inhibiting tumor progression in lung adenocarcinoma [32]. These findings suggested that *SLC2A1*, *ADARB1*, and *COL10A1* can be potential candidates for target therapy to inhibit metastasis in lung adenocarcinoma.

In our drug sensitivity analysis, we utilized two datasets (GSE31210 and GSE50081) consisting of patients who had not received any adjuvant therapies. Using the oncoPredict method, we predicted the response to 12 clinical drugs based on prognostic outcomes [33–35]. In the GSE31210 dataset, high-risk patients exhibited greater sensitivity to three drugs, including afatinib, crizotinib and gefitinib. In the GSE50081 dataset, high-risk patients were more sensitive to an extended list of 8 drugs, such as afatinib, alectinib, crizotinib, docetaxel, gefitinib, gemcitabine, paclitaxel and pemetrexed. In Yang's study, high-risk lung adenocarcinoma patients showed sensitivity to drugs docetaxel and gefitinib, which was relatively consistent with our findings in the high-risk groups [34]. Another study also identified an 8-gene signature predicting sensitivity to the afatinib, and gefitinib, most of which also showed sensitivity in our high-risk groups [36]. Although this study focused on a limited number of drugs due to computational constraints, this work highlights the value of prioritizing clinically approved drugs in the process of therapeutic decision-making. Nevertheless, our drug sensitivity predictions need to be clinically validated, as the lack of clinical response data restricts the translational potential of these findings. Future larger and independent studies should aim to verify the clinical utility of the 3-gene signature in distinguishing risk groups and guiding drug selection. These validations will be crucial to ensuring the applicability of our 3-gene signature to optimize the treatment outcomes in NSCLC.

5 Conclusion

This research successfully established a solid 3-gene signature with promising prospects for the early diagnosis and prognosis of NSCLC patients. The gene signature achieved high sensitivity and specificity in different datasets, indicating its potential to serve as an effective diagnostic model. Furthermore, our findings highlighted a significant association between the 3-gene signature and patient survival outcomes, supporting its application as a prognostic factor. In addition to the diagnostic and prognostic significance, the 3-gene signature also explained potential therapeutic approaches, suggesting specific drugs that may benefit high-risk patients. Taken together, these findings emphasized the clinical significance of the 3-gene signature in the improvement of personalized treatment approaches for NSCLC.

Conflict of Interest Statement. All authors have read the journal's policy on disclosure of potential conflicts of interest and declare no potential conflicts of interest with respect to the research, authorship, and/or publication of this article.

Data Availability Statement. Only publicly available data were used in this study, and data sources and handling of these data are described in the Materials. Further information is available from the corresponding author upon request.

Author Contributions. MNN conceptualized the study. MNN and DPN designed the study and planed the experiments. MNN administrated the project. MNN supervised the project. DPN, NDN, and TDN collected and cleaned the data. DPN, NDN, and TDN performed the analysis. DPN prepared all figures. DPN wrote the original draft. MNN and TDN revised the manuscript. All authors have read the journal's authorship agreement and the manuscript has been reviewed and approved by all named authors. All authors have read and approved the final manuscript.

Appendix

Supplementary Table S1. Clinical information in collected datasets for diagnosis

Variables		GSE19188	GSE19804	GSE18842	GSE33532
Gender	Female	34	120	–	20
	Male	100	0	–	80
	NA	22	–	–	–
Age	< 65	–	76	–	55

(*continued*)

Supplementary Table S1. (*continued*)

Variables		GSE19188	GSE19804	GSE18842	GSE33532
	≥ 65	–	44	–	45
Tissue	Normal	65	60	45	20
	Cancer	91	60	46	80
Stage	I	–	35	–	56
	II	–	12	–	24
	III	–	12	–	–
	IV	–	1	–	–
	NA	–	60	–	20

Supplementary Table S2. Clinical information in collected datasets for prognosis

Variables		GSE37745	GSE14814	GSE29013	GSE30219	GSE31210	GSE50081
Gender	Female	89	42	17	43	121	83
	Male	107	91	38	250	105	98
Age	< 65	94	87	24	166	164	59
	≥ 65	102	46	31	126	62	122
	NA	–	–	–	1	–	–
Stage	I	130	73	24	–	168	127
	II	35	60	14	–	58	54
	III	27	–	17	–	–	–
	IV	4	–	–	–	–	–
Smoking	Yes	–	–	53	262 Current: 134 Before: 128	115	136 Current: 57 Before: 79
	No	–	–	2	25	111	24
	NA	–	–	–	6	–	21
Chemotherapy	Yes	29	71	34	29	–	–
	No	71	62	21	258	–	–
	NA	96	–	–	6	–	–
Radiotherapy	Yes	–	–	–	56	–	–
	No	–	–	–	233	–	–
	NA	–	–	–	4	–	–

Supplementary Table S3. Univariate and multivariate Cox proportional hazard regression analyses in GSE37745 dataset, based on 168 probesets

Variables		Univariate		Multivariate	
		HR (95% CI)	*p*	HR (95% CI)	*p*
Age	≥ 65 versus <65	1.438 (1.033–2.001)	0.032	1.332 (0.951–1.866)	0.096
Gender	Male versus female	1.096 (0.788–1.523)	0.585	0.928 (0.661–1.304)	0.667
Stage	I-II versus III-IV	0.586 (0.385–0.891)	0.013	0.605 (0.396–0.926)	0.02
Risk score	High versus low	2.110 (1.504–2.960)	1.52e-05	2.001 (1.406–2.846)	1.15e-04

Stage is classified according to TNM classification

Supplementary Table S4. Univariate and multivariate Cox proportional hazard regression analyses in GSE14814 dataset

Variables		Univariate		Multivariate	
		HR (95% CI)	*p*	HR (95% CI)	*p*
Age	≥ 65 versus <65	1.839 (1.102–3.071)	0.02	1.924 (1.138–3.250)	0.015
Gender	Female versus male	0.617 (0.339–1.123)	0.114	0.629 (0.339–1.167)	0.141
Stage	II vs. IB	1.734 (1.042–2.885)	0.034	1.964 (1.171–3.296)	0.011
Chemotherapy	Yes versus No	0.796 (0.479–1.321)	0.377	0.715 (0.426–1.199)	0.203
Risk score	High versus low	0.616 (0.364–1.043)	0.072	0.477 (0.277–0.821)	0.008

Stage is classified according to TNM classification

Supplementary Table S5. Univariate and multivariate Cox proportional hazard regression analyses in GSE29013 dataset

Variables		Univariate		Multivariate	
		HR (95% CI)	*p*	HR (95% CI)	*p*
Age	≥ 65 versus <65	1.068 (0.421–2.709)	0.89	1.806 (0.655–4.982)	0.254

(*continued*)

Supplementary Table S5. (*continued*)

Variables		Univariate		Multivariate	
		HR (95% CI)	*p*	HR (95% CI)	*p*
Gender	Female versus male	1.176 (0.437–3.167)	0.749	1.848 (0.584–5.851)	0.296
Stage	III versus I-II	3.471 (1.366–8.816)	0.009	9.138 (2.688–31.072)	0.0004
Smoking	Yes versus no	0.903 (0.119–6.848)	0.921	0.064 (0.003–1.359)	0.078
Chemotherapy	Yes versus no	1.629 (0.592–4.479)	0.345	2.234 (0.706–7.070)	0.172
Risk score	High versus low	5.712 (1.312–24.87)	0.020	26.440 (2.978–234.758)	0.003

Stage is classified according to TNM classification

Supplementary Table S6. Univariate and multivariate Cox proportional hazard regression analyses of disease-free survival in GSE30219 dataset

Variables		Univariate		Multivariate	
		HR (95% CI)	*p*	HR (95% CI)	*p*
Age	≥ 65 versus <65	2.021 (1.511–2.703)	2.14e-06	1.915 (1.410–2.601)	3.23e-05
Gender	Female versus male	0.568 (0.357–0.904)	0.017	0.644 (0.392–1.056)	0.081
Smoking	Never versus A, F	0.407 (0.212–0.781)	0.007	0.666 (0.331–0.338)	0.253
Chemotherapy	Yes versus no	1.534 (0.972–2.419)	0.066	1.099 (0.658–1.836)	0.718
Radiotherapy	Yes versus no	1.867 (1.339–2.604)	0.0002	1.405 (0.966–2.044)	0.075
Risk score	High versus low	2.164 (1.596–2.936)	6.92e-07	1.847 (1.348–2.531)	0.0001

A. Active, F. Former

Supplementary Table S7. Univariate and multivariate Cox proportional hazard regression analyses of overall survival in GSE30219 dataset

Variables		Univariate		Multivariate	
		HR (95% CI)	*p*	HR (95% CI)	*p*
Age	≥ 65 versus <65	2.001 (1.51–2.651)	1.38e-06	1.905 (1.415–2.565)	2.18e-05
Gender	Female versus male	0.588 (0.374–0.925)	0.022	0.669 (0.410–1.091)	0.107
Smoking	Never versus A, F	0.399 (0.208–0.764)	0.006	0.621 (0.308–1.254)	0.184
Chemotherapy	Yes versus No	1.362 (0.874–2.124)	0.173	1.023 (0.622–1.681)	0.929
Radiotherapy	Yes versus No	1.747 (1.263–2.415)	0.0007	1.401 (0.967–2.031)	0.075
Risk score	High versus Low	1.977 (1.475–2.65)	5.18e-06	1.679 (1.242–2.270)	0.0007

A. Active, F. Former

Supplementary Table S8. Univariate and multivariate Cox proportional hazard regression analyses in GSE31210 dataset

Variables		Univariate		Multivariate	
		HR (95% CI)	*p*	HR (95% CI)	*p*
Age	≥ 65 versus <65	1.888 (0.960–3.714)	0.066	2.562 (1.288–5.095)	0.007
Gender	Female versus male	0.659 (0.338–1.281)	0.219	0.853 (0.324–2.243)	0.747
Stage	II versus I	4.232 (2.175–8.236)	2.17e-05	2.826 (1.403–5.692)	0.004
Smoking	No versus Yes	0.611 (0.3124–1.195)	0.15	0.700 (0.263–1.865)	0.476
Risk score	High versus Low	5.311 (2.059–13.700)	0.0005	4.216 (1.567–11.348)	0.004

Stage is classified according to TNM classification

Supplementary Table S9. Univariate and multivariate Cox proportional hazard regression analyses in GSE50081 dataset

Variables		Univariate		Multivariate	
		HR (95% CI)	*p*	HR (95% CI)	*p*
Age	≥ 65 versus <65	1.559 (0.932–2.608)	0.09	1.702 (1.011–2.867)	0.045
Gender	Female versus male	0.517 (0.318–0.841)	0.008	0.510 (0.314–0.830)	0.007
Stage	II versus I	1.689 (1.049–2.718)	0.031	1.652 (1.019–2.680)	0.042
Risk score	High versus low	2.542 (1.37–4.716)	0.003	2.489 (1.332–4.648)	0.004

Stage is classified according to TNM classification

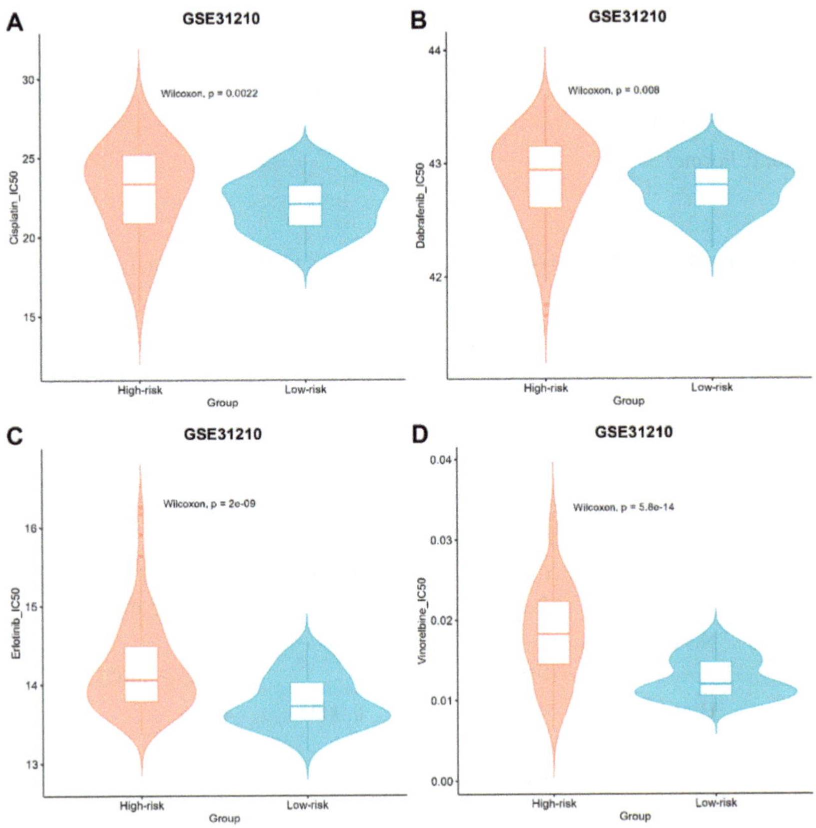

Supplementary Fig. S1. Response prediction of 4 sensitive drugs in the low-risk in GSE31210 dataset (high-risk, n = 126; low-risk, n = 100), *p*-value calculated by the Wilcoxon rank-sum test.

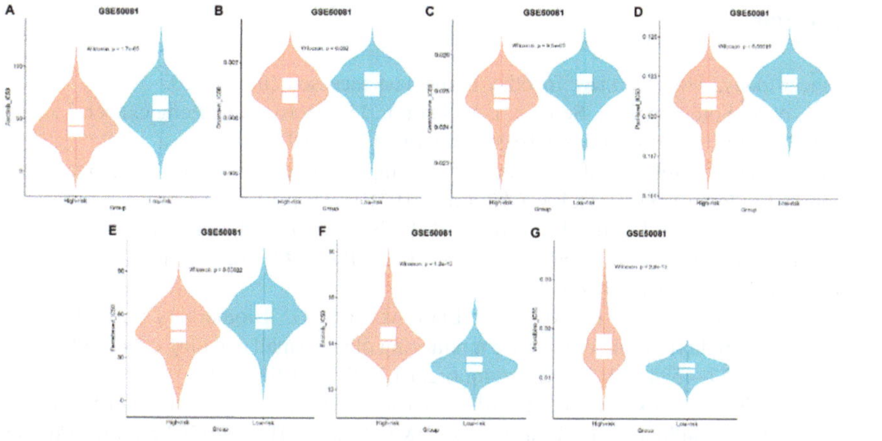

Supplementary Fig. S2. Response prediction of 5 sensitive drugs in the high-risk (**A-E**) and 2 sensitive drugs in the low-risk (**F, G**) in GSE50081 dataset (high-risk, n = 130; low-risk, n = 51), *p*-value calculated by the Wilcoxon rank-sum test.

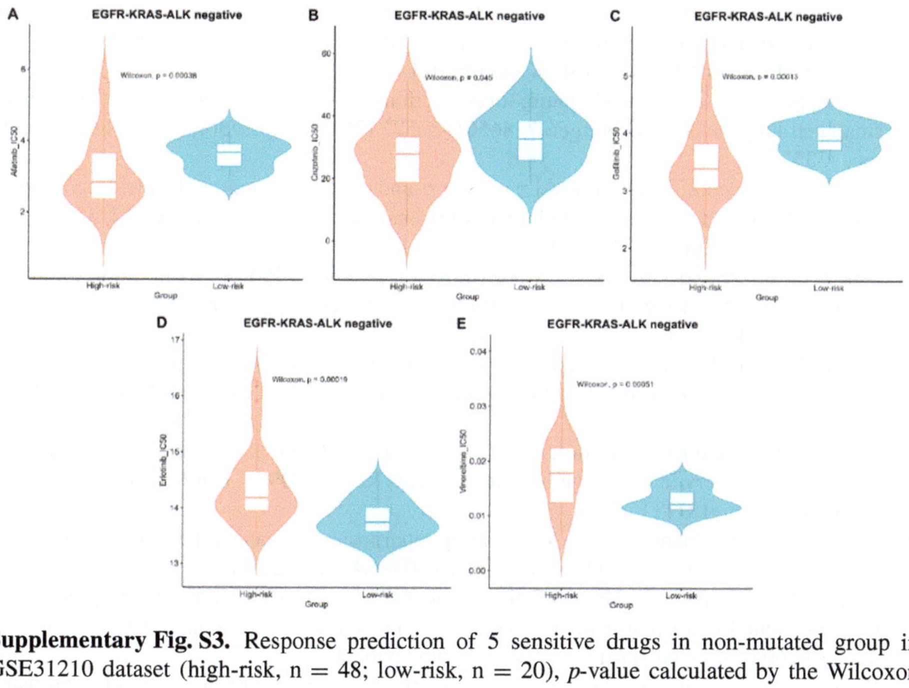

Supplementary Fig. S3. Response prediction of 5 sensitive drugs in non-mutated group in GSE31210 dataset (high-risk, n = 48; low-risk, n = 20), *p*-value calculated by the Wilcoxon rank-sum test.

References

1. Cancer today., http://gco.iarc.fr/today/home, last accessed 2023/12/20.
2. Thai, A.A., Solomon, B.J., Sequist, L.V., Gainor, J.F., Heist, R.S.: Lung cancer. Lancet. **398**, 535–554 (2021). https://doi.org/10.1016/S0140-6736(21)00312-3
3. Zhang, J. et al.: Global burden of lung cancer: implications from current evidence. Annals of. Cancer Epidemiol. **5** (2021). https://doi.org/10.21037/ace-20-31
4. Duma, N., Santana-Davila, R., Molina, J.R.: Non-small cell lung cancer: epidemiology, screening, diagnosis, and treatment. Mayo Clin. Proc. **94**, 1623–1640 (2019). https://doi.org/10.1016/j.mayocp.2019.01.013
5. Tang, Y., Qiao, G., Xu, E., Xuan, Y., Liao, M., Yin, G.: Biomarkers for early diagnosis, prognosis, prediction, and recurrence monitoring of non-small cell lung cancer. Onco. Targets Ther. **10**, 4527–4534 (2017). https://doi.org/10.2147/OTT.S142149
6. Wood, R., Taylor-Stokes, G.: Cost burden associated with advanced non-small cell lung cancer in Europe and influence of disease stage. BMC Cancer. **19**, 214 (2019). https://doi.org/10.1186/s12885-019-5428-4
7. Goebel, C., Louden, C.L., McKenna, R., Onugha, O., Wachtel, A., Long, T.: Diagnosis of non-small cell lung cancer for early stage asymptomatic patients. Cancer Genomics Proteomics. **16**, 229–244 (2019). https://doi.org/10.21873/cgp.20128
8. Seijo, L.M. et al.: Biomarkers in lung cancer screening: achievements, promises, and challenges. J. Thorac. Oncol. **14**, 343–357 (2019). https://doi.org/10.1016/j.jtho.2018.11.023
9. Ma, N., Si, L., Yang, M., Li, M., He, Z.: A highly expressed mRNA signature for predicting survival in patients with stage I/II non-small-cell lung cancer after operation. Sci. Rep. **11**, 5855 (2021). https://doi.org/10.1038/s41598-021-85246-x
10. Shahid, M. et al.: An 8-gene signature for prediction of prognosis and chemoresponse in non-small cell lung cancer. Oncotarget. **7**, 86561–86572 (2016). https://doi.org/10.18632/oncotarget.13357
11. Wu, Q. et al.: Identification of novel biomarkers and candidate small molecule drugs in non-small-cell lung cancer by integrated microarray analysis. Onco. Targets Ther. **12**, 3545–3563 (2019). https://doi.org/10.2147/OTT.S198621
12. Pan, J. et al.: A two-miRNA signature (miR-33a-5p and miR-128-3p) in whole blood as potential biomarker for early diagnosis of lung cancer. Sci. Rep. **8**, 16699 (2018). https://doi.org/10.1038/s41598-018-35139-3
13. Šutić, M. et al.: Diagnostic, predictive, and prognostic biomarkers in non-small cell lung cancer (NSCLC) management. J. Pers. Medicine. **11**, 1102 (2021). https://doi.org/10.3390/jpm11111102
14. Yang, W. et al.: Genomics of drug sensitivity in cancer (GDSC): a resource for therapeutic biomarker discovery in cancer cells. Nucleic Acids Res. **41**, D955–D961 (2013). https://doi.org/10.1093/nar/gks1111
15. Xia, Y., Sun, M., Huang, H., Jin, W.-L.: Drug repurposing for cancer therapy. Sig Transduct Target Ther. **9**, 1–33 (2024). https://doi.org/10.1038/s41392-024-01808-1
16. Maeser, D., Gruener, R.F., Huang, R.S.: OncoPredict: an R package for predicting in vivo or cancer patient drug response and biomarkers from cell line screening data. Brief. Bioinform. **22**, bbab260 (2021). https://doi.org/10.1093/bib/bbab260
17. Klaus, B., Reisenauer, S.: An end to end workflow for differential gene expression using Affymetrix microarrays. F1000Res. **5**, 1384 (2018). https://doi.org/10.12688/f1000research.8967.2
18. Bair, E., Hastie, T., Paul, D., Tibshirani, R.: Prediction by supervised principal components. J. Am. Stat. Assoc. **101**, 119–137 (2006)

19. Bair, E., Tibshirani, R.: Semi-supervised methods to predict patient survival from gene expression data. PLoS Biol. **2**, e108 (2004). https://doi.org/10.1371/journal.pbio.0020108
20. Fenlon, C., O'Grady, L., Doherty, M.L., Dunnion, J.: A discussion of calibration techniques for evaluating binary and categorical predictive models. Prev. Vet. Med. **149**, 107–114 (2018). https://doi.org/10.1016/j.prevetmed.2017.11.018
21. Wang, X. et al.: Function of low ADARB1 expression in lung adenocarcinoma. PLoS One. **14**, e0222298 (2019). https://doi.org/10.1371/journal.pone.0222298
22. Wang, X. et al.: Role of downregulated ADARB1 in lung squamous cell carcinoma. Mol. Med. Rep. **21**, 1517–1526 (2020). https://doi.org/10.3892/mmr.2020.10958
23. Wang, K. et al.: Identification of differentially expressed genes in non-small cell lung cancer. Aging (Albany NY). **11**, 11170–11185 (2019). https://doi.org/10.18632/aging.102521
24. Andriani, F. et al.: Diagnostic role of circulating extracellular matrix-related proteins in non-small cell lung cancer. BMC Cancer. **18**, 899 (2018). https://doi.org/10.1186/s12885-018-4772-0
25. Wei, Z. et al.: TTYH3, a potential prognosis biomarker associated with immune infiltration and immunotherapy response in lung cancer. Int. Immunopharmacol. **110**, 108999 (2022). https://doi.org/10.1016/j.intimp.2022.108999
26. Wu, Y. et al.: Uridine-cytidine kinase 2 (UCK2): a potential diagnostic and prognostic biomarker for lung cancer. Cancer Sci. **110**, 2734–2747 (2019). https://doi.org/10.1111/cas.14125
27. Zhou, H.-X. et al.: Plasma LUNX mRNA, a non-invasive specific biomarker for diagnosis and prognostic prediction of non-small cell lung cancer. Am. J. Cancer Res. **6**, 452–458 (2016)
28. Wang, Y. et al.: Metabolic reprogramming induced by inhibition of SLC2A1 suppresses tumor progression in lung adenocarcinoma. Int. J. Clin. Exp. Pathol. **10**, 10759–10769 (2017)
29. Liang, Y. et al.: Upregulated collagen COL10A1 remodels the extracellular matrix and promotes malignant progression in lung adenocarcinoma. Front. Oncol. **10**, 573534 (2020). https://doi.org/10.3389/fonc.2020.573534
30. Wu, J. et al.: METTL16 promotes stability of SYNPO2L mRNA and leading to cancer cell lung metastasis by secretion of COL10A1 and attract the cancer-associated fibroblasts. Int. J. Biol. Sci. **20**, 4128–4145 (2024). https://doi.org/10.7150/ijbs.95375
31. Zhou, Z. et al.: GLUT1 promotes cell proliferation via binds and stabilizes phosphorylated EGFR in lung adenocarcinoma. Cell Commun. Signal. **22**, 303 (2024). https://doi.org/10.1186/s12964-024-01678-8
32. Qian, D., Zha, D., Sang, Y., Tao, J., Zhuang, B., Cheng, Y.: A-to-I RNA edited miR-3167 restrains malignant Behaviors of lung adenocarcinoma by influencing SSR2-meditated hippo Signaling. Mol. Carcinog. **64**, 1552–1563 (2025). https://doi.org/10.1002/mc.70011
33. Pan, R. et al.: A novel immune cell signature for predicting osteosarcoma prognosis and guiding therapy. Front. Immunol. **13** (2022)
34. Yang, C. et al.: Prognostic risk signature and comprehensive analyses of endoplasmic reticulum stress-related genes in lung adenocarcinoma. J Immunol Res. **2022**, e6567916 (2022). https://doi.org/10.1155/2022/6567916
35. Zhang, J., Chen, M., Fang, C., Luo, P.: A cancer-associated fibroblast gene signature predicts prognosis and therapy response in patients with pancreatic cancer. Front. Oncol. **12**, 1052132 (2022). https://doi.org/10.3389/fonc.2022.1052132
36. Cheng, T. et al.: Development of a ferroptosis-based model to predict prognosis, tumor microenvironment, and drug response for lung adenocarcinoma with weighted genes co-expression network analysis. Front. Pharmacol. **13**, 1072589 (2022). https://doi.org/10.3389/fphar.2022.1072589

Prevalence of *Toxocara* Spp. Eggs in Soil From Public Parks in Tra Vinh City, Tra Vinh Province, Vietnam in 2024

Ly Khanh Linh[1], Huynh Nhut Thuy[1], Nguyen Van Thong[1], Nguyen Ngoc The[1], Tao Gia Phu[1], and Minh-Anh Dang Trinh[2,3](✉)

[1] Tra Vinh University, Tra Vinh, Vietnam
[2] School of Biotechnology, International University, Ho Chi Minh City, Vietnam
dtmanh@hcmiu.edu.vn
[3] Viet Nam National University, Ho Chi Minh City, Vietnam

Abstract. *Toxocara* spp., prevalent globally, primarily infects dogs (*T. canis*) and cats (*T. cati*). Humans acquire infection by accidentally ingesting embryonated eggs from contaminated soil, raw vegetables, or water, or via direct contact with infected pets. In tropical regions like Vietnam, factors such as free-roaming animals and suboptimal sanitation elevate transmission risk, especially among children. Once ingested, *Toxocara*-larvae migrate through human tissues (liver, lungs, eyes, central nervous system), causing silent but severe chronic damage. Detecting *Toxocara*-eggs in soil is crucial for assessing environmental contamination, identifying community-level infection risks, and guiding public health interventions, particularly in children's play areas like parks. The goal of this study was to determine the presence of *Toxocara*-eggs in soil samples from public parks in Tra Vinh City and identify associated environmental factors. This cross-sectional descriptive study, conducted from February to April 2024, surveyed five public parks in Tra Vinh City: Le Thanh Ton, Vo Nguyen Giap, Nguyen Dang, Pham Ngu Lao, and Kien Thi Nhan. Soil samples were collected from tree bases, and pet defecation spots. Standardized sampling protocols ensured consistency. *Toxocara* egg prevalence in soil was 5.6%. Contamination rates varied by park: Nguyen Dang (0%), Pham Ngu Lao (3.4%), Le Thanh Ton (1.1%), Kien Thi Nhan (1.1%), and Vo Nguyen Giap (0%). A significant association was found between *Toxocara*-egg presence and free-roaming pets. Parks where animals roamed freely had 10.17 times higher than in parks where this behavior was absent (95% confidence interval [CI]: 1.08–95.66, $p = 0.031$). The contamination was significantly linked to residents allowing pets to roam freely in parks.

Keywords: *Toxocara* spp., · public parks in Tra Vinh province · soil parasites

1 Introduction

Human toxocariasis caused by *Toxocara canis* or *Toxocara cati* is currently a significant public health concern worldwide [1]. The definitive hosts of *Toxocara* spp. are dogs and cats, in which the parasites develop into adult worms living in the small intestine

L. Vu et al. (Eds.): PBC 2025, 2026.
https://doi.org/10.1007/978-3-032-29579-8_11

of their definitive hosts. Dogs and cats become infected by ingesting embryonated eggs from contaminated sources such as soil and food, or via vertical transmission from infected mothers to offspring through the placenta (in dogs) or through nursing (in cats). Under favorable environmental conditions (humidity and temperatures between 10–35 °C), the eggs develop through embryonic stages to form first-stage larvae and subsequently second-stage larvae within the eggshell. Eggs containing second-stage larvae are infectious and can survive in the soil for months to several years.

Humans are accidental hosts infected by ingesting *Toxocara* spp. eggs containing larvae. Since humans are not the specific hosts, the eggs hatch in the intestine releasing larvae that penetrate the intestinal mucosa, enter the bloodstream, and migrate to the liver, lungs, and other tissues, causing localized inflammatory reactions and tissue damage. Clinical manifestations vary depending on the affected organs and infection intensity, ranging from asymptomatic cases to severe syndromes such as hepatitis, pneumonia, meningitis, or ocular damage leading to blindness. In humans, *Toxocara* spp. larvae become "paratenic hosts" or cause "incomplete parasitism". The larvae can evade the host immune system and survive in the human body for months to years [2].

To gain a deeper understanding of the status of *Toxocara* spp. larvae in soil and to provide baseline data for future comparative studies, thereby raising public awareness of the silent yet significant risks posed by *Toxocara* spp., we conducted this research.

2 Study Subjects and Methods

Study subjects This cross-sectional descriptive study was conducted in all five public parks within Tra Vinh City, including Le Thanh Ton Park (Ward 1), Vo Nguyen Giap Park (Ward 8), Nguyen Dang Park (Ward 7), Pham Ngu Lao Park (Ward 1), and Kien Thi Nhan Park (Ward 7). They are most frequently used recreational areas in Tra Vinh City and are known to have regular dog- and cat-walking activities. Soil samples were collected from high-risk contamination sites, includings the bases of trees, dense bushes, and areas frequently visited by dogs and cats. The survey targeted residents living in and around the vicinity of these five parks.

Sample size The sample size was calculated based on the formula

$$n = Z^2_{1-\frac{\alpha}{2}} \frac{p(1-p)}{d^2}$$

- n study sample size
- Z value derived from the standard normal distribution at a 95% confidence level
- p expected prevalence rate, with p = 0.358 [3]
- d desired margin of error, with d = 0.1

Applying these parameters to the formula, the calculated sample size was 88 samples.

The sample size was calculated using a single-proportion formula with an expected contamination prevalence (p) of 35.8% based on a previous Vietnamese study of N. Thi Lan Anh. Within each park, samples were collected from locations considered at higher risk of fecal contamination, including six sample from tree bases, six from shaded grass areas, and six from zones frequently used by pets, to obtain a total of 18 samples.

Sampling method Convenient sampling was used to select the samples.

Procedure

The survey procedure for detecting *Toxocara* spp. eggs in soil was conducted in three main steps. First, soil samples were collected in the morning by taking the surface soil layer (approximately 1 cm deep) within an area of 1 m^2, avoiding soil that is too deep or too dry. Each sample weighed about 100 g and was placed into a zipper bag labeled clearly with sample ID, location, and collection date.

Next, samples were processed in the laboratory using the Romanenko method. Each 100 g sample was divided into two 50 g portions to simultaneously apply two egg isolation techniques: (1) NaOH–$NaNO_3$ flotation and (2) saturated salt flotation. In the NaOH–$NaNO_3$ method, soil samples were mixed with 0.5% NaOH, centrifuged twice, then supplemented with saturated $NaNO_3$ solution to float the eggs. For the saturated salt method, samples were thoroughly shaken, allowed to settle, centrifuged, and the sediment was directly examined microscopically by drop; the remaining suspension was allowed to float in a glass tube and observed after 45 min.

Finally, *Toxocara* spp. eggs were detected under a light microscope at 10X and 40X magnifications, based on their characteristic shape, thick pitted shell, and size range of approximately 65–85 μm in length. Because *T. canis* and *T. cati* eggs overlap in morphology and size, we report findings as *Toxocara* spp. eggs without attempting species-level differentiation. All slides were examined independently by two trained parasitologists. Discrepant results were reviewed jointly until a consensus was reached. The results were carefully recorded.

Data Management and Analysis Data were updated and managed using Microsoft Excel software. Statistical analysis was performed using Stata version 14.0.

Data Analysis Descriptive statistics were used to summarize the prevalence of *Toxocara* spp. egg contamination, information on residents bringing dogs and cats to the parks, and the characteristics, knowledge, and behaviors of the participants.

Analytical statistics involved the Chi-square test or Fisher's Exact test to determine associations between variables, with a significance level set at $p < 0.05$. Epidemiological measures, specifically prevalence ratios (PR) with 95% confidence intervals (CI), were used to quantify associations. The criteria for identifying significant associations included $p < 0.05$ and 95% CI not containing the value 1.

Ethical Considerations The study was approved by the Ethics Committee of Tra Vinh University according to Decision No. 229/GCT-HĐĐĐ dated May 15, 2024 (Fig. 1).

3 Results and Discussion

3.1 Prevalence of *Toxocara* Spp. Egg Contamination in Soil

Figure 2 illustrated that the study, which involved 90 soil samples, found a *Toxocara* egg contamination prevalence of 5.6% in park soil. This rate was lower than findings from other studies in Vietnam, such as research by Nguyen Thi Lan Anh et al. (2016), which reported a 35.8% contamination rate in soil samples [3], and Bui Van Tuan (2018), who observed a 25% contamination rate in soil from public playgrounds [4]. The prevalence

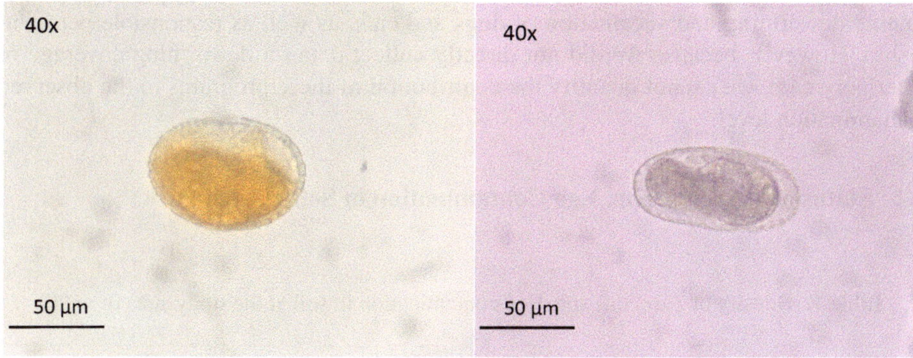

Fig. 1. *Toxocara* spp. eggs were observed under the microscope

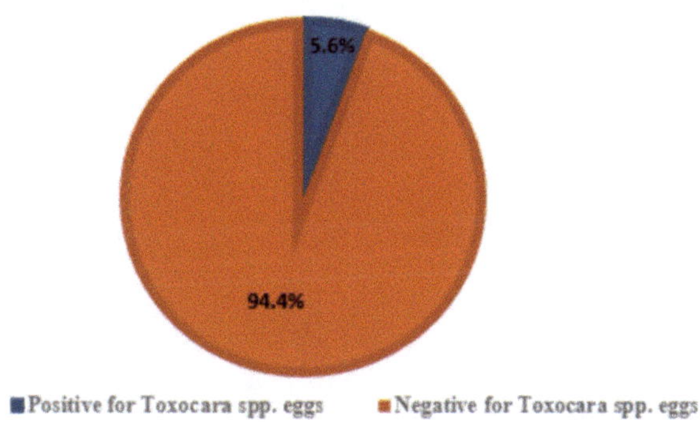

Fig. 2. Percentage (%) of *Toxocara* spp. egg contamination in soil

was also lower compared to studies conducted in other countries, including Sudhakar N.R (2013) in India with a 17.02% rate [5], M. Choobineh et al. (2019) in Iran at 16% [6], Alexis Israel Vargas Nava et al. (2020) at 16.7% [7], and Lukashev et al. (2020) in Russia at 40% [8].

The lower prevalence observed in our study may be due to differences in study locations, sampling periods, methodologies (e.g., sample depths, techniques, and egg recovery protocols). Previous studies conducted in different places and at different times may yield varying results because of environmental, climatic, and living condition differences for dogs and cats. Moreover, variations in soil sampling techniques and analytical methods can also affect the outcomes of *Toxocara* spp. egg detection. Additionally, fluctuations in dog and cat populations and the effectiveness of disease control measures can lead to changes in the prevalence of *Toxocara* spp. eggs over time. Improvements in sanitation and environmental management conditions may also contribute to the reduction in contamination rates.

The relatively low prevalence of *Toxocara* spp. eggs (5.6%) may reflect, at least in part, improved urban pet management and deworming practices in Tra Vinh City in recent years. Local veterinary authorities and public health programs have promoted routine deworming and vaccination of dogs and cats, as well as responsible pet ownership. However, because we did not directly collect data on deworming coverage or veterinary care, we cannot quantify the contribution of these programs to the observed contamination level.

3.2 Status of *Toxocara* Spp. Egg Contamination in Soil at each Park

Table 1. Density of *Toxocara* spp. Egg contamination in soil at the study sites (n = 90)

Sample collection site	Number of samples tested	Number of positive samples	Percentage (%)
Nguyen dang park, ward 7	18	0	0
Pham ngu lao park, ward 1	18	3	3.4 (0.04–0.42)
Le thanh ton park, ward 1	18	1	1.1 (0.001–0.27)
Kien thi nhan park, ward 7	18	1	1.1 (0.001–0.27)
Vo Nguyen giap park, ward 8	18	0	0
Total	90	5	5.6 (0.02–0.13)

Table 1 showed that 18 soil samples were tested at each park. No *Toxocara* spp. eggs were found in any of the 18 soil samples from Nguyen Dang Park and Vo Nguyen Giap Park (0%), resulting in an contamination rate of 0% for both. Le Thanh Ton Park and Kien Thi Nhan Park each had 1 soil sample positive for *Toxocara* spp. eggs out of 18 samples tested at each park (1.1%). Pham Ngu Lao Park recorded the highest rate, with 3 soil samples positive for *Toxocara* spp. eggs out of 18 samples tested (3.4%).

According to the study results, the technique using NaOH-$NaNO_3$ to detect *Toxocara* spp. eggs appeared sensitive than was superior to the method using saturated salt solution, as it detected all five positive samples. NaOH is a strong base capable of breaking down organic tissues and cleaning the sample, while $NaNO_3$ has strong oxidative properties that can help decompose unwanted organic substances, making it easier to detect the worm eggs. Upon thermal decomposition, $NaNO_3$ releases oxygen, providing a strong oxidative environment for reactions, which can enhance the effectiveness of breaking down organic matter and thereby highlight the worm eggs. Both NaOH and $NaNO_3$ have high solubility in water, allowing them to disperse evenly in the solution and interact effectively with the components of the sample, whereas saturated salt solution may have limitations in solubility and dispersion .

However, we did not perform formal validation experiments to quantify the sensitivity or inter-assay variability of these methods. This represents a methodological

Table 2. Results of soil testing for *Toxocara* spp. eggs using NaOH–$NaNO_3$ and saturated salt solution methods at each park

Sampling site	Number of positive samples	NaOH-$NaNO_3$	Saturated salt solution
Nguyen dang park, ward 7	0	0	0
Pham ngu lao park, ward 1	3	2	1
Le thanh ton park, ward 1	1	1	0
Kien thi nhan park, ward 7	1	1	0
Vo nguyen giap park, ward 8	0	0	0

limitation that may have led to underestimation of the true contamination prevalence (Table.2).

3.3 Frequency and Rate of Knowledge and Behavior of the People

Table 3. Knowledge and behavior of the people

Knowledge and behaviors of the people	Frequency (n)	Percentage (%)
Knowledge about dog/cat roundworm disease		
Yes	84	93.3
No	6	6.7
Allowing dogs/cats to roam freely		
Yes	28	31.1
No	62	68.9
Disposing of dog/cat feces		
Use tissue and dispose of in trash	19	21.1
Dispose in a pet waste bag	38	42.2
Other	33	36.7

The proportion of people who knew or had heard about dog/cat roundworm infection was 93.3%, which was higher than the 48.2% rate reported by Bui Van Tuan's study (2018) [4]. This suggested that public awareness about dog and cat roundworm larvae infection had been increasing, possibly due to widespread media coverage and individuals learning about the disease from relatives who had undergone medical examinations or tests at healthcare facilities.

However, the survey results on correct knowledge about the risk of *Toxocara* spp. egg infection were only 26.7%, lower than Bui Van Tuan's study (2018) which found a 45.6% rate [4]. Several reasons could explain this lack of accurate knowledge: many individuals may not have had access to sufficient and precise information; some might have mistakenly believed the disease was not dangerous or only occurred in rural areas, despite its potential presence anywhere; and there might have been a scarcity of effective community health education programs aimed at raising awareness about this illness (Table 3).

According to the study, 31.1% of people allowed their dogs/cats to roam freely, a rate lower than Bui Van Tuan's study (2018), where 83.5% of dogs roamed freely [4]. The reduced rate of free-roaming dogs/cats could be attributed to several factors: the risk of dog/cat bites and rabies transmission, which posed a severe community danger; the pollution of streets with feces and urine, negatively impacting the environment and human health; and the potential for conflicts between households or neighbors due to uncontrolled animals.

Regarding interview results, 21.1% of individuals used tissue paper and disposed of it in a trash bin. Another 42.2% managed dog/cat feces by placing it into feces bags. The remaining 36.7% reported using other disposal methods, such as throwing feces into vacant lots, using ash to dry feces before sweeping it onto the ground, or flushing it down the toilet. Disposing of feces by putting it in feces bags or using tissue paper and throwing it in the trash were considered correct disposal methods, accounting for 63.3% of cases. Safe collection and disposal of dog/cat feces were essential to prevent the spread of worm eggs into the environment. To protect the environment and health, everyone needed to handle feces responsibly, carefully, and hygienically.

3.4 The Relationship Between People's Knowledge and Behavior and the Contamination of *Toxocara* Spp. Eggs in Soil at Parks in Tra Vinh City

The study did not find a statistically significant association between the presence of *Toxocara* spp. eggs in soil and people's awareness of canine/feline roundworm. Survey results showed that individuals who were aware of the disease had a 0.28 times lower likelihood of encountering contaminated soil compared to those who were not aware ($p > 0.05$). Canine/feline roundworm was a public health issue that could occur anywhere, regardless of whether the setting was rural or urban. In Tra Vinh province, residents likely gained knowledge about the disease through mass media, healthcare facilities, public health education campaigns, or information from the Tra Vinh Center for Disease Control (Table 4).

The study found a statistically significant relationship between the presence of *Toxocara* spp. eggs in soil and the practice of allowing dogs and cats to roam freely. Specifically, the odds of soil contamination with *Toxocara* spp. eggs were 10.17 times higher when residents allowed their pets to roam freely compared to when pets were kept indoors ($OR = 10.17$; 95% CI: 1.08–95.66, $p = 0.031$). This finding indicates that allowing pets

Table 4. The relationship between people's knowledge and behavior and the contamination of *Toxocara* spp. eggs in parks

Knowledge and Behavior	Infection with *Toxocara* spp. eggs		OR (CI 95%)	p*
	Yes (%)	No (%)		
Knowledge about dog/cat roundworm disease				
Yes	4 (4.8)	80 (95.2)	0.24 (0.02–2.68)	0.298
No	1(16.7)	5 (83.3)	1	
Allowing dogs/cats to roam freely				
Yes	4 (14.3)	24 (85.7)	10.17 (1.08–95.66)	0.031
No	1 (1.6)	61 (98.4)	1	
Disposing of dog/cat feces				
Use tissue and dispose of in trash	1 (5.3)	18 (94.7)	1	0.67
Dispose in a pet waste bag	3 (7.9)	35 (92.1)	1.54 (0.15–15.91)	
Other	1 (3.0)	32 (97.0)	0.56 (0.03–9.54)	

p Fisher exact test*

to roam freely increases the likelihood of environmental contamination with *Toxocara* eggs. There is no existing research with comparable findings on this factor, so comparisons could not be made. When dogs or cats are allowed to roam, they may defecate anywhere, increasing the risk of parasite eggs spreading into the environment. *Toxocara* spp. eggs can survive in soil for months and even across seasons, increasing the risk of infection to both humans and other animals. People may inadvertently ingest these eggs through contact with contaminated soil, dirty hands, or contaminated food. The larvae can then enter the intestines, migrate to internal organs, and survive in the body. Roaming animals may also contaminate water sources, further increasing human exposure and infection risk. Therefore, controlling free-roaming dogs and cats is an important measure to protect both human and environmental health.

The study also examined methods of pet feces disposal but found no statistically significant relationship between soil contamination and the use of tissue paper to collect and dispose of feces in trash bins. Similarly, no significant relationship was found with the practice of disposing of feces in poop bags (1.54 times more likely than tissue disposal, $p > 0.05$) or alternative disposal methods (0.56 times less likely than tissue disposal, $p > 0.05$). According to the literature available, no studies have examined these disposal methods in relation to *Toxocara* spp. egg presence in soil, so comparisons could not be made. If pet waste is not disposed of properly, parasite eggs in feces can contaminate soil and pose a risk to public health. We did not collect detailed quantitative data on environmental factors such as soil moisture, pH, shade, or park sanitation indices. As a result, we could only estimate crude associations between behavioral variables and contamination, without adjustment for potential confounders. This is an important limitation, and future studies should incorporate standardized environmental measurements

4 Conclusion

The prevalence of *Toxocara* spp. eggs in soil at public parks in Tra Vinh city was 5.6%. The study found a statistically significant association between the presence of *Toxocara* spp. eggs in soil and the behavior of allowing dogs/cats to roam freely. This highlights the importance of pet management and raising community awareness in preventing zoonotic parasitic diseases.

Acknowledgments. We would like to thank Tra Vinh University and International University, Viet Nam National University-HCM for support of time and facilities for this study.

References

1. Rostami, A. et al.: Human toxocariasis—a look at a neglected disease through an epidemiological 'prism'. Infect. Genet. Evol. **74**, 104002 (2019)
2. Liravizadeh, S., Khabisi, S.A., Khorashad, A.S., Mirahmadi, H.: Molecular determination of *Toxocara* spp. eggs isolated from public parks and playgrounds in Zahedan, Southeast Iran. J. Parasitol. Res. **2024**, 2132696 (2024)
3. Anh, N.T.L., Thuy, D.T.T., Hoan, D.H., Hop, N.T., Dung, D.T.: Levels of *Toxocara* infections in dogs and cats from urban Vietnam together with associated risk factors for transmission. J. Helminthol. **90**(4), 508–510 (2016)
4. Bui, V.T.: Current Situation and Related Factors of Toxocara Canis Larval Infection in Humans in Mo Duc District, Quang Ngai Province in 2016 and Effectiveness of some Interventions, Ph.D. dissertation, Institute of Hygiene and Epidemiology, (2018)
5. Sudhakar, N.R., Kumar, A.B., Sharma, P.K., Singh, M.N.: Prevalence of *Toxocara* species eggs in soil samples of public health importance in and around Bareilly, Uttar Pradesh, India. Vet.Y World. **6**(2), 87–90 (2013)
6. Choobineh, M., Mikaeili, F., Sadjjadi, S., Ebrahimi, S., Iranmanesh, S.J.: Molecular characterization of *Toxocara* spp. eggs isolated from public parks and playgrounds in shiraz, Iran. J. Helminthol. **93**(3), 306–312 (2019)
7. Nava, A.I.V., Del Campo, N.C., Verdugo, I.E., Loera, J.J.P., Tizoc, C.L.B., Camacho, S.M.G.J.I.J.P.: Prevalence and viability of *Toxocara* spp. eggs in soil of public parks in northwestern Mexico. Iran. J. Parasitol. **15**(2), 196–203 (2020)
8. Lukashev, A.N., Ruzina, M.N., Akhmadishina, L.V.: *Toxocara* prevalence in dogs, cats and the environment in Russia. Adv. Parasitol. **109**, 801–817 (2020)

Optimization of Genomic DNA Extraction from Animal Tissue Using Response Surface Methodology

Song-Tan Nguyen, Huu-Khanh Le, Kim-Ngan Nguyen Ngoc, Thanh-Huong Do Thi, Nhat-Uyen Pham Khuong, and Dinh-Chuong Pham(✉)

Faculty of Applied Sciences, Ton Duc Thang University, Ho Chi Minh City, Vietnam
phamdinhchuong@tdtu.edu.vn

Abstract. The main aim of the present research was to optimize genomic DNA extraction from animal tissue using Response Surface Methodology (RSM). The effect of animal tissue weight (10–90 mg), the final concentration of Proteinase K (0.6–2.2 mg/mL), and GuSCN concentration in Binding Buffer ST2 (3.5–5.5 M) were investigated. Genomic DNA concentration is the main criterion to help evaluate the process of optimizing the genomic DNA extraction from animal tissue. In addition, other criteria to assess the extraction process include purity, stability, etc. After optimization, the animal tissue weight is 57 mg, the final concentration of Proteinase K is 1.4 mg/mL, and 4.6111 M GuSCN in Binding Buffer ST2 gives an experimental genomic DNA concentration of 210.08 ± 10.91 μg /mL, 7.37% larger than the model's predicted result, this parameter was compared with experiment and the results showed no significant error. This proves that the RSM model has good repeatability, and optimal adjustment ability and plays an important role in optimization survey parameters. The genomic DNA extraction chemicals have a shelf life of at least 9 months and can extract DNA with a variety of input samples. The simple extraction process and recovery efficiency are not much different from commercialized kits, purity is always within an acceptable range (A260/280: 1.8 – 2.0), and can extract tissue samples with a small weight (from 10 mg).

Keywords: Optimization · Response surface methodology · Genomic DNA · Animal tissue

1 Introduction

Animals are multicellular organisms with complex biological systems, each carrying genetic information crucial for growth, development, and functioning. This genetic information is encoded in DNA, located in every cell of an animal's body. DNA serves as the blueprint for all biological functions, containing instructions that direct cellular activities, protein synthesis, and inheritance of traits [7].

Genomic DNA (gDNA) in animals refers to the complete set of DNA within an organism, encompassing all the genes and non-coding sequences across chromosomes

L. Vu et al. (Eds.): PBC 2025, 2026.
https://doi.org/10.1007/978-3-032-29579-8_12

within the nucleus. Genomic DNA is highly structured and organized to ensure stability and accurate transmission of genetic information across generations. It is also central to studies in genetics, molecular biology, and biotechnology, enabling scientists to understand genetic diversity, evolutionary relationships, and mechanisms of heredity in animals. Analyzing genomic DNA is therefore fundamental in fields like animal breeding, disease research, and conservation biology, as it provides insights into genetic health, adaptation, and resilience of animal species.

DNA is the "starting" molecule for any research, especially in molecular biology, because it is the smallest molecule that encodes the expression of any characteristic of any organism. DNA extraction is time-consuming, and DNA extraction products are raw materials for many molecular biology methods, including Southern blot, SNP, PCR, Real-time PCR, and gene sequencing. Therefore, this is a very important role; DNA extraction is the premise for genetic research. DNA extraction is a time-consuming and chemical-intensive process, and extraction efficiency and purity are important factors in evaluating the process. Therefore, optimizing the DNA extraction process is necessary to shorten the extraction time and use the right amount of chemicals to obtain a stable, effective, and pure amount of DNA.

The goal of a DNA extraction process that does not require expensive equipment does not use toxic chemicals, is economical, and is simple and fast is to optimize DNA extraction from animal tissue using the silica column membrane method. This method allows the collection of DNA with high purity and efficiency in a short time [3]. Although the silica column-based DNA extraction method has a several drawback such as must use of many supporting equipment (centrifuge, thermal incubation tank, vortex machine), the amount of DNA obtained is limited by the membrane area and the ability to separate samples with high viscosity is poor, there are many advantages and suitable for almost extracted conditions in laboratory, with simple process, high recovery efficiency, low execution time, high purity compared to traditional methods, low implementation costs, safe chemicals [10].

Product after optimization of DNA extraction process from animal tissue is necessary and important in many fields. It can especially serve experimental subjects at school that have limited time, such as molecular genomic experiments, or research and theses of students in the education environment. Serves for medical testing and qualitative tests to see whether animals are infected or not. Researching disease-causing genes, thereby developing diagnosis and treatment methods. Creating vaccines helps animals fight infectious diseases, protecting the health of animals and humans. Helps with animal origin tests for animals, foods of animal origin, etc. for customs procedures, certificates, and licensing. Contributed greatly to biodiversity research, helping to extract and clone DNA of endangered and hard to reach species, thereby building phylogenetic trees and assessing the level of genetic diversity [9].

Normally, the DNA extraction kit is based on four basic steps in the DNA extraction process are cell lysis, protein removal, DNA precipitation and DNA purification. Kit manufacturers will mix chemicals to perform the above steps, depending on the manufacturer that the chemicals are mixed with a different number of tubes. The DNA generated from the kit must be quality-checked for DNA content, DNA purity, DNA integrity (DNA length). In addition, one must assess the stability of the kit after a period of storage [4].

In addition, to optimize important factors that directly affect the extraction process such as sensitivity and chemical concentration, response surface methodology (RSM) is a suitable method to optimize this kit. Response Surface Methodology is an efficient, quick, and cost-effective technique for assessing the influence of influencing factors on any system having input variables and output objective functions, and it uses data from reality experiments [6].

Response surface methodology (RSM) is a group of statistical and mathematical techniques, which are based on the fit of a polynomial equation to the experimental value. RSM is used for analyzing and modeling complex interactions between multiple independent variables and the responses they produce on the dependent variable based on experimental values. It has to describe a data set's behavior to make statistical predictions [2].

2 Methods

2.1 Chemicals and Materials

Animal tissue material. Pig samples (muscle, intestine, liver, spleen) – pig (*Sus scrofa*) and shrimp (*Litopenaeus vannamei*) were collected from the southern area of Vietnam. Catfish (*Pangasianodon hypophthalmus*) were collected from Hai Fish Hatchery Farm, Ho Chi Minh City. After pre-treatment, fish and shrimp dissection, and sample washing, samples were stored at −20 °C in the refrigerator before being used for research. The main object for the single factor and RSM experiment used in the study is a pig muscle sample.

Reagents. The chemicals used in the study were purchased from different companies, including Tris-HCl (Sigma-Aldrich, Merck, Germany), NaCl (Labscan, Thailand), Na_2EDTA (Sigma-Aldrich, Merck, Germany), SDS (Sigma-Aldrich, Merck, Germany), GuSCN (Bio basic, Canada), Urea (Xilong Scientific, China), Triton X-100 (Xilong Scientific, China), Ethanol absolute (Merck, Germany), Proteinase K (ABT, Vietnam) and silica column (Biosharp, China), pellet pestle (ABT, Vietnam).

Control kit: ISOLATE II Genomic DNA Kit (Bioline, owned by Meridian Bioscience Inc, USA).

2.2 Genomic DNA Extraction from Animal Tissue Procedures

In this study, genomic DNA was extracted from animal tissue using silica column-based DNA extraction method. Briefly, the animal tissue sample was prepared and pre-treated to transfer to 1.5 mL Eppendorf. Then, the sample was ground using a pestle in 200 μL of Lysis Buffer ST1 (50 mM Tris-HCl, 20 mM NaCl, 20 mM Na2EDTA, 2% SDS; pH 8.0) before adding 25 μL of 20 mg/mL Proteinase K solution and incubating the sample at 55 °C for 1 h. After that, 200 μL of Binding Buffer ST2 (GuSCN—concentration survey, 10 mM Urea, 10 mM Tris-HCl, 20% Triton X-100; pH 4.4) was added to the sample and continued to incubate the sample at 70 °C for 15 min. Further, add 200 μL Ethanol absolute and transfer the sample (lysed) to the column (prepared to place the silica column into a 2 mL collection tube), followed by centrifugation at 13,000 rpm for 1 min. The fluid was removed, and the collection tube was reused.

Following the DNA in the silica column, wash with 500 μL of wash buffer ST3a (4 M GuSCN, 20 mM Tris-HCl; pH 6.5) at 13,000 rpm for 1 min, repeat removing fluid, and keep the collection tube for the next step. In the next wash step, the silica membrane was washed with 500 μL of wash buffer ST3b (100 mM NaCl, 10 mM Tris-HCl; pH 7.0) at 13,000 rpm for 1 min and discarded flow-through and reused collection tube. After that, the silica membrane was dried at 13,000 rpm for 1 min. Finally, the silica column was placed in 1.5 mL Eppendorf (used collection tube was removed), after that add 100 μL of elution buffer ST4 (10 mM Tris-HCl, 1 mM Na2EDTA; pH 8.0) onto the center of silica membrane, before incubating at temperature room for 1 min and centrifugation at 13,000 rpm for 1 min, and stored at 4 °C.

Binding Buffer ST2 was formulated to ensure effective lysis and promote DNA-silica binding in the chaotropic environment. GuSCN serves as the main chaotropic agent that rapidly denatures proteins, inactivates nucleases, and removes the hydration layer around DNA and silica. Urea enhances protein unfolding, while Triton X-100 supports membrane solubilization. The mildly acidic pH further strengthens DNA adsorption to the silica membrane. This combination enables efficient lysis and maximizes nucleic acid capture.

2.3 Single Factor Experiments

Investigate the effect of animal tissue weight on the efficiency of DNA extraction from animal tissue. The experiment had six treatments, and each was carried out twice. Firstly, prepare 10 mg, 30 mg, 50 mg, 70 mg, 90 mg, 110 mg tissue samples respectively, and transfer them into 1.5 mL Eppendorf. Perform sample extraction according to the protocol (Sect. 2.2) under the condition that other factors are kept constant: 2.2 mg/mL Proteinase K (add 25 μL stock 20 mg/mL Proteinase K), 6 M GuSCN (in Binding Buffer ST2).

Investigate the effect of Proteinase K concentration on the efficiency of DNA extraction from animal tissue. The experiment had six treatments, and each was carried out twice. Following the procedure (Sect. 2.2), 50 mg of tissue sample was prepared and transferred to 1.5 mL Eppendorf. Conduct sample extraction according to the protocol with 20 mg/mL Proteinase K stock suction volumes of 5 μL, 10 μL, 15 μL, 20 μL, 25 μL, 30 μL (final Proteinase K concentration changes to 0.6 mg/mL, 1 mg/mL, 1.4 mg/mL, 1.8 mg/mL, 2.2 mg/mL, and 2.6 mg/mL, respectively) while other factors were kept constant in 6 M GuSCN in Binding Buffer ST2.

Investigate the effect of GuSCN concentration on the efficiency of DNA extraction from animal tissue. Firstly, 50 mg of tissue sample was prepared and transferred into 1.5 mL Eppendorf. Carry out sample extraction according to the protocol (Sect. 2.2) using binding buffer ST2 with different concentrations of GuSCN (3.5 M, 4 M, 4.5 M, 5 M, 5.5 M, 6 M, respectively) under the condition that other factors are kept constant at

2.2 μg/mL Proteinase K (absorb 25 μL of Proteinase K stock 20 mg/mL). The experiment had six treatments, and each was carried out twice.

2.4 Determination of Genomic DNA Concentration and DNA Purity

After extraction, the sample was diluted 15 times in elution buffer or double distilled water (pH 7), and 280 μL of elution buffer was mixed with 20 μL of sample in a 1.5 mL Eppendorf. The diluted sample was determined by determining the optical density at 260 nm and the quality of DNA was evaluated by measuring the ratio of A260/A280 on an ultraviolet spectrophotometer (Evolution 220 UV-VIS, Thermo Scientific—USA). After that, to examine DNA purity and quantification by calculating, pure DNA has an OD260/OD280 ratio of ~1.8 [1].

$$\text{dsDNA concentration} = 50\ \mu\text{g/mL} \times \text{OD260} \times \text{dilution factor} \tag{1}$$

2.5 Optimization of Genomic DNA Extraction from Animal Tissue Using RSM

Experimental design. Based on the results obtained from experiments investigating the effect of single factors, the response surface methodology was used to optimize the extraction conditions to obtain effective DNA extraction from animal tissue. Three-level Box-Behnken design (BBD) for three-factor was employed to evaluate the influence of genomic DNA extraction factors. Genomic DNA extraction process from animal tissue was conducted under different animal tissue weights, the final Proteinase K concentration, and GuSCN concentration in Binding Buffer ST2. The relationship between DNA concentration and extraction conditions is expressed through a second-order polynomial equation as follows [5]:

$$\begin{aligned} Y = {} & b_0 + b_1X_1 + b_2X_2 + b_3X_3 + b_{12}X_1X_2 + b_{13}X_1X_3 + b_{23}X_2X_3 \\ & + b_{11}X_1^2 + b_{22}X_2^2 + b_{33}X_3^2 \end{aligned} \tag{2}$$

where Y is the genomic DNA concentration (μg/mL) obtained after extraction; X_1, X_2, and X_3 are the weight of animal tissue (mg), Proteinase K concentration (mg/mL), and GuSCN concentration (M), respectively; β_0 is a constant, $\beta_1, \beta_2, \beta_3$ are first-order coefficients; $\beta_{11}, \beta_{22}, \beta_{33}$ are second-order coefficients; $\beta_{12}, \beta_{13}, \beta_{23}$ are interaction coefficients between factors.

The BBD model for three-factor suggests 15 treatments, corresponding to 15 groups of extraction conditions; these groups include the same factors but different survey parameters. Extract the animal tissue sample with each proposed condition, then measure the absorbance at 260 nm and determine the DNA concentration in the sample.

Statistical analysis and obtaining the optimal extraction process. Minitab 16 (Minitab Inc., PA, USA) was employed to conduct an analysis of variance (ANOVA), establish the regression model, and determine the optimal extraction conditions for maximizing the genomic DNA concentration [8]. Interactive graph of factors drawn with Origin 2024 software. Values in research were expressed as mean ± standard deviation (SD). Differences in values expressed with different letters (a, b, c…) represent statistically significant differences ($P < 0.05$).

2.6 Stability of Chemical

To evaluate the shelf life and stability of the optimized reagents and genomic DNA extraction procedures, based on the optimized procedures, storing the reagents at room temperature (except Proteinase K stored at −20 °C) genomic DNA extractions were performed at different time points (immediately after preparation—0 month, after 3, 6, and 9 months).

2.7 Tested on Different Types of Animal Tissues

Different subjects have different amounts of DNA, to evaluate the optimized genomic DNA extraction chemicals and procedures, and the ability to extract different subjects, based on the optimized procedures, perform genomic DNA extraction of different subjects such as pigs, shrimps, and catfish in different organs.

2.8 Comparison with Commercial Kit

Perform animal tissue sample extraction with optimized procedure. With the control kit, the genomic DNA extractions from all animal tissue samples were performed by ISOLATE II Genomic DNA kit (Meridian Bioscience Inc., USA), according to the manufacturer's protocols.

3 Results and Discussion

3.1 Single Factor Experiments

Investigate the effect of animal tissue weight on the efficiency of DNA extraction from animal tissue. The study found that genomic DNA (gDNA) concentration from animal tissue increases with tissue weight, reaching a maximum of approximately 400 μg/mL. Starting from 10 mg tissue with a DNA yield of 50.25 ± 3.18 μg/mL, the concentration rises to 412.88 ± 12.2 μg/mL at 90 mg. However, increasing tissue weight further does not boost DNA yield under the same extraction conditions (1 h with 2.2 mg/mL Proteinase K) (Fig. 1).

Investigate the effect of Proteinase K concentration on the efficiency of DNA extraction from animal tissue. Data from Fig. 2 shows that DNA concentration increases with Proteinase K concentration, reaching a peak at 1.4 mg/mL (277.88 ± 17.50 μg/mL), 1.79 times higher than the initial level. Beyond this concentration, DNA yield declines, with the lowest yield at 2.6 mg/mL (123 ± 16.97 μg/mL), a 2.26-fold decrease from the peak. This indicates that Proteinase K concentration impacts DNA extraction efficiency, with 1.4 mg/mL yielding the highest DNA from 50 mg of tissue in 1 h. Higher enzyme concentrations may disrupt DNA attachment to the silica membrane, causing reduced DNA concentration.

Investigate the effect of GuSCN concentration on the efficiency of DNA extraction from animal tissue. Data from Fig. 3 indicates that GuSCN concentration in ST2 buffer

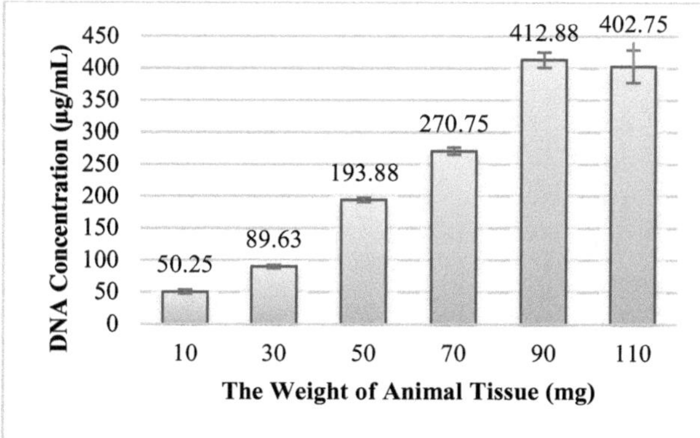

Fig. 1. The effect of animal tissue weight on gDNA concentration.

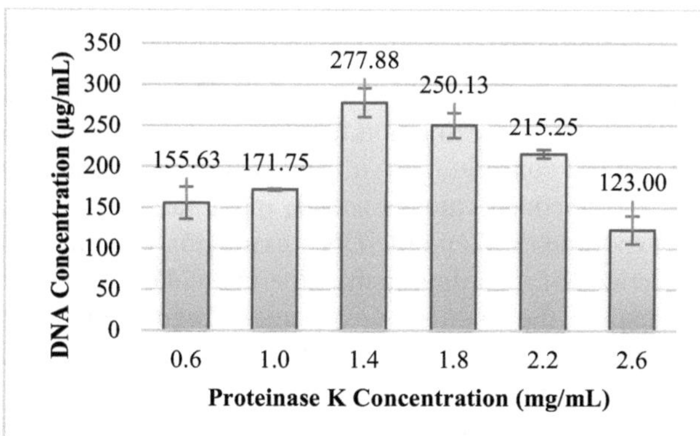

Fig. 2. The effect of Proteinase K concentration on gDNA concentration.

significantly affects DNA extraction from animal tissue. At 3.5 M GuSCN, DNA concentration is low (123.75 ± 5.3 μg/mL), but it increases as GuSCN concentration rises, peaking at 4.5 M with 287.63 ± 13.26 μg/mL—an increase of 2.32 times. Beyond 4.5 M, DNA concentration remains steady at around 200 μg/mL. GuSCN aids in cell membrane lysis, protein denaturation, nuclease inactivation, and DNA binding to silica membranes. Thus, 4.5 M GuSCN optimizes DNA extraction efficiency at high concentrations.

The optimal concentration of 4.5 M GuSCN reflects the balance required for efficient chaotropic action. At lower concentrations, protein denaturation and dehydration of DNA and silica are insufficient, resulting in poor binding. As GuSCN increases, denaturation and nuclease inactivation improve, enhancing DNA recovery. Beyond 4.5–5 M, the effect reaches saturation and solution viscosity may hinder flow-through, explaining the plateau in DNA yield.

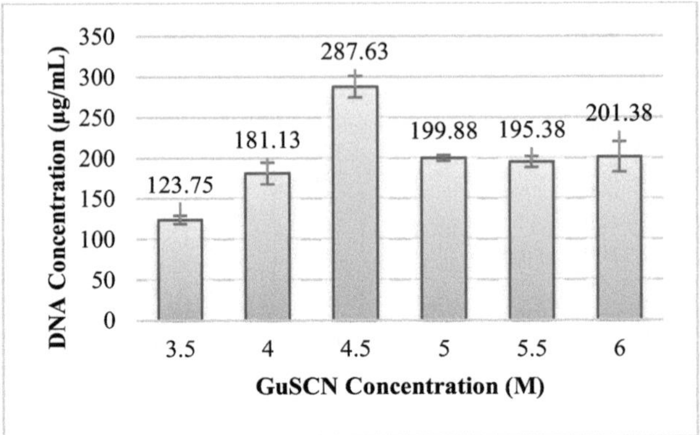

Fig. 3. The effect of GuSCN concentration on gDNA concentration.

3.2 RSM Model Development

The RSM method, utilizing a Box-Behnken model with three input variables at three levels, was used to examine how extraction factors (animal tissue weight, Proteinase K concentration, and GuSCN concentration) impact genomic DNA concentration. Table 1 presents the coded and uncoded values of these input variables, summarizing results from single-factor experiments. Genomic DNA extraction conditions using the RSM-BBD model were carried out according to the experimental design to determine optimal conditions. Table 2 displays the treatments, experimental extraction results, and predicted values, where the central experiments (run numbers 5, 9, 11) demonstrated highlighting reproducibility and high accuracy.

This allowed the relationship between input variables and measured responses to be effectively modeled; the regression equation describes the correlation between DNA concentration and coding variables as follows:

$$Y = -363.8 + 6.124X_1 - 45.1X_2 + 179.2X_3 - 0.05947X_1 * X_1 - 25.13X_2 * X_2$$
$$-24.01X_3 * X_3 + 0.0563X_1 * X_2 + 0.1474X_1 * X_3 + 24.32X_2 * X_3 \quad (3)$$

Response variable in there:
+Y: gDNA concentration (µg/mL).
+X_1: Animal tissue weight (mg).
+X_2: Proteinase K concentration (mg/mL).
+X_3: GuSCN concentration (M).

The model predictions and actual values show good agreement, indicating that the established model provides accurate and satisfactory results. The overall impact of input variables on gDNA concentration was analyzed using T-test (Table 3). Mostly low P-values (<0.05) were obtained, indicating that the corresponding parameters were significant. As a result, the reaction's critical components include the initial term, all of the quadratic terms, the three linear terms (X_1-X_3), and the interaction term. Consequently, the response surface curves may be plotted and the perfect conditions for maximizing the response can be predicted using the created model (Table 4).

Table 1. Results of range limits and variability of factors affecting the process of DNA extraction from animal tissue.

Variables	Symbols	Variable levels		
		- 1	0	1
Animal tissue weight (mg)	X_1	10	50	90
Proteinase K concentration (mg/mL)	X_2	0.6	1.4	2.2
GuSCN concentration (M)	X_3	3.5	4.5	5.5

Table 2. Experimental results and predictions from the RSM—BBD model.

Treatments	Variables			DNA Concentration (μg/mL)	
	X1	*X2*	*X3*	Experimental value	Predicted value
1	90	2.2	4.5	$119.167^{cd} \pm 15.685$	116.552
2	50	0.6	5.5	$141.000^{bc} \pm 5.414$	139.297
3	10	1.4	5.5	$35.333^{e} \pm 2.082$	34.421
4	50	0.6	3.5	$172.333^{ab} \pm 6.146$	169.193
5*	50	1.4	4.5	$194.083^{a} \pm 12.141$	191.764
6	50	2.2	3.5	$123.417^{cd} \pm 20.249$	125.120
7	90	1.4	5.5	$120.333^{cd} \pm 10.408$	119.807
8	10	1.4	3.5	$36.667^{e} \pm 1.893$	37.193
9*	50	1.4	4.5	$187.375^{a} \pm 6.541$	191.764
10	90	0.6	4.5	$115.875^{cd} \pm 11.137$	118.104
11*	50	1.4	4.5	$193.833^{a} \pm 6.331$	191.764
12	50	2.2	5.5	$169.917^{ab} \pm 19.609$	173.057
13	10	0.6	4.5	$45.500^{e} \pm 1.803$	48.115
14	10	2.2	4.5	$41.583^{e} \pm 1.283$	39.354
15	90	1.4	3.5	$98.083^{d} \pm 3.503$	98.995

*: center point, X_1: The weight of animal tissue (mg), X_2: Proteinase K concentration (mg/mL), X_3: GuSCN concentration (M); values are expressed as mean ± standard deviation (n = 3). Values with different letters (a, b, c…) represent statistically significant differences ($P < 0.05$).

Comparing the correlation between experimental gDNA concentrations and predictions from the model gives results as shown in Fig. 4. Through the high coefficient of determination ($R^2 = 99.83\%$), it is found that the model can be used for experimental prediction. Thereby, it can be seen that the regression equation accurately describes the data obtained from the experiment.

RSM was plotted using Origin 2024 software to study the influence of factors on genomic DNA concentrations as well as the interactions between these factors. Based on

Table 3. Significance of the coefficients in the empirical model.

Term	Coef	SE Coef	T value	P value
Constant	191.764	2.313	82.896	0.000[a]
X1	36.797	1.417	25.976	0.000[a]
X2	−2.578	1.417	−1.820	0.128
X3	4.510	1.417	3.184	0.024[a]
X1*X1	−95.148	2.085	−45.631	0.000[a]
X2*X2	−16.085	2.085	−7.714	0.001[a]
X3*X3	−24.012	2.085	−11.516	0.000[a]
X1*X2	1.802	2.003	0.900	0.410
X1*X3	5.896	2.003	2.943	0.032[a]
X2*X3	19.458	2.003	9.713	0.000[a]

[a]: $P < 0.05$ indicates that the model terms are significant

Table 4. Variance analysis for RSM model of BBD design.

Source	DF	Seq SS	Adj SS	Adj MS	F	P
Regression	9	47390.4	47390.4	5265.6	327.99	0.000
Linear	3	11048.0	11048.0	3682.7	229.39	0.000
X1	1	10832.1	10832.1	10832.1	674.73	0.000
X2	1	53.2	53.2	53.2	3.31	0.128
X3	1	162.7	162.7	162.7	10.14	0.024
Square	3	34675.8	34675.8	11558.6	719.99	0.000
X1*X1	1	31794.0	33426.7	33426.7	2082.15	0.000
X2*X2	1	752.9	955.3	955.3	59.50	0.001
X3*X3	1	2128.9	2128.9	2128.9	132.61	0.000
Interaction	3	1666.5	1666.5	555.5	34.60	0.001
X1*X2	1	13.0	13.0	13.0	0.81	0.410
X1*X3	1	139.1	139.1	139.1	8.66	0.032
X2*X3	1	1514.5	1514.5	1514.5	94.34	0.000
Residual Error	5	80.3	80.3	16.1		
Lack-of-Fit	3	51.3	51.3	17.1	1.18	0.488
Pure Error	2	28.9	28.9	14.5		
Total	14	47470.6				
		S = 4.00674		PRESS = [illegible]		
$R^2 = 99.83\%$		Predicted $R^2 = 98.13\%$		Adjusted $R^2 = 99.53$		

the regression equation, when fixing one factor and changing the value of the remaining 2 factors, a matrix of points will be created, 3D response surface cells will be drawn from that matrix, it is very simple and useful for assessing the interactive effects of factors on the response function. These types of graphs show the effects of two factors

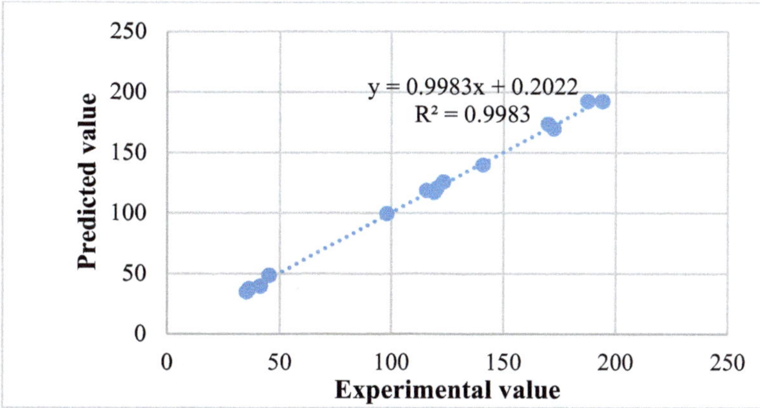

Fig. 4. Correlation between experimental and predicted DNA concentration.

on one reaction at a time. In all graphs, the remaining two elements are kept at the center. Elliptical contours are obtained when perfect interactions between independent variables occur.

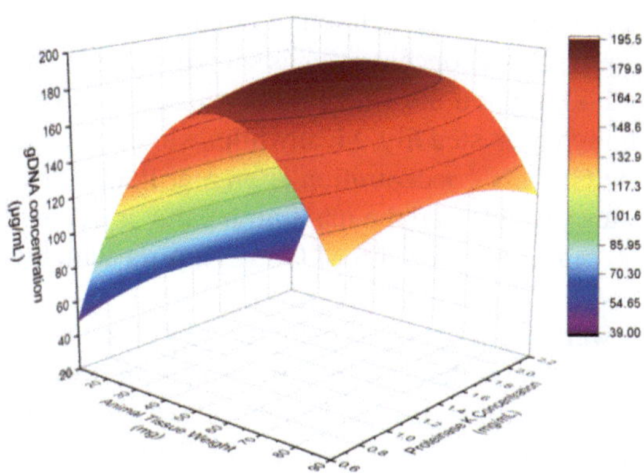

Fig. 5. Response surface plot showing the interaction between animal tissue weight (mg) and Proteinase K concentration (mg/mL) to gDNA concentration.

Figure 5 shows the combined effect of animal tissue weight and Proteinase K concentration on the gDNA when the remaining factors were maintained at their constant level. When extracting gDNA from animal tissue with 4.5 M GuSCN in ST2 buffer, tissue weight is in the range of 10–40 mg and Proteinase K concentration is in the range of 0.6–2.2 mg/mL, the DNA concentration is genome increases steadily. And when the tissue weight is in the range of 40–70 mg, the gDNA concentration reaches a large value and the maximum value reaches about 195.5 µg/mL when extracting 60 mg of tissue with 1.4 mg/mL Proteinase K (aspirate 15 µL 2.2 mg/mL Proteinase K).

As animal tissue weight continues to increase, gDNA concentration will continue to decrease steadily. This shows that gDNA extraction is an irreversible and unbalanced process, it is necessary to find the optimal weight of extracted animal tissue and Proteinase K to achieve maximum extraction efficiency of the tissue sample.

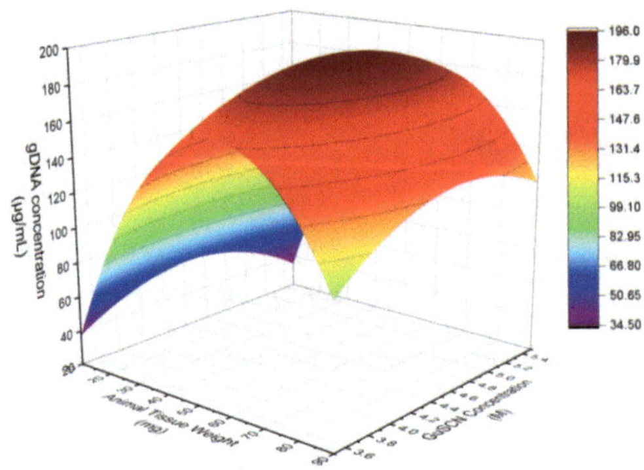

Fig. 6. Response surface plot showing the interaction between animal tissue weight (mg) and GuSCN concentration (M) to genomic DNA concentration.

The interaction between the two extraction conditions, tissue weight and GuSCN concentration, to genomic DNA concentration is shown in Fig. 6. When the tissue weight is between 10 mg and 40 mg and the GuSCN concentration is between 3.5 and 3.7 M, gDNA concentration increases steadily. Meanwhile, when the tissue weight is between 40 and 70 mg and the GuSCN concentration is greater than 3.7 M, the genomic DNA concentration fluctuates strongly and the maximum value is about 196 µg/mL when the extraction is about 60 mg tissue with 4.5 M GuSCN, after which gDNA concentration begins to decrease.

As animal tissue weight continues to increase, gDNA concentration will continue to decrease steadily. This shows that gDNA extraction is an irreversible and unbalanced process, it is necessary to find the optimal weight of extracted animal tissue to achieve maximum extraction efficiency of the tissue sample. The gDNA concentration showed signs of low extraction when the GuSCN concentration was low, but when starting from 4 M GuSCN, the gDNA concentration increased but did not show signs of much change and was stable. Therefore, it is necessary to find the optimal point of both animal tissue volume and GuSCN concentration.

Figure 7 illustrates the interaction between tissue weight and GuSCN concentration in relation to genomic DNA concentration. When Proteinase K concentration is between 0.6 and 1 mg/mL and GuSCN concentration is between 3.5 and 4.2 M, genomic DNA concentration steadily increases. However, when Proteinase K is increased to between 1 and 1.6 mg/mL and GuSCN to 4.2 to 4.8 M, the DNA concentration fluctuates, peaking at about 192.2 µg/mL with 1.4 mg/mL Proteinase K and 4.5 M GuSCN, before declining. At higher concentrations (1.6 to 2.2 mg/mL Proteinase K and 4.8 to 5.5 M GuSCN), genomic DNA concentration again rises steadily. This indicates that gDNA extraction is an irreversible and unbalanced process, highlighting the importance of optimizing both Proteinase K and GuSCN concentrations for maximum extraction efficiency.

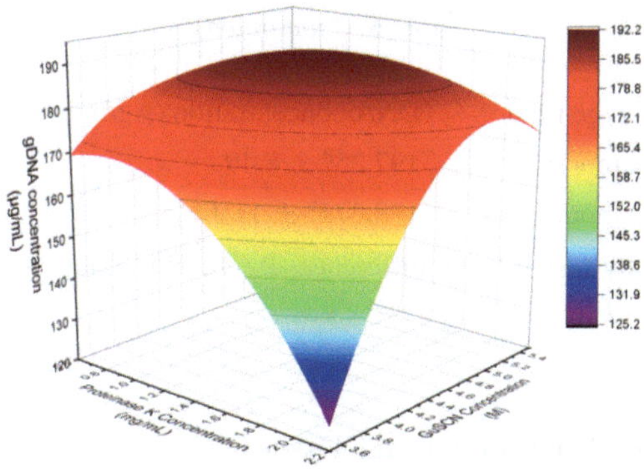

Fig. 7. Response surface plot showing the interaction between Proteinase K concentration (mg/mL) and GuSCN concentration (M) to gDNA concentration.

Obtaining optimal extraction conditions. By solving the RSM model based on the regression equation, the maximal genomic DNA concentration was predicted to be 195.6516 µg/mL using 57.6768 mg of animal tissue, 1.4 mg/mL of final concentration Proteinase K, and 4.6111 M GuSCN in Binding Buffer ST2. To suit the reality conditions in the laboratory, the animal weight sample was adjusted to 57 mg. Besides that, to confirm the prediction, experiments were conducted under optimal conditions. The actual gDNA concentration obtained was 210.08 ± 10.91 µg/mL, which was 7.37% higher than the model's predicted result, aligning well with the model's forecast. This result suggests that the developed RSM model is effective and can accurately represent the relationship between extraction factors and gDNA concentration in genomic DNA extraction from animal tissue.

3.3 Stability of Chemicals

After storing chemicals at room temperature (except Proteinase K stored at −20 °C), at time points such as 0, 3 months, 6 months, and 9 months proceed to extract DNA. The results obtained in Table 5 show differences in DNA concentrations at different time points, but this difference is not large and insignificant. At the 9-month mark, the standard deviation is higher than the previous milestones, showing that as time passes, the stability of the kit will be affected but not significantly (2.22% compared to the average value). From the values obtained from the experiment, it can be concluded that within 9 months from the date of buffer production, the ability to extract DNA will not be affected and the shelf life of chemicals can be at least 9 months (Table 6).

Table 5. Stability of chemicals.

Storage time	DNA concentration (µg/mL)
0	$202.875^{a} \pm 2.65$
3rd month	$196.125^{a} \pm 2.65$

(continued)

Table 5. *(continued)*

Storage time	DNA concentration (μg/mL)
6th month	$197.25^{a} \pm 8.49$
9th month	$198.375^{a} \pm 11.14$

[a]: represents a statistically insignificant difference ($P < 0.05$). Values are expressed as mean ± standard deviation ($n = 3$)

3.4 Test on Different Types of Animal Tissues

Table 6. Results of DNA extraction from different organs.

No	Species	Organ	Amount	DNA yield (μg)	OD260/280
1	Pig	Muscle	50 mg	19.8375 ± 1.11	1.92 ± 0.009
2		Intestine	50 mg	48 ± 2.55	1.98 ± 0.040
3		Liver	50 mg	139.8375 ± 5.25	1.87 ± 0.005
4		Spleen	50 mg	71.25 ± 3.08	1.85 ± 0.006
5	Shrimp	Muscle	50 mg	33.525 ± 0.95	1.98 ± 0.019
6		Feet	50 mg	49.2 ± 2.97	2.01 ± 0.031
7	Catfish	Liver	50 mg	100.35 ± 5.20	1.80 ± 0.009
8		Muscle	50 mg	19.65 ± 4.99	1.99 ± 0.007

Values are expressed as mean ± standard deviation ($n = 2$).

With the DNA extraction kit from animal tissue, the above experiment shows that the kit's ability to extract and purify is diverse, almost all animal tissue samples can be extracted from different species of animals, different animals such as pigs, shrimp, catfish to different organs from those species such as muscles, internal organs, etc. Each organ from different species has a different ability to collect DNA, the liver almost always has more DNA than the muscle groups, and DNA collected from the catfish liver (100.35 ± 5.20 μg) also has a difference compared to pig liver (139.8375 ± 5.25 μg). Using this kit for experiments from different species is feasible, depending on the purpose of use, choose the appropriate input object.

Ratio of OD_{260} and OD_{280} represents the purity of DNA. All samples have $OD_{260}/_{280}$ in the range of 1.8–2.0, proving the kit's ability to effectively extract and purify DNA. However, there are tissue samples with $OD_{260}/_{280}$ values near 2.0 and greater, those samples show signs of RNA contamination (muscle, intestine), to improve that, RNase can be used. Therefore, the genomic DNA extraction kit, in addition to separating many samples, also has reliable DNA purity.

3.5 Comparison with Commercial Kit

Table 7. Comparison of 2 kits.

No	Sample	Amount	DNA concentration (μg/mL)	
			Bioline	Research kit
1	Muscle (pig)	50 mg	$246.375^{b} \pm 7.95$	$202.5^{c} \pm 3.18$
2	Liver (catfish)	50 mg	$895^{a} \pm 7.07$	$836.5^{a} \pm 19.09$

Values are expressed as mean ± standard deviation (n = 2). Values with different letters (a, b, c…) represent statistically significant differences ($P < 0.05$).

Table 7 shows that when extracting DNA from the same pig muscle tissue sample, the optimal chemical research yields DNA with a concentration of 202.5 ± 3.18 μg/mL. However, if the DNA is extracted with the ISOLATE II Genomic DNA kit, the resulting concentration is approximately 246.375 ± 7.95 μg/mL, 1.22 times higher than that of the research chemical.

Besides, when extracting DNA from catfish liver, the DNA concentration of the research chemical is 836.5 ± 19.09 μg/mL, and that of the commercial kit is 895 ± 7.07 μg/mL; the difference in DNA concentration is not significant. The available data shows that the ISOLATE II Genomic DNA kit has better DNA recovery ability than the research kit. However, there are samples where the difference is insignificant, and there are samples with more differences but still within the allowable range.

The difference may be due to factors such as Proteinase K acting differently, the origin of the difference. The columns and DNA recovery capabilities of the silica columns of the two kits are completely different. Sample handling during the experiment is not the same. The chemicals of the ISOLATE II Genomic DNA kit are more optimal and stable than the research kit.

4 Conclusion

This research reports genomic DNA was extracted from animal tissue using silica column-based DNA extraction method. The study used an RSM optimization model to establish a statistically significant quadratic model for gDNA extraction from animal tissue. With a high correlation of 99.83% between experimental and predicted results, the model optimized key factors weight, Proteinase K concentration, and GuSCN concentration. Successful optimization of genomic DNA extraction process from animal tissue with optimal conditions such as using 57 mg animal tissue, 1.4 mg/mL Proteinase K, 4.6111 M GuSCN. Therefore, the RSM model has been successfully implemented to improve gDNA extraction conditions from animal tissues with high extraction efficiency, small input weight, optimizing the extraction process in a short time, stable, purity.

The chemical can extract various samples from different organs, demonstrating its versatility. When stored at room temperature (Proteinase K at −20 °C), and remains effective for stable gDNA extraction even after 9 months.

This research serves as a foundation for developing high-quality kits aimed at detecting bacteria. By establishing effective methods and optimal conditions for DNA extraction, the study provides essential insights into producing reliable diagnostic kits. These kits can facilitate early detection of bacterial infections, enhance laboratory efficiency, and support applications in both clinical and environmental testing, ultimately contributing to improved public health and disease management.

References

1. Barbas, C.F., Burton, D.R., Scott, J.K., Silverman, G.J.: Quantitation of DNA and RNA. Cold Spring Harb. Protoc. **2007**(11), pdb-ip47 (2007)
2. Bezerra, M.A., Santelli, R.E., Oliveira, E.P., Villar, L.S., Escaleira, L.A.: Response surface methodology (RSM) as a tool for optimization in analytical chemistry. Talanta **76**(5), 965–977 (2008)
3. Dairawan, M., Shetty, P.J.: The evolution of DNA extraction methods. Am. J. Biomed. Sci. Res **8**(1), 39–45 (2020)
4. Denhart, M., Doraiswamy, V.: Advantages of PCR Master Mix (2001)
5. Hassan, M.A., Bakhiet, E.K., Hussein, H.R., Ali, S.G.: Statistical optimization studies for polyhydroxybutyrate (PHB) production by novel Bacillus subtilis using agricultural and industrial wastes. Int. J. Environ. Sci. Technol. **16**, 3497–3512 (2019)
6. Khuri, A.I., Mukhopadhyay, S.: Response surface methodology. Wiley Interdiscip. Rev.: Comput. Stat. **2**(2), 128–149 (2010)
7. McFall-Ngai, M., et al.: Animals in a bacterial world, a new imperative for the life sciences. Proc. Natl. Acad. Sci. **110**(9), 3229–3236 (2013)
8. Pham, D.C., et al.: Optimization of ultrasound-assisted extraction of flavonoids from celastrus hindsii leaves using response surface methodology and evaluation of their antioxidant and antitumor activities. Biomed. Res. Int. **2020**(1), 3497107 (2020)
9. Piggott, M.P., Taylor, A.C.: Remote collection of animal DNA and its applications in conservation management and understanding the population biology of rare and cryptic species. Wildl. Res. **30**(1), 1–13 (2003)
10. Woodard, D.L., Howard, A.J., Down, J.A.: Process for purifying DNA on hydrated silica. United States (1993)

Detection of Bacteriocin-Producing *Bacillus Subtilis* by Multiplex PCR

Thao-Nhi Thi Nguyen and Dinh-Chuong Pham(✉)

Faculty of Applied Sciences, Ton Duc Thang University, Ho Chi Minh City, Vietnam
phamdinhchuong@tdtu.edu.vn

Abstract. Bacteriocins, a class of bioactive antimicrobial peptides synthesized in various bacteria, offer potential solutions to combine with or alter traditional antibiotics in treating some multi-drug resistance pathogens; still, detecting bacteriocin-producing strains is laborious. In this study, *Bacillus subtilis* ATCC 6633 strains, which have been demonstrated to generate bacteriocins, are used to develop a rapid method based on Multiplex PCR to identify strains able to produce two popular bacteriocins, subtilin and subtilosin A. The crude bacteriocin obtained from *B. subtilis* exhibited antimicrobial activity against *Staphylococcus aureus*, with an inhibition zone of 11.0 ± 0.5 mm after 48 h of incubation. PCR profiles revealed Multiplex PCR, employing three primer pairs targeting spaS, ytcP, and albA genes, successfully amplified specific DNA fragments from *B. subtilis* ATCC 6633 with amplicon size 221 bp, 336 bp, and 556 bp, respectively. The specificity of the primers was validated against other Gram-positive bacterial species. This study also highlights the applicability of findings in soil samples as a rapid and reliable tool for screening *B. subtilis* strains from natural environments, paving the way for the discovery of novel bacteriocin-producing strains.

Keywords: *Bacillus subtilis* · *Bacillus subtilis* ATCC 6633 · Bacteriocins · Multiplex PCR · Multi-drug resistance · Subtilin · Subtilosin A

1 Introduction

Bacteriocins, a minuscule antimicrobial peptide ranging from 30 to 60 amino acids, are synthesized and released extracellularly in various bacteria, can solve some of the most challenging problems caused by multi-drug resistant pathogens [1, 2]. According to previous reports, significant animal and plant pathogens such as enterotoxigenic *Escherichia coli* (ETEC), methicillin-resistant *Staphylococcus aureus* (MRSA), vancomycin-resistant Enterococcus (VRE), etc. are all inhibited by bacteriocins [3, 4]. The bactericidal action is mainly based on their ability to attach to receptors on the surfaces of bacteria, including cytotoxicity on the bacteria [2]. Since bacteriocins are distinguished from traditional broad-spectrum antibiotics in terms of their antimicrobial mechanisms and due to their unique properties such as low molecular weights, thermal and protease stability, no or very low toxicity, low tendency to generate resistance, and low immunogenicity, this makes them promising candidates to alter traditional antibiotics to treat multi-drug resistant pathogens in the future [5, 6].

L. Vu et al. (Eds.): PBC 2025, 2026.
https://doi.org/10.1007/978-3-032-29579-8_13

B. subtilis is a fast-growing, Gram-positive, aerobic bacterium, with applications in various fields, and as a model organism for studying endospore formation and biofilm development [7]. Known to be non-pathogenic to people and animals, *B. subtilis* strains are trusted and safe probiotic supplements and produce bacteriocins with enormous potential for use in place of antibiotics [8]. Numerous bacteriocin-producing species are present in the genus *Bacillus*, such as *B. subtilis*, which generates subtilin and subtilosin A [9]. However, not all *B. subtilis* strains are capable of producing these two types of bacteriocins. For example, *B. subtilis* 168 is not capable of producing subtilin, while *B. subtilis* ATCC 6633 has the subtilin coding sequence in its genome [10]. As of late, identifying new antimicrobial substances or bacteriocin–producing strains has been a time-consuming and imprecise task, and replicating or conducting additional experiments is necessary to ensure the accuracy of the outcome [6]. Therefore, using molecular approaches such as PCR to amplify bacteriocin-encoding genes for bacteriocins is a desirable, rapid, and efficient method to identify bacteriocin-producing strains.

2 Materials and Methods

2.1 Searching Sequences and Designing Primer Pairs

Information about the specific sequence of *B. subtilis* and bacteriocin-encoding sequences is available on the NCBI website (National Center for Biotechnology Information, website: https://blast.ncbi.nlm.nih.gov/Blast.cgi) based on previously published studies and confirmed by the BLAST tool. The sequence specificity was determined based on the Per-Ident and Query cover parameters. The GenBank ID of selected sequences was recorded.

2.2 Searching Sequences and Designing Primer Pairs

Table 1 lists the target sequences for the development of the primer set, including ytcP, spaS, and albA. Based on the selected sequences, Primer3Plus (https://www.bioinformatics.nl/cgi-bin/primer3plus/primer3plus.cgi) was used to design the primer pairs for the multiplex PCR reaction. The specificity of the primer was checked with the Primer-BLAST tool.

Table 1. Target sequences for amplification

Sequence	GenBank ID	Product
ytcP	EU669668.1	An ABC–type transporter–like protein
spaS	DQ452514.1	Precursor subtilin peptide
albA	AJ430547.1	Precursor subtilosin A peptide

2.3 Culturing and Proliferation of *B. Subtilis*

B. subtilis strain ATCC 6633 was purchased from the Institute of Drug Quality Control Ho Chi Minh City and kept in TSA medium at 4 °C. *B. subtilis* was then subcultured on LB agar plates, and incubated for 24 h at 37 °C. Next, a bacterial colony from LB agar plates was collected into a falcon containing 10 ml of LB broth media and cultured at 37 °C with continuous shaking at 180 rpm for 24 h. After 24 h, 1 mL of the inoculum was transferred into a flask with 100 mL of the liquid LB medium for the second enrichment. The flask was cultured for another 24 h at 37 °C with 180 rpm of steady shaking.

2.4 Antibacterial Activity of Crude Bacteriocins

B. subtilis ATCC 6633 was cultured in LB media for 12, 24, 36, and 48 h. Cell-free supernatants were collected by centrifugation at 4 °C for 15 min at 6,000 rpm, then filter-sterilized using a 0.22 µm filter. Similarly, *Vibrio vulnificus* was cultured in ASPW media as a negative control. The antibacterial activity of the crude bacteriocins sample was assessed against *S. aureus* using the agar well diffusion method. Briefly, the overnight culture of *S. aureus* was adjusted to a concentration of 10^8 CFU/mL and spread on MHA plates. Wells were created on the plates and filled with the following: blank control (LB broth), positive control (ampicillin), negative control (*V. vulnificus* supernatant), and *B. subtilis* supernatants The plates were left undisturbed for a few hours to allow the supernatant to diffuse into the agar and then cultured at 37 °C for 18 h. The experiment was repeated three times to ensure reproducibility.

2.5 Extracting and Purifying of *B. Subtilis* Genome

The second-enrichment *B. subtilis* was transferred into labeled 1.5 mL Eppendorf tubes and centrifuged at 6,000 rpm for 5 min to collect the bacterial pellet. The genomic DNA of *B. subtilis* and other control bacteria were extracted from bacterial cells using an ISOLATE II Genomic DNA kit. The process involved cell lysis by proteinase K and lysis buffer, DNA binding to a silica membrane then washing and eluting bacterial genome. The process was followed strictly to the manufacturer's instructions. Spectrophotometric analysis was used to assess the concentration and quality of the purified genome.

2.6 Singleplex PCR

The Singplex PCR was performed to check the specificity of the designed primer pairs listed in Table 2 on the *B. subtilis* genome.

Table 2. Primer pairs informations

Primers	Sequence 5′–3′	Product length	Tm (°C)	References
BS–ytcP–F	TCG ATT CTT ACT GGG CGC TC	336 bp	62	Kwon *et al.* (2009)

(continued)

Table 2. (*continued*)

Primers	Sequence 5′–3′	Product length	Tm (°C)	References
BS–ytcP–R	CCG ACC CCA TTT CAG ACA TAT		62	
BS–spaS–F	TGA AAG GAG GTG ACC AAT ATG	221 bp	60	This study
BS–spaS–R	GAG TCA GAG CAA GGT GAT GC		60	
BS–albA–F	CTG AGG GTG GCG TGT TAG AA	556 bp	62	This study
BS–albA–R	CGC ATG AGT GTT CCG TTT GT		60	
IC–FP	ACA TAC TAC AGG TCC AAG TG	102 bp	53	This study
IC–RP	AGG GAG AAC TGG TTC TTG G		53	

Singleplex PCR reactions were prepared using a primer mix consisting of 80 μL of ultrapure water, 10 μL of forward primer (100 μM), and 10 μL of reverse primer (100 μM). The final volume of the primer mix was 100 μL. For each PCR reaction, a 10 μL reaction mixture was prepared containing 3.53 μL of ultrapure water, 5 μL of MyTaq™ Mix (Meridian Bioscience, Germany), 0.5 μL of primer mix, and 0.97 μL of extracted genomic DNA (at a concentration of 51.75 μg/mL). According to the program indicated in Table 3, amplification was done in a thermocycler. The amplified products were cooled to 4 °C.

Table 3. Thermal cycle for PCR reaction

Temperatures	Times	Cycles
94 °C	5 min	1
94 °C	30 s	30
57 °C	25 s	
72 °C	25 s	
72 °C	7 min	1
4 °C	∞	

2.7 Multiplex PCR

A primer mix for Multiplex PCR was prepared by combining 20 μL of ultrapure water with 10 μL of each forward and reverse primer listed in Table 2 (at a concentration of 100 μM). The final volume of the primer mix was 100 μL. For each PCR reaction, a 10 μL reaction mixture was prepared containing 2.53 μL of ultrapure water, 5 μL of MyTaq™ Mix (Meridian Bioscience, Germany), 0.5 μL of primer mix, 1 μL of DNA IC, and 0.97 μL of extracted genomic DNA (at a concentration of 51.75 μg/mL). According to the program indicated in Table 3, amplification was done in a thermocycler. The amplified products were cooled to 4 °C.

2.8 Double-Check Results with Multiplex PCR on Control Bacteria

Three mixtures of primer including BS–ytcP with IC, BS–spaS with IC, and BS–alb with IC, respectively, were prepared with 60 μL of ultrapure water with 10 μL of each forward and reverse primer listed in Table 2 (at a concentration of 100 μM). The final volume of the primer mix was 100 μL. The PCR samples were prepared into a 0.2 mL PCR tube with a DNA template of genome extract from control bacteria: *Bacillus cereus, Clostridium perfringens*, and *Lactobacillus acidophilus*. According to the program indicated in Table 3, amplification was done in a thermocycler. The amplified products were cooled to 4 °C.

2.9 Electrophoresis

To prepare samples for agarose gel electrophoresis, 5 μL of each extracted genome sample was mixed with 1 μL of 6X GelRed Loading Buffer with TriColor (ABT, Vietnam). Similarly, PCR products were prepared by combining 3 μL of ultrapure water, 2 μL of the PCR product, and 1 μL of 6X GelRed Loading Buffer with TriColor (ABT, Vietnam). All samples were then analyzed by 1% agarose gel electrophoresis at 100 V for 15–20 min. A 1000 bp molecular size marker was used for extracted bacterial genomes, while a 100 bp marker was used for PCR products. Bands were shown under ultraviolet light.

2.10 Collecting and Processing Soil Samples

The soil sample is collected from orchard soil in Vinh Long Province at a depth of at least 30 cm with a sterile spoon. The sample is stored in sterile containers, labeled appropriately, and transported to the laboratory for further use. The collected sample was divided into 4 treatments as shown in Table 4. The treated samples were incubated for 48 h at room temperature.

Table 4. Experimental treatment layout

No.	Experimental layout	Soil sample (g)	Solution volume (mL)
1	Sterile water	5	5
2	LB and MRS medium	5	5
3	Culturing medium of *L. acidophilus*	5	5
4	Culturing medium of *B. subtilis*	5	5

Following the protocol outlined by Stein *et al.* (2002), samples were suspended in sterile distilled water at a ratio of 1:9 (w/v) and homogenized at room temperature for 30 min at 150 rpm [10]. The suspension was then heat-treated at 80 °C for 1 h to inactivate non-spore-forming organisms, followed by cooling and filtration through a sieve to remove large debris. A 1 mL aliquot of the suspension was transferred to a 10 mL flask containing LB broth supplemented with 50 mg/L nystatin to inhibit fungal growth. The flask was incubated at 37 °C with shaking at 180 rpm for 24 h. After 24-h incubation, 1 mL of the culture was transferred to a 100 mL flask containing LB broth supplemented with 50 mg/L nystatin. The flask was then incubated at 37 °C with shaking at 180 rpm for another 24 h.

2.11 Extracting and Purifying the Bacterial Genome from Soil Samples

The bacteria isolated from soil samples were prepared from overnight culture in LB media by the methods described above. The inoculum was transferred into labeled 1.5 mL Eppendorf tubes and centrifuged at 6,000 rpm for 5 min to collect the bacterial pellet. The genome extraction was conducted using an ISOLATE II Genomic DNA kit, following the instructions of the manufacturer.

2.12 Multiplex PCR with Treated Samples

The PCR primer mix and sample were prepared according to the following step described in Sect. 2.6. The sample was mixed well and put into the PCR machine. The Multiplex PCR reaction was performed according to the thermal cycle set in Table 3 and then examined by agarose gel electrophoresis (1% w/v) in the voltage condition of 100 V for 15 to 20 min with a molecular size maker of 100 bp.

3 Results and Discussion

3.1 Searching *B. Subtilis*–Specific and Bacteriocin–Encoding Sequences

The ytcP sequence (GenBank ID: EU669668.1) encodes a putative ABC-type transporter protein in *B. subtilis*. Kwon *et al.* (2009) designed ytcP primers based on this sequence to specifically amplify *B. subtilis* DNA. These primers successfully amplified the expected-sized fragment from 14 *B. subtilis* reference strains and 3 isolates, but not from other *Bacillus* species [11]. Meanwhile, spaS and albA genes were targeted for the detection of bacteriocin-producing *B. subtilis* strains, specifically those producing subtilin and subtilosin A. The spaS gene (GenBank ID: DQ452514.1) encodes the precursor peptide for subtilin. Banerjee and Hansen (1988) cloned the subtilin-encoding gene cluster from *B. subtilis* ATCC 6633 [12]. The albA gene (GenBank ID: AJ430547.1) is located downstream of the subtilosin A structural gene, sbo, and is essential for subtilosin A biosynthesis. This makes the albA gene a useful marker for identifying subtilosin A-producing *B. subtilis* strains [13].

3.2 Designing Primers

Three primer pairs were designed and optimized for PCR. The BS-ytcP primer pair (from Kwon *et al.*, 2009) was used as a reference primer for *B. subtilis* detection. To optimize the Multiplex PCR reaction, the forward primer of the BS-ytcP pair was modified to produce a shorter 336 bp amplicon. Additionally, two novel primer pairs were designed to amplify the genes encoding subtilin and subtilosin A, with expected amplicon sizes of 221 bp and 556 bp, respectively, with T_m ranges from 60 and 62 °C. Simultaneously, the amplification products of three primer pairs were calculated to be 100–220 bp separated to ensure optimal band separation during electrophoresis. Furthermore, internal control primers were included to monitor DNA extraction and PCR.

3.3 Culturing and Proliferating of *B. Subtilis*

After *B. subtilis* ATCC 6633 was subcultured from TSA media and incubated at 37 °C for 24 h, similar–looking white colonies appeared on the surface of the LB agar plate. The colonies resemble spherical with moist textures and fimbriate margins with smooth and viscous surfaces. When the *B. subtilis* colony was transferred from the agar plate to LB broth media and incubated for 24 h at 37 °C in an incubator, the bacteria proliferated and clouded the medium, and part of the white bacterial pellets may be observed sinking to the bottom of the flask.

3.4 Antibacterial Activity of Crude Bacteriocins

The crude bacteriocins obtained after 24 h of culture begin to have an inhibition zone. The diameter of the inhibition zone gradually increased and reached 11.0 ± 0.5 mm at 48 h-obtained culture. It can be suggested that strain *B. subtilis* ATCC 6633 is capable of producing bacteriocins and confirming the occurrence of the bacteriocin encoding gene in this bacterial genome by comparing it with the wells containing crude solution obtained from the *V. vulnificus* strain that can not produce bacteriocins (Fig. 1).

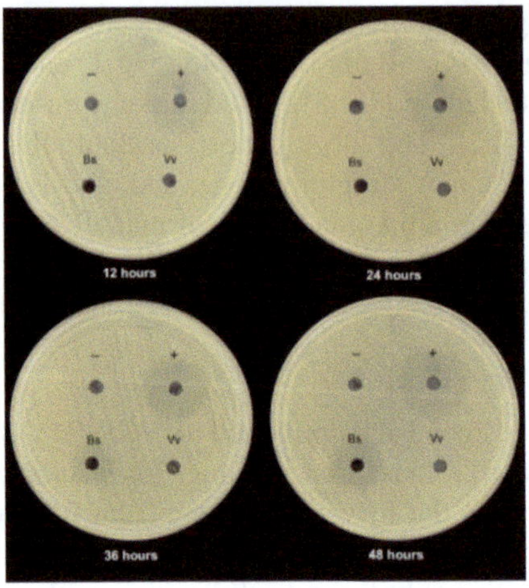

Fig. 1. Antibacterial activity of crude bacteriocins. (−) LB broth, (+) Ampicillin (100 μg/mL), Bs crude bacteriocins from *B. subtilis*, Vv crude solution from *V. vulnificus*

3.5 Extracting and Purifying of *B. Subtilis* Genome

Figure 2 demonstrates an enormous, bright band that appears in lane 1, close to the well mouth, above the 10 kb line of the 1000 bp ladder. This indicates that the DNA obtained through purification remains the whole bacterial genome.

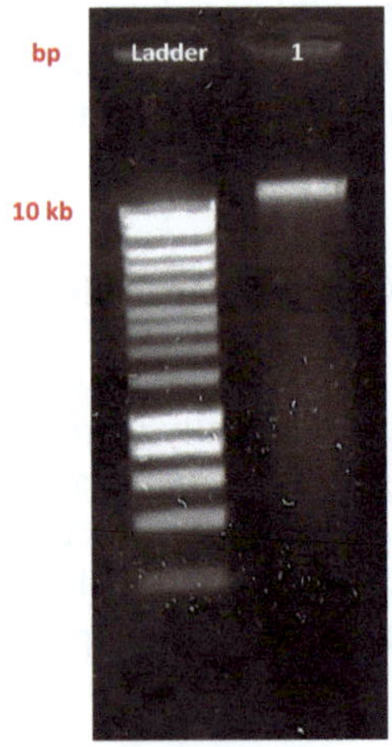

Fig. 2. Electrophoretic profile of genome extraction from *B. subtilis*. lane Ladder ladder DNA 1000 bp, lane IC Genome of *B. subtilis*

The DNA concentration was also determined by measuring the optical density (OD) at 260 nm. The DNA concentration in the sample calculated is 51.75 μg/mL.

3.6 Singleplex and Multiplex PCR

As shown in Fig. 3, the amplified products of Singleplex PCR exhibited bright bands at the expected sizes. The internal control (IC) band at 102 bp confirmed the successful PCR process. The BS-ytcP primer pair amplified a 336 bp product, confirming the presence of *B. subtilis* DNA in the sample. The BS-spaS and BS-albA primer pairs amplified products of approximately 221 bp and 556 bp, respectively, as expected. The clear and distinct bands obtained for all three primer pairs indicate successful amplification of the target genes in the *B. subtilis* genome. The Multiplex PCR results (Multi-lane) show that all four primer pairs (IC, BS-ytcP, BS-spaS, and BS-albA) amplified their target sequences, with clear and well-separated bands. A faint band below the 100 bp marker is likely a residual primer and encounters no impact on the results of the experiment.

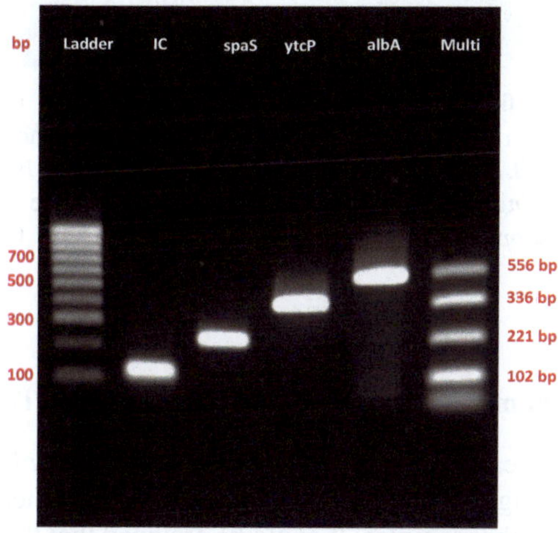

Fig. 3. Electrophoretic profile of singleplex PCR and multiplex PCR. lane ladder ladder DNA 100 bp, lane IC exogenous IC—102 bp, lane spaS: BS–spaS—221 bp, lane ytcP: BS–ytcP—336 bp, lane albA: BS–albA—556 bp, lane multi mPCR—*B. subtilis*

3.7 Double–Check the Specificity of Primers on Control Bacteria

The specificity of the primer pairs was verified using the genomes *B. cereus, L. acidophilus*, and *C. perfringens*. The BS-ytcP primer pair was found to be specific to *B. subtilis*, while the BS-spaS and BS-albA primers were specific to *B. subtilis* strains producing subtilin and subtilosin A, respectively. The BS-spaS primers exhibited low specificity, amplifying multiple DNA fragments in *B. cereus* and *L. acidophilus*. In contrast, only the IC band was observed for *C. perfringens*. Similarly, the BS-albA primers produced multiple, non-specific bands for *B. cereus* and *L. acidophilus*. However, a clear band was observed for *B. subtilis*, suggesting differences in the subtilosin gene sequence or the production of subtilin-like peptides in these species. As expected, only the IC band was observed for *C. perfringens*, which does not produce subtilosin A (Fig. 4).

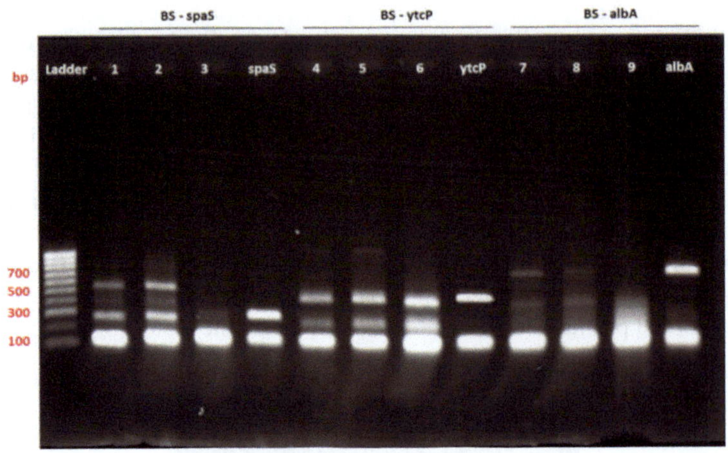

Fig. 4. Electrophoretic profile of multiplex PCR on control bacteria. lane ladder ladder DNA 100 bp, lane 1 BS-spaS—*B. cereus*, lane 2 BS-spaS—*L. acidophilus*, lane 3 BS-spaS—*C. perfringens*, lane spaS BS-spaS—*B. subtilis*, lane 4: BS-ytcP—*B. cereus*, lane 5 BS-ytcP—*L. acidophilus*, lane 6: BS-ytcP—*C. perfringens*, lane ytcP: BS-ytcP—*B. subtilis*, lane 7: BS- albA—*B. cereus*, lane 8: BS-albA—*L. acidophilus*, lane 9: BS-albA—*C. perfringens*, lane ytcP: BS-albA—*B. subtilis*

3.8 Processing Soil Samples and Extracting Genome from Treated Samples

After 48-h incubation, treatment 4, which was treated with *B. subtilis* culture medium, showed visible orange pigments on the surface of the soil. Further, the fluid was partly absorbed by the soil in other samples. It could be assumed that the genome purification procedure was successful in getting genomes from treated samples according to the results in Fig. 5, bands can be observed in lanes 1 to 4, above the highest mark of the 1000 bp DNA ladder.

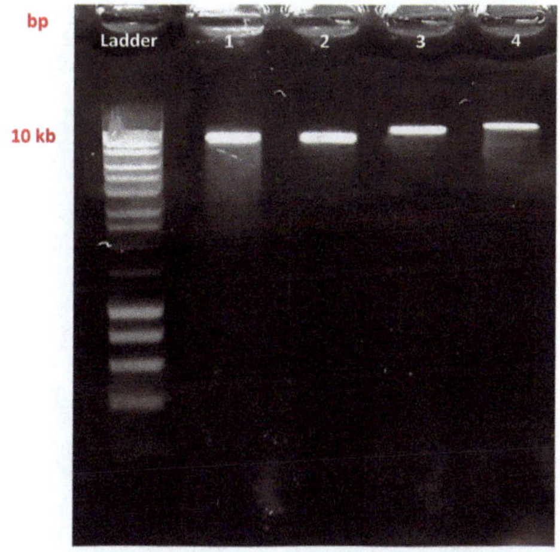

Fig. 5. Electrophoretic profile of genome extraction from treated samples. Lane Ladder: Ladder DNA 1000 bp, lane 1 genome—sample 1, lane 2 genome—sample 2, lane 3 genome—sample 3, lane 4 genome—sample 4

3.9 Multiplex PCR on Genome Bacterial Isolated from the Soil Sample

The amplification product of the IC internal control primer is represented as a band with a size of 102 bp which appears in lanes 1 to 3 of the results displayed in Fig. 6. Meanwhile, 4 evenly separated bright bands that matched the size of the expected amplification resulting from the specified primers appeared in lane 4, which corresponded to the sample treated with *B. subtilis* bacterial culture. The impurity band in 4 lanes, which is 100 bp below the DNA ladder, is formed via residual primers from the PCR process and has no impact on the final result.

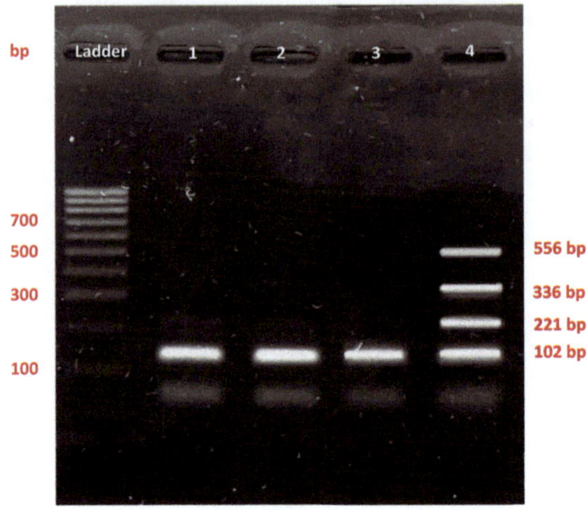

Fig. 6. Electrophoretic profile of multiplex PCR with genome from treated samples. lane ladder ladder DNA 1000 bp, lane 1 mPCR—sample 1, lane 2 mPCR—sample 2, lane 3 mPCR—sample 3, lane 4 mPCR—sample 4.

4 Conclusion

Three sets of primers were effectively constructed through the research to amplify gene fragments limited to *B. subtilis* as well as gene fragments encoding for bacteriocins subtilosin A and subtilin, respectively. Implementing 1% agarose gel electrophoresis, the primer pairs BS–spaS (221 bp), BS–ytcP (336 bp), and BS–albA (556 bp) display distinct, clear bands that correspond to their expected sizes. To control the PCR processes and genome purification while simultaneously amplifying the 102 bp product, an internal control primer pair was also created. The research also showed that a gene fragment specific to *B. subtilis* was amplified by the primer pair BS–ytcP, which was not observed in other Gram-positive bacteria, including strains of the same genus as *B. cereus*. Additionally, two primer pairs, BS–spaS and BS–albA, displayed specific results for genes found in the gene cluster that encodes subtilin and subtilosin A, respectively. The study could be successfully implemented in application with natural and biological samples, as evidenced by the outcomes of multiplex PCR on bacterial genome samples obtained from soil samples treated with *B. subtilis* bacterium. The study exposed may be broadened to encompass the production of recombinant proteins or the identification of several species of bacteria that produce subtilin and subtilosin A by using a Multiplex PCR reaction.

References

1. Gradisteanu, P.G., Popa, L.I.: Bacteriocins in the era of antibiotic resistance: rising to the challenge. Pharmaceutics. **13**(2), 196 (2021)
2. Yang, S.-C., Lin, C.-H., Sung, C.T., Fang, J.-Y.: Antibacterial activities of bacteriocins: application in foods and pharmaceuticals. Front. Microbiol. **5**, 241 (2014)

3. Cotter, P.D., Ross, R.P., Hill, C.: Bacteriocins—a viable alternative to antibiotics? Nat. Rev. Microbiol. **11**(2), 95–105 (2013)
4. Grinter, R., Milner, J., Walker, D.: Bacteriocins Active against Plant Pathogenic Bacteria. Portland Press Ltd (2012)
5. Zhang, Q., Kobras, C.M., Gebhard, S., Mascher, T., Wolf, D.: Regulation of heterologous subtilin production in *Bacillus subtilis* W168. Microb. Cell Factories. **21**(1), 1–16 (2022)
6. Zou, J., Jiang, H., Cheng, H., Fang, J., Huang, G.: Strategies for screening, purification and characterization of bacteriocins. Int. J. Biol. Macromol. **117**, 781–789 (2018)
7. Errington, J., Aart, L.T.V.D.: Microbe profile: *Bacillus subtilis*: model organism for cellular development, and industrial workhorse. Microbiology. **166**(5), 425–427 (2020)
8. Lu, Z., Guo, W., Liu, C.: Isolation, identification and characterization of novel *Bacillus subtilis*. J. Vet. Med. Sci. **80**(3), 427–433 (2018)
9. Liu, Q., Gao, G., Xu, H., Qiao, M.: Identification of the bacteriocin subtilosin a and loss of purL results in its high-level production in *bacillus* amyloliquefaciens. Res. Microbiol. **163**(6–7), 470–478 (2012)
10. Stein, T., Borchert, S., and others: Dual control of subtilin biosynthesis and immunity in *Bacillus subtilis*. Mol. Microbiol., **44**(2), 403–416 (2002).
11. Kwon, G.-H., Lee, H.-A.: And others: development of a RAPD-PCR method for identification of *bacillus* species isolated from Cheonggukjang. Int. J. Food Microbiol. **129**(3), 282–287 (2009)
12. Banerjee, S., Hansen, J.N.: Structure and expression of a gene encoding the precursor of subtilin, a small protein antibiotic. J. Biol. Chem. **263**(19), 9508–9514 (1988)
13. Thennarasu, S., Lee, D.-K., and others: Membrane permeabilization, orientation, and antimicrobial mechanism of subtilosin a. Chem. Phys. Lipids, **137**(1–2), 38–51 (2005).

Chemical Composition, Antimicrobial Activity, and Antidiabetic Potential of Dichloromethane Extract from *Cryptotympana Mandarina*

Ton That Huu Dat(✉), Vu Thi Thanh Tam, Le Canh Viet Cuong, and Pham Hong Thai

Mientrung Institute for Scientific Research, Vietnam National Museum of Nature, Vietnam Academy of Science and Technology, Hue, Vietnam
tthdat@vnmn.vast.vn

Abstract. This study investigates the chemical composition and biological activities of a dichloromethane (DCM) extract from the cicada *Cryptotympana mandarina.* Gas chromatography–mass spectrometry (GC-MS) analysis identified 16 major compounds, comprising 69.68% of the total extract. The predominant constituents included (9*E*)-9-octadecenoic acid, benzeneacetic acid, 1*H*-indole-3-carboxaldehyde, and *n*-hexadecanoic acid. *In vitro* assays revealed that the extract showed antidiabetic activity through inhibiting α-amylase and α-glucosidase with IC_{50} values of 3.68 ± 0.24 mg/mL and 6.75 ± 0.02 mg/mL, respectively. Furthermore, the extract also exhibited antimicrobial activity against a panel of microorganisms with MICs ranging from 256 to 1024 μg/mL. These results highlight the potential of the *C. mandarina* DCM extract as a natural source of compounds with antidiabetic and antimicrobial properties.

Keywords: Antimicrobial activity · Antidiabetic activity · *Cryptotympana mandarina* · Dichloromethane

1 Introduction

Insects represent the most diverse group of animals on Earth, occupying a wide range of terrestrial and aquatic habitats. While approximately one million insect species have been formally described, estimates suggest that many more remain undiscovered, with projections ranging from 563 million to as high as 2.2 billion species [1]. Among these, cicadas, members of the family *Cicadidae*, are a prominent component of Vietnam's insect fauna. At least 131 cicada species, spanning 45 genera, have been documented in Vietnam [2].

In traditional East Asian medicine, cicadas and their exuviae are highly valued for their therapeutic properties. Particularly, cicada shells, commonly referred to as "cicada periostracum" or "cicada exuviae", have been traditionally used to treat various ailments, including fever, seizures, skin rashes, and eye disorders [3–5]. Recent studies have identified several bioactive compounds in cicada-derived extracts, such as ergosterol and

L. Vu et al. (Eds.): PBC 2025, 2026.
https://doi.org/10.1007/978-3-032-29579-8_14

its peroxides, cyclic heptapeptides like bassianin and beauvericin, as well as polysaccharides. These constituents have demonstrated diverse biological activities, including antioxidant, anti-inflammatory, antimicrobial, and anticancer effects [6–9].

Although natural products derived from insects, particularly cicadas, have garnered increasing scientific interest due to their potential in traditional medicine, bioprospecting efforts focused on cicadas remain largely underexplored. This study aims to investigate the chemical composition and biological properties of a dichloromethane (DCM) extract from the cicada *Cryptotympana mandarina*. By bridging traditional medicinal knowledge with modern scientific approaches, this research contributes to a deeper understanding of the biochemical diversity found in cicadas and highlights their potential as a source of novel bioactive compounds.

2 Materials and Methods

2.1 Biological Materials

The cicada *C. mandarina* Distant was collected in Bach Ma National Park, Vietnam, in May 2024. Species identification was confirmed by Prof. Pham Hong Thai (Mientrung Institute for Scientific Research, Vietnam National Museum of Nature, Vietnam Academy of Science and Technology).

2.2 *C. Mandarina* DCM Extract Preparation

Dried and powdered *C. mandarina* (100 g) was subjected to ultrasonic-assisted extraction with ethanol (3 × 1000 mL) for 2 h at 50–60 °C each time. The combined ethanol extracts were concentrated under reduced pressure to yield a crude residue (10.4 g). This residue was suspended in warm water (200 mL) and successively partitioned with dichloromethane (3 × 200 mL) to obtain the dichloromethane extract (DCM), yielding 6.8 g.

2.3 Gas Chromatography-Mass Spectrometry (GC-MS) Analysis

GC–MS analysis was performed using an Agilent 7890B gas chromatograph coupled with an Agilent 5977A mass detector (Agilent Technologies, Santa Clara, CA, USA) [10]. Separation was achieved with an HP-5MS capillary column (30 m × 0.250 mm × 0.25 μm film thickness; Agilent Technologies, Santa Clara, CA, USA). The sample (1 μL) was injected in splitless mode with a flow rate of 1 mL/min. Helium (99.999% purity) was used as the carrier gas at 1 mL/min. The oven temperature was initially set at 60 °C for 2 min, then increased to 260 °C at 5 °C/min and held for 1 min. The inlet and ionization source temperatures were set at 260 °C and 280 °C, respectively. The solvent delay was 3.00 min. The MS detector was operated in electron ionization (EI) mode at 70 eV, scanning the mass-to-charge (m/z) range of 50–550 in full scan mode. Data were processed using Agilent ChemStation software C.01.10 (Agilent Technologies, Santa Clara, CA, USA). Compound identification was achieved by comparing the spectra to a stored MS library (W8N08 and NIST08). The relative percentage of each compound was calculated based on GC peak areas.

2.4 Antidiabetic Assay

2.4.1 α-Amylase Inhibitory Assay

The α-amylase (Sigma-Aldrich, St. Louis, MO, USA) enzyme inhibitory activity of the *C. mandarina* DCM extract was determined according to the described method by Dat et al. [11]. In brief, the reaction mixture consisting of 50 μL sample was incubated with 50 μL starch azure solution and 50 μL α-amylase (1.0 U/mL) in Tris-HCl buffer (pH 6.9) in 96-well plates at 37 °C for 10 min. The reaction was stopped by adding 50 μL of acetic acid 50%, and then the absorbance of reactions was recorded at 650 nm using an ELx800 absorbance microplate reader (BioTek Instruments, Winooski, VT, USA). The inhibition activity was calculated as follows: inhibition (%) = 100 × [1—(As—Abs)/(Ac—Acb)]. Where: As is the absorbance of the sample, Asb is the absorbance of the sample blank, Ac is the absorbance of the control, and Acb is the absorbance of the control blank. Acarbose was used as a positive control.

2.4.2 α-Glucosidase Inhibitory Assay

α-Glucosidase (Sigma-Aldrich, St. Louis, MO, USA) enzyme inhibitory activity of the *C. mandarina* DCM extract was determined according to the described method by Dat et al. [11] In brief, the reaction mixture consisting of 50 μL sample was incubated with 50 μL α-glucosidase (0.5 U/mL) in 0.1 M potassium phosphate buffer (pH 6.8) in 96-well plates at 37 °C for 10 min. The reaction was started by adding 50 μL of 5 mM 4-Nitrophenyl β-D-glucopyranoside (pNPG), followed by incubation at 37 °C for 30 min. The reaction was stopped by adding 50 μL of 0.2 M Na_2CO_3, and then the absorbance of reactions was recorded at 405 nm using an ELx800 absorbance microplate reader (BioTek Instruments, Winooski, VT, USA). The inhibition activity was calculated as follows: inhibition (%) = 100 × [1—(As—Abs)/(Ac—Acb)]. Where: As is the absorbance of the sample, Asb is the absorbance of the sample blank, Ac is the absorbance of the control, and Acb is the absorbance of the control blank. Acarbose was used as a positive control.

2.5 Antimicrobial Assay

Antimicrobial activity of the *C. mandarina* DCM extract was tested against seven reference microorganisms i.e., *Salmonella enterica* ATCC 13076, *Escherichia coli* ATCC 25922, *Pseudomonas aeruginosa* ATCC 27853, *Enterococcus faecalis* ATCC 29212, *Bacillus cereus* ATCC 14579, *Staphylococcus aureus* ATCC 25923, *Candida albicans* ATCC 10231 using the broth microdilution method as described by Dat et al. [12]. Briefly, 100 μL of the bacterial inoculum (1 × 10^6 CFU/mL) was added to wells containing 100 μL of the extracts at a range of different concentrations in 96-well plates. The plate was incubated at 37 °C for 24 h, the absorbance at 630 nm was then measured using ELx800 absorbance microplate reader (BioTek Instruments, Winooski, VT, USA). MICs of the antibacterial extracts were determined as the lowest concentration, at which there was no growth of the bacteria. For the yeast, 100 μL of the inoculum (2 to 5 × 10^5 CFU/mL for yeast) was added to wells containing 100 μL of the extracts at a range of different concentrations in 96-well plates. The plate was incubated at 28 °C for 48 h. MICs of the anti-yeast extracts were determined as the lowest concentration, at which there was no growth of the yeast by the absorbance at 530 nm using an ELx800 absorbance microplate reader (BioTek Instruments, Winooski, VT, USA).

3 Results and Discussion

3.1 Chemical Constituents Present in the *C. Mandarina* DCM Extract

The GC-MS chromatogram of the *C. mandarina* DCM extract (Fig. 1) revealed multiple peaks, indicating the presence of diverse chemical constituents. Subsequent analysis identified a total of 16 compounds, which accounted for 69.68% of the total composition of the extract (Table 1 and Fig. 2). These identified compounds belong to several chemical classes, including fatty acids, amides, alkanes, aromatic derivatives, and diketopiperazines. Among these, fatty acids and indole derivatives were the most abundant, with prominent compounds such as (9*E*)-9-octadecenoic acid (27.65%), benzeneacetic acid (6.44%), 1*H*-indole-3-carboxaldehyde (5.94%), and *n*-hexadecanoic acid (5.04%).

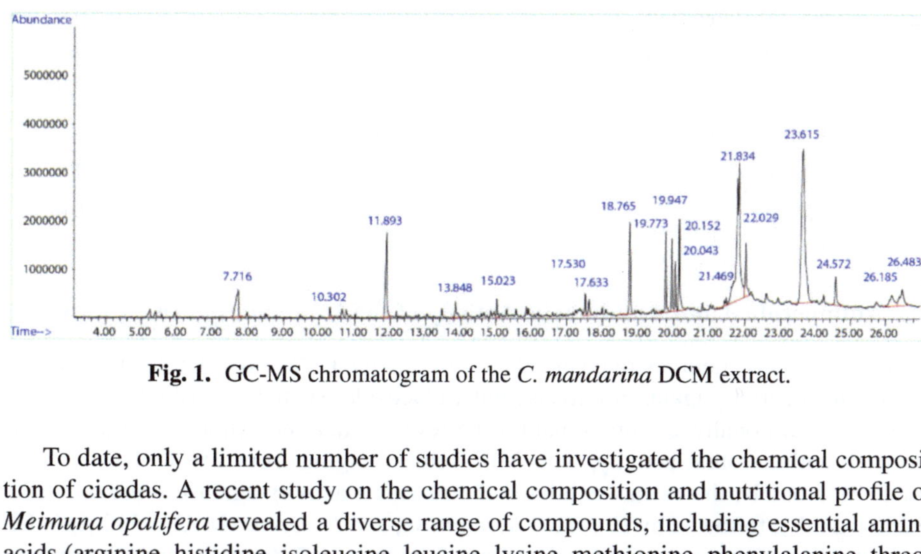

Fig. 1. GC-MS chromatogram of the *C. mandarina* DCM extract.

To date, only a limited number of studies have investigated the chemical composition of cicadas. A recent study on the chemical composition and nutritional profile of *Meimuna opalifera* revealed a diverse range of compounds, including essential amino acids (arginine, histidine, isoleucine, leucine, lysine, methionine, phenylalanine, threonine, tryptophan, valine) and non-essential amino acids (alanine, aspartic acid, cysteine, glutamine, glutamic acid, glycine, proline, serine, tyrosine). The species also contained saturated fatty acids (e.g., palmitic acid, stearic acid), monounsaturated fatty acids (e.g., *cis*-oleic acid), and polyunsaturated fatty acids (e.g., *cis*-linoleic acid). Additionally, phenolic acids such as gallic acid, protocatechuic acid, chlorogenic acid, vanillic acid, *p*-coumaric acid, ferulic acid, and sinapic acid were identified, along with flavonoids including rutin, quercetin, apigenin, kaempferol, and myricetin. However, the chemical profile of cicadas is not static and changes dynamically throughout their lifecycle [7]. Several studies have also highlighted that cicada exoskeletons are a valuable source of structural biopolymers such as chitin and chitosan [13–15].

Table 1. Chemical constituents present in the *C. mandarina* DCM extract

No	Compounds	%
1	Hexanoic acid	3.79
2	Hexanamide	0.55
3	Benzeneacetic acid	6.44
4	Benzeneacetamide	1.31
5	Tricosane	0.89
6	Tetradecane	1.10
7	3-Pyrrolidin-2-yl-propionic acid	1,06
8	1*H*-indole-3-carboxaldehyde	5.94
9	3-Propan-2-yl-2,3,6,7,8,8a-hexahydropyrrolo[1,2-a]pyrazine-1,4-dione	4.08
10	3-Isobutylhexahydropyrrolo[1,2-a]pyrazine-1,4-dione	3.25
11	Octahydro-5h,10h-dipyrrolo[1,2-a:1',2'-d]pyrazine-5,10-dione	2.39
12	*n*-Hexadecanoic acid	5.04
13	*Trans*-13-octadecenoic acid, methyl ester	0.48
14	(9*E*)-9-octadecenoic acid	27.65
15	3-Benzylhexahydropyrrolo[1,2-a]pyrazine-1,4-dione	2.48
16	Cycloeicosane	3.23
	Total	**69.68**

3.2 Antimicrobial and Antidiabetic Activity of the *C. Mandarina* DCM Extract

Antidiabetic assays demonstrated that the *C. mandarina* DCM extract exhibited promising inhibitory activity against α-amylase and α-glucosidase with IC_{50} of 3.68–6.75 mg/ml (Table 2). Additionally, antimicrobial testing revealed that the extract exhibited broad-spectrum antimicrobial activity, with MIC values ranging from 256 to 1024 μg/mL (Table 3). Notably, the extract exhibited the strongest activity against *S. aureus* (MIC = 256 μg/mL), followed by *E. coli*, *P. aeruginosa*, and *B. cereus* (MIC = 512 μg/mL), while weaker effects was observed against *S. enterica*, *E. faecalis*, and *C. albicans* (MIC = 1024 μg/mL). These results suggest the presence of potential antimicrobial and antidiabetic compounds in the *C. mandarina* DCM extract.

GC-MS analysis (Table 1) identified several compounds with known bioactivities in the *C. mandarina* DCM extract. For example, fatty acids such as compounds **1**, **3**, **12–14** have demonstrated antimicrobial, anticancer, and anti-inflammatory properties [16–20]. Compound **1** also showed antidiabetic effects by improving hyperglycemia and hyperinsulinemia in high-fat diet (HFD)-fed mice [21]. Furthermore, cyclic dipeptides, including compounds **9–11** and **15**, have been reported to exhibit broad bioactivities such as antimicrobial, antibiofilm, anticancer, and antioxidant activities [22–25]. These findings highlight the potential of the *C. mandarina* DCM extract as a valuable source of bioactive compounds.

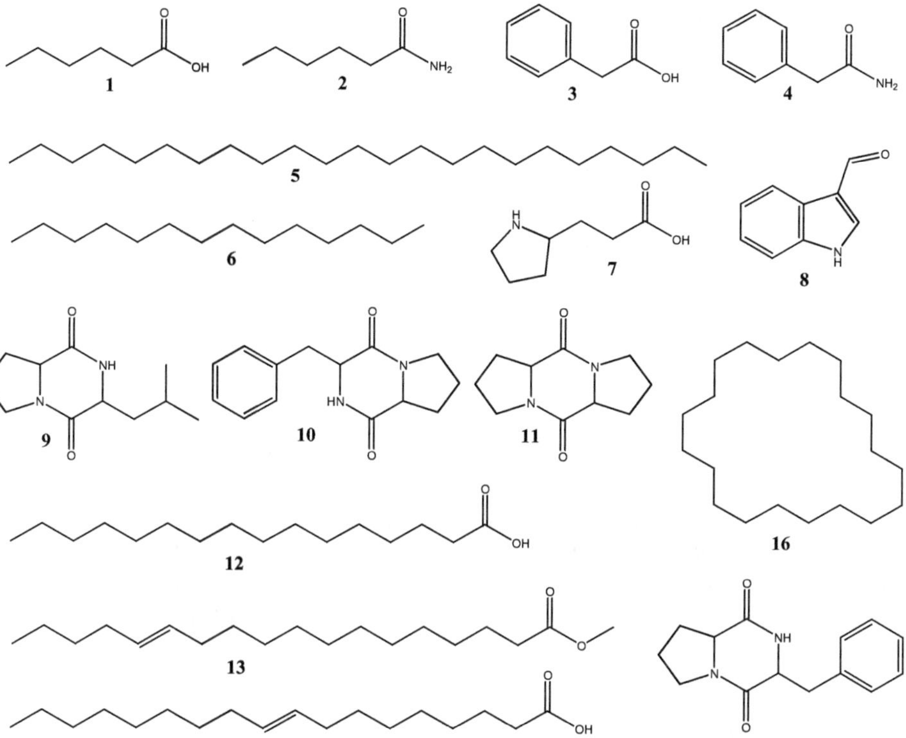

Fig. 2. Chemical structure of compounds identified in the *C. mandarina* DCM extract

Table 2. Antidiabetic activity of the *C. mandarina* DCM extract

Sample	IC_{50}, mg/mL	
	α-amylase inhibition	α-glucosidase inhibition
C. mandarina DCM extract	3.68 ± 0.24	6.75 ± 0.02
Acarbose	0.105 ± 0.003	0.121 ± 0.004

Although cicadas are recognized for their potential in traditional medicine, studies investigating their biological activities remain limited, with most research focused on the anti-inflammatory and antibacterial properties of cicada shells. The anti-inflammatory effects of cicada slough are attributed to the reduction of pro-inflammatory cytokines such as TNF-α, IFN-γ, and IL-6 [26]. N-acetyldopamine dimers derived from cicada slough have demonstrated antioxidant and anti-inflammatory activities by scavenging 1,1-diphenyl-2-picrylhydrazyl (DPPH) radicals, inhibiting reactive oxygen species (ROS) generation and nitric oxide (NO) production, suppressing nuclear factor-κB (NF-κB) activation, and downregulating the expression of pro-inflammatory mediators

Table 3. Antimicrobial activity of the *C. mandarina* DCM extract

Microorganisms	MIC (μg/ml)
Escherichia coli	512
Salmonella enterica	1024
Pseudomonas aeruginosa	512
Enterococcus faecalis	1024
Staphylococcus aureus	256
Bacillus cereus	512
Candida albicans	1024

including inducible nitric oxide synthase (iNOS), IL-6, TNF-α, and cyclooxygenase-2 (COX-2) in LPS-induced RAW264.7 cells [27]. Additionally, chitooligosaccharides extracted from cicada slough exhibited strong antibacterial activity against *B. subtilis*, *S. aureus*, and *E. coli* [6]. A dichloromethane-ethanol extract of the shell of *Cicada pustulata* showed antimicrobial activity against *S. aureus* and *Mycobacterium tuberculosis*, producing inhibition zones of 13 mm and 16 mm, respectively [28]. Notably, this study is the first to report the antidiabetic activity of cicadas.

4 Conclusions

This study explores the chemical profile and biological activities of a dichloromethane (DCM) extract derived from the cicada *Cryptotympana mandarina.* Gas chromatography–mass spectrometry (GC-MS) analysis identified 16 major compounds with main components such as (9*E*)-9-octadecenoic acid, benzeneacetic acid, 1*H*-indole-3-carboxaldehyde, and *n*-hexadecanoic acid. *In vitro* experiments demonstrated that the extract inhibited α-amylase and α-glucosidase with IC_{50} values of 3.68 ± 0.24 mg/mL and 6.75 ± 0.02 mg/mL, respectively. Additionally, the extract showed antimicrobial activity against all tested microorganisms with MICs ranging from 256 to 1024 μg/mL. These findings suggest that the DCM extract of *C. mandarina* may serve as a promising natural source of antidiabetic and antimicrobial agents.

Acknowledgements. Ton That Huu Dat was funded by the Postdoctoral Scholarship Programme of Vingroup Innovation Foundation (VINIF), code VINIF.2024.STS.58.

References

1. Thai, P.H., Yang, J.-T.: A contribution to the Cicadidae fauna of Vietnam (Hemiptera: Auchenorrhyncha), with one new species and twenty new records. Zootaxa **2249**(1), 1–19 (2009). https://doi.org/10.11646/zootaxa.2249.1.1
2. Li, X., Wiens, J.J.: Estimating global biodiversity: the role of cryptic insect species. Syst. Biol. **72**(2), 391–403 (2023). https://doi.org/10.1093/sysbio/syac069

3. Park, J., Choi, G.: Review on herbal medicinal materials in the korean pharmacopoeia and the Korean herbal pharmacopoeia. Korean Herb. Med. Inform. **4**(2), 9–35 (2016)
4. Noh, P., Kim, W.J., Song, J.H., Park, I., Choi, G., Moon, B.C.: Rapid and simple species identification of cicada exuviae using COI-based SCAR assay. Insects **11**(3), 168 (2020). https://doi.org/10.3390/insects11030168
5. Song, B.K., Won, J.H., Kim, S.: Historical medical value of Donguibogam. J. Pharmacopuncture **19**(1), 16–20 (2016). https://doi.org/10.3831/KPI.2016.19.002
6. Wu, S.J., Pan, S.K., Wang, H. Bin, Wu, J.H.: Preparation of chitooligosaccharides from cicada slough and their antibacterial activity. Int. J. Biol. Macromol. **62**, 348–351 (2013). https://doi.org/10.1016/j.ijbiomac.2013.09.042
7. Li, H., Chumroenphat, T., Bunyatratchata, A., Boonarsa, P., Wrigley, C., Siriamornpun, S.: Chemical composition and nutritional profile of cicada (*Meimuna opalifera* Walker) at different developmental stages: Implications for functional food applications. Food Chem X. **21**, (2024). https://doi.org/10.1016/j.fochx.2023.101081
8. Zhang, E., et al.: A minireview of the medicinal and edible insects from the traditional Chinese medicine (TCM). Front. Pharmacol. **14**, 1125600 (2023). https://doi.org/10.3389/fphar.2023.1125600
9. Xie, X., Guo, H., Liu, J., Wang, J., Li, H., Deng, Z.: Edible and medicinal progress of *Cryptotympana atrata* (Fabricius) in China. Nutrients **15**(19), 4266 (2023). https://doi.org/10.3390/nu15194266
10. Dat, T.T.H., et al.: Pharmacological properties, volatile organic compounds, and genome sequences of bacterial endophytes from the mangrove plant *Rhizophora apiculata* Blume. Antibiotics **10**(12), 1491 (2021). https://doi.org/10.3390/antibiotics10121491
11. Dat, T.T.H., Cuong, L.C.V., Ha, D.V., Oanh, P.T.T., Nhi, N.P.K., Anh, H.L.T., Quy, P.T., Bui, T.Q., Triet, N.T., Nhung, N.T.A.: The study on biological activity and molecular docking of secondary metabolites from *Bacillus* sp. isolated from the mangrove plant *Rhizophora apiculata* Blume. Reg. Stud. Mar. Sci. **55**, 102583 (2022). https://doi.org/10.1016/j.rsma.2022.102583
12. Dat, T.T.H., Cuc, N.T.K., Cuong, P.V., Smidt, H., Sipkema, D.: Diversity and antimicrobial activity of vietnamese sponge-associated bacteria. Mar. Drugs **19**(7), 353 (2021). https://doi.org/10.3390/md19070353
13. Al Shaqsi, N.H.K., Al Hoqani, H.A.S., Hossain, M.A., Al Sibani, M.A.: Optimization of the demineralization process for the extraction of chitin from *Omani Portunidae* segnis. Biochem Biophys Rep. **23**, 100779 (2020). https://doi.org/10.1016/j.bbrep.2020.100779
14. Poerio, A., Petit, C., Jehl, J.P., Arab-Tehrany, E., Mano, J.F., Cleymand, F.: Extraction and physicochemical characterization of chitin from cicada orni sloughs of the south-eastern French mediterranean basin. Molecules **25**(11), 2543 (2020). https://doi.org/10.3390/molecules25112543
15. Izadi, H., Asadi, H., Bemani, M.: Chitin: a comparison between its main sources. Front Mater. **12**, 1537067 (2025). https://doi.org/10.3389/fmats.2025.1537067
16. Pierre, K., Thapa, N., Liu, Q., Jibrin, M.O., Jones, J.B., Zhang, S.: Effectiveness of hexanoic acid for the management of bacterial spot on tomato caused by *Xanthomonas perforans*. Agriculture **15**(7), 695 (2025). https://doi.org/10.3390/agriculture15070695
17. Wang, X., Zhang, C., Bao, N.: Molecular mechanism of palmitic acid and its derivatives in tumor progression. Front. Oncol. **13**, 1224125 (2023). https://doi.org/10.3389/fonc.2023.1224125
18. Kwon, K.M., et al.: Phenylacetic acid, an anti-vaginitis metabolite produced by the vaginal symbiotic bacterium *Chryseobacterium gleum*. Sci. Rep. **14**, 12226 (2024). https://doi.org/10.1038/s41598-024-62947-7

19. Pan, H., et al.: The antibacterial mechanism of phenylacetic acid isolated from *Bacillus megaterium* L2 against *Agrobacterium tumefaciens.* PeerJ **10**, e14304 (2022). https://doi.org/10.7717/peerj.14304
20. Huang, C.B., Alimova, Y., Myers, T.M., Ebersole, J.L.: Short- and medium-chain fatty acids exhibit antimicrobial activity for oral microorganisms. Arch. Oral Biol. **56**(7), 650–654 (2011). https://doi.org/10.1016/j.archoralbio.2011.01.011
21. Ikeda, T., Nishimoto, Y., Ichikawa, D., Matsunaga, T., Kawauchi, A., Kimura, I.: Hexanoic acid improves metabolic health in mice fed high-fat diet. bioRxiv. 2025.02.20.639216 (2025). https://doi.org/10.1101/2025.02.20.639216
22. Singh, V.K., Mishra, A., Jha, B.: 3-benzyl-hexahydro-pyrrolo[1,2-a]pyrazine-1,4-dione extracted from *Exiguobacterium indicum* showed anti-biofilm activity against *Pseudomonas aeruginosa* by attenuating quorum sensing. Front. Microbiol. **10**, 1269 (2019). https://doi.org/10.3389/fmicb.2019.01269
23. Jia, J., et al.: 2,5-Diketopiperazines: a review of source, synthesis, bioactivity, structure, and MS fragmentation. Curr. Med. Chem. **30**(9), 1060–1085 (2022). https://doi.org/10.2174/0929867329666220801143650
24. Yu, B., Zeng, W., Zhou, Y., Li, N., Liang, Z.: Characterization and bioactive metabolite profiling of *Streptomyces* sp. Y009: a mangrove-derived actinomycetia with anticancer and antioxidant potential. microorganisms **12**(11), 2300 (2024). https://doi.org/10.3390/microorganisms12112300
25. Al-Askar, A., Al-Otibi, F.O., Abo-Zaid, G.A., Abdelkhalek, A.: Pyrrolo[1,2-a]pyrazine-1,4-dione, hexahydro-3-(2-methylpropyl), as the primary secondary metabolite of *Bacillus* spp., could be an effective antifungal agent against the soil-borne fungus, *Sclerotium bataticola.* Egypt. J. Chem. **67**(13), 1009–1022 (2024). https://doi.org/10.21608/ejchem.2024.325664.10571
26. Kim, H., Cho, S., Kim, M., Kim, H., Ryu, J., Jo, S., Lee, G., Ryu, M.: Anti-inflammatory effects of *Cryptotympana atrata* Fabricius slough shed on contact dermatitis induced by dinitrofluorobenzene in mice. Pharmacogn Mag. **10**(Suppl 2), S377–S382 (2014). https://doi.org/10.4103/0973-1296.133288
27. Xu, M.Z., et al.: Antioxidant and anti-inflammatory activities of N-acetyldopamine dimers from *Periostracum Cicadae.* Bioorg. Med. Chem. **14**(23), 7826–7834 (2006). https://doi.org/10.1016/j.bmc.2006.07.063
28. Ma, G., Wu, L., Shao, F., Zhang, C., Wan, H.: Antimicrobial activity of 11 insects extracts against multi drug resistant (MDR) strains of bacteria and fungus. In: IOP Conference Series: Earth and Environmental Science **252**, 022132 (2019). https://doi.org/10.1088/1755-1315/252/2/022132

Toward Sustainable Diabetes Treatment: Mechanistic Insights into *Pandanus Amaryllifolius*

Huu Thang Tran[1,3], Hoang Phuong Vu Pham[1,3], Ngoc Dien Nguyen[1,3], Kim Nhuong Thi Nguyen[2,3], and Minh Nam Nguyen[2,3,4](✉)

[1] Faculty of Pharmacy, University of Health Sciences, Ho Chi Minh City, Vietnam
[2] Center for Genetics and Reproductive Health (CGRH), University of Health Sciences, Ho Chi Minh City, Vietnam
nmnam@uhsvnu.edu.vn
[3] Vietnam National University Ho Chi Minh City, Ho Chi Minh City, Vietnam
[4] Department of Biomedical Engineering, Faculty of Medicine, University of Health Sciences, Ho Chi Minh City, Vietnam

Abstract. To construct and analyze the interaction network of compounds in *Pandanus amaryllifolius* leaves (PAL) and protein targets related to Type 2 Diabetes Mellitus (T2DM), this study applied a Network Pharmacology approach. We utilized DAVID for biochemical pathway analysis and visualization, and Cytoscape to visualize the impact of PAL compounds on T2DM-related targets. This was combined with molecular docking analysis using AutoDock Vina. Our results identified 14 compounds from PAL that met bioavailability and drug-likeness criteria, affecting 97 proteins associated with T2DM. Network pharmacological analysis highlighted six hit compounds-Luteolin, Apigenin, Quercetin, Genistein, Fisetin, and β-tocopherol—which interact with ten potential targets: AKT1, EGFR, ESR1, MMP2, MMP9, CCNB1, CDK1, PTGS2, KDR, and SRC. Molecular Docking studies further indicated that these compounds possess promising ability binding potential with these ten targets, showing stable interactions with binding energy ranging from −10.0 to −5.3 kcal/mol. The integrated approach, combining compound-protein target-biochemical pathway analysis with molecular docking, clarifies the mechanisms of potential compounds in PAL for treating T2DM.

Keywords: *Pandanus amaryllifolius* · Type 2 diabetes mellitus (T2DM) · Network Pharmacology

1 Introduction

Type 2 diabetes mellitus (T2DM) is a widespread chronic metabolic disorder. Current treatment often targets single pathways and requires lifelong medication, which may cause side effects and create financial burdens for patients [1]. Recently, natural compounds from medicinal plants have gained attention because they can act on multiple targets, and may offer more holistic therapeutic benefits [2].

L. Vu et al. (Eds.): PBC 2025, 2026.
https://doi.org/10.1007/978-3-032-29579-8_15

Pandanus amaryllifolius leaves (PAL), commonly known as pandan leaf, are traditionally used in folk medicine due to their wide range of health benefits. Although their pharmacological properties are well recognized, the mechanisms by which its bioactive compounds may influence T2DM remain poorly understood [3].

Advances in Network Pharmacology (NP) and Molecular Docking (MD) have opened new avenues for exploring complex disease mechanisms and multi-target interactions. NP allows us to clarify how active compounds in PAL may interact with key proteins, and signaling pathways involved in T2DM [4]. Complementarily, MD simulates how these compounds bind to potential protein targets, providing insight into their potential therapeutic effects [5].

In this study, we combine NP and MD approaches using publicly available databases of T2DM-related proteins, and phytochemicals in PAL to identify promising multi-target compounds. Our goal was to clarify the molecular mechanisms through which PAL may exert antidiabetic effects.

2 Materials and Methods

2.1 Study Design

This was an *in silico* study applying Network Pharmacology and Molecular Docking approaches to investigate the multi-target mechanisms of bioactive compounds from *Pandanus amaryllifolius* leaves in the treatment of T2DM.

2.2 Study Setting and Duration

All computational analyses were conducted on a Gigabyte Gaming G5 GD 51S1123SO laptop from May 18th, 2023, to September 15th, 2024.

2.3 Study Subjects

This study focused on identifying bioactive compounds extracted from PAL that achieved drug-likeness and oral bioavailability criteria, as well as T2DM-related target proteins.

2.4 Sample Size and Sampling Method

All available data of phytochemicals in PAL and T2DM-associated proteins were collected from open-access databases up to September 15, 2024.

2.5 Study Procedures

- Collection and screening of bioactive compounds from PAL.
- Prediction of potential T2DM-related targets for the selected bioactive compounds.
- Enrichment analysis of biochemical pathways associated with PAL compounds in the context of T2DM.
- Construction and analysis of a pharmacological network linking potential compounds, targets, and pathways.
- Molecular docking between core compounds and core target proteins.

2.6 Techniques, Tools and Data Collection Process

Collection and Screening of Bioactive Compounds in PAL

Advanced search tools on Google Scholar and PubMed were used to identify compounds reported in *Pandanus amaryllifolius* leaves (PAL) using combinations of keywords including "phytochemical", or "constituents", and "*Pandanus amaryllifolius* leaves". Only reputable, peer-reviewed articles synthesizing PAL compounds experimentally identified in prior phytochemical studies were included (accessed August 2023) [6]. The resulting compounds were then screened based on two key criteria, bioavailability > 0.3 and drug-likeness > 0.18, using SwissADME and Molsoft.

Prediction of Targets Related to T2DM for Bioactive Compounds in PAL

Targets of the screened compounds were predicted using SwissTargetPrediction and retained only when the probability score was ≥ 0.7. These predicted targets were then intersected with T2DM-related targets collected from GeneCards (accessed 18 August 2023) and relevant review studies to generate a set of disease-specific targets for compounds found in PAL. GeneCards is a widely used, curated human proteins compendium integrating peer-reviewed experimental evidence and multi-omics resources, providing a reliable foundation for target nomination in *in silico* studies (https://www.genecards.org/).

Pathway Enrichment Analysis

The T2DM-related targets were input into DAVID for Gene Ontology (GO) and KEGG pathway enrichment analysis. Key pathways of the compounds in PAL were selected based on a p-value < 0.05 and ranked by count values in descending order. These potential pathways were then combined with the T2DM-related targets to construct a pharmacological network.

Construction and Analysis of the Potential Compound–Target–Pathway Interaction Network

The interactions of potential T2DM-related targets in PAL were analyzed using the STRING online tool. The visualization of Network was performed using Cytoscape, and topological analysis (Degree, Closeness, and Betweenness) was conducted with CytoHubba to identify the most prominent targets. Cytoscape software was used to visualize the network of potential compounds, targets, and pathways. The hit compounds contributing to the anti-T2DM activity of PAL were identified based on their interactions with central targets of disease.

Molecular Docking

Interactions between the key compounds in PAL and the central T2DM targets were evaluated using molecular docking. The 3D structures of key protein targets were collected from the Protein Data Bank (PDB) and met the following criteria which are determined by X-ray crystallography with resolution less than 4 Å [7], co-crystallized with a ligand, no mutations, and without missing amino acids at the ligand-binding site. Discovery Studio 2019 was used to separate the protein and co-crystallized ligand into two distinct files. Hydrogen atoms were then added to the protein, a grid box was generated using AutoDock Tools, and the protein file was saved in .pdbqt format. The 2D structures of PAL compounds were downloaded from PubChem in .sdf format. Ubuntu operating system was used for converting all ligand files from .sdf to .pdbqt format. Molecular docking with multiple ligands was performed using AutoDock Vina with Ubuntu support. Docking results were evaluated based on two criteria that are binding affinity (kcal/mol) and interactions with amino acids at the binding site.

2.7 Data Processing and Analysis

Data were processed, analyzed, and visualized using Python programming language on the Kaggle platform, Excel 2016, Cytoscape 3.10.2, and SRPlot.

3 Results

3.1 Screening of Bioactive Compounds in PAL

A total of 96 compounds were initially identified from PAL through the data collection process. Following the screening for drug-likeness and bioavailability criteria, 14 compounds were selected for further analysis, as presented in Table 1.

Table 1. Bioactive compounds in PAL

Numbers	PubChem ID	IUPAC	Classification	Drug-likeness	Bioavailability
1	71255	Flufylline	Alkaloids	1.96	0.55
2	131752803	Pandamarine	Alkaloids	0.78	0.55
3	457801	Clionasterol	Phytosterols	0.78	0.55
4	72276	Epicatechin	Flavonoids	0.64	0.55
5	73160	Catechin	Flavonoids	0.64	0.55
6	5280794	Stigmasterol	Phytosterols	0.62	0.55
7	5280343	Quercetin	Flavonoids	0.52	0.55
8	5280863	Kaempferol	Flavonoids	0.5	0.55
9	14985	Vitamin E	Tocopherols	0.48	0.55
10	6857447	β-tocopherol	Tocopherols	0.47	0.55

(continued)

Table 1. *(continued)*

Numbers	PubChem ID	IUPAC	Classification	Drug-likeness	Bioavailability
11	5281614	Fisetin	Flavonoids	0.46	0.55
12	5280961	Genistein	Flavonoids	0.44	0.55
13	5280443	Apigenin	Flavonoids	0.39	0.55
14	5280445	Luteolin	Flavonoids	0.38	0.55

3.2 Prediction of T2DM-Related Targets of Compounds in PAL

The analysis using SwissTargetPrediction indicated targets for the 14 PAL compounds that met the criteria for drug-likeness and bioavailability. This initial prediction identified a total of 104 targets. However, no protein targets were predicted for two compounds, Epicatechin and Catechin. To expand and validate the potential T2DM-related targets, several review studies and the GeneCards database were utilized, identifying 4,736 and 11,621 related targets, respectively. After removing duplicate proteins from these two sources, a total of 97 overlapping targets were identified, which represent the final set of T2DM-related targets for further analysis.

3.3 Biochemical Pathway Enrichment Analysis

The pathway of enrichment analysis for the 97 overlapping targets is illustrated in Fig. 1. Based on Gene Ontology (GO) analysis, the potential targets of the compounds in PAL are involved in critical processes such as protein autophosphorylation and regulation of pancreatic β-cell apoptosis. KEGG pathway analysis also revealed that these targets are involved in several key signaling pathways such as PI3K/AKT, estrogen signaling, FoxO, and Rap1 pathways, which are closely associated with the development and progression of T2DM and its complications. These findings suggest that the compounds in PAL may exert their effects through multiple biochemical pathways to regulate T2DM-related pathophysiological processes, highlighting their potential therapeutic effects against T2DM.

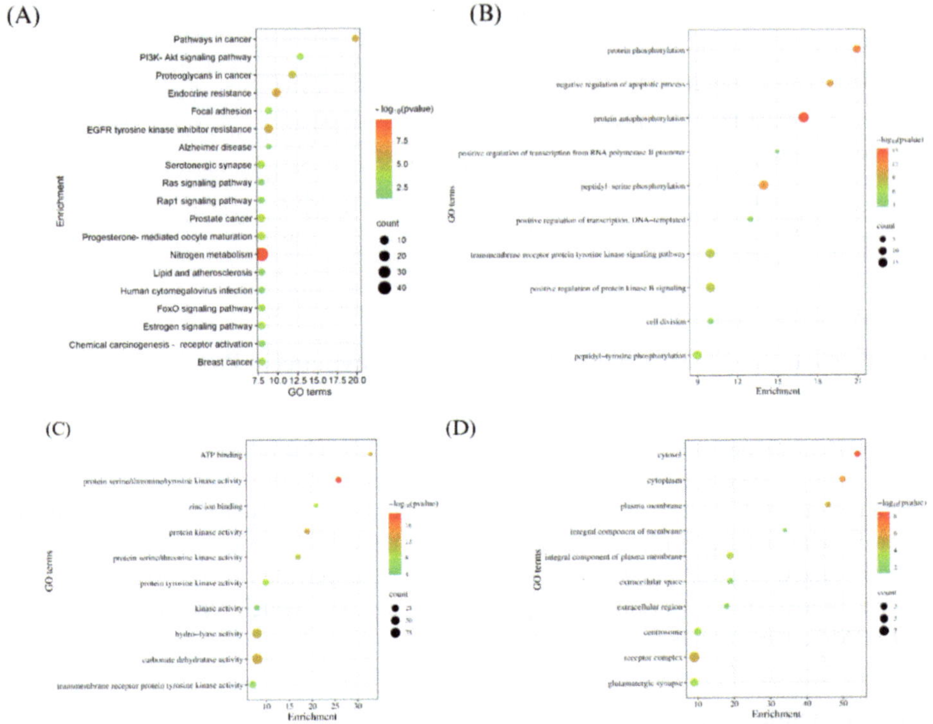

Fig. 1. GO/KEGG enrichment of PAL T2DM-related targets. Input: 97 overlapping targets. Enrichment was performed in DAVID. Panels: **a** Kyoto Encyclopedia of Genes and Genomes (KEGG) pathways; **b** GO Biological Process (BP); **c** GO Molecular Function (MF); **d** GO Cellular Component (CC).

3.4 Pharmacological Network Analysis

A protein–protein interaction (PPI) network of the 97 overlapping targets was constructed using the STRING database, resulting in a network with 93 nodes and 1.058 edges, indicating strong interactions among these targets. From this network, the top 10 central targets with the highest scores were identified, including AKT1, MMP2, MMP9, CDK1, CCNB1, ESR1, EGFR, SRC, PTGS2, and KDR. These proteins play key roles in regulating biochemical pathways related to T2DM.

The compound–target interaction network consisted of 22 nodes and 23 edges (Fig. 2a), highlighting differences in the levels of interaction among compounds. Specifically, six compounds exhibited no interactions with the 10 central targets, whereas the other six compounds were predicted to have at least one connection. Among them, Quercetin is the notable compound with eight interactions, suggesting its potential therapeutic effects. In contrast, β-tocopherol and Genistein showed only one connection each. Additionally, a more comprehensive network of Potential Compound—Potential Target—Potential Pathway was constructed, consisting of 34 nodes and 92 edges (Fig. 2b), providing deeper insights into the relationships between central targets and T2DM-related signaling pathways.

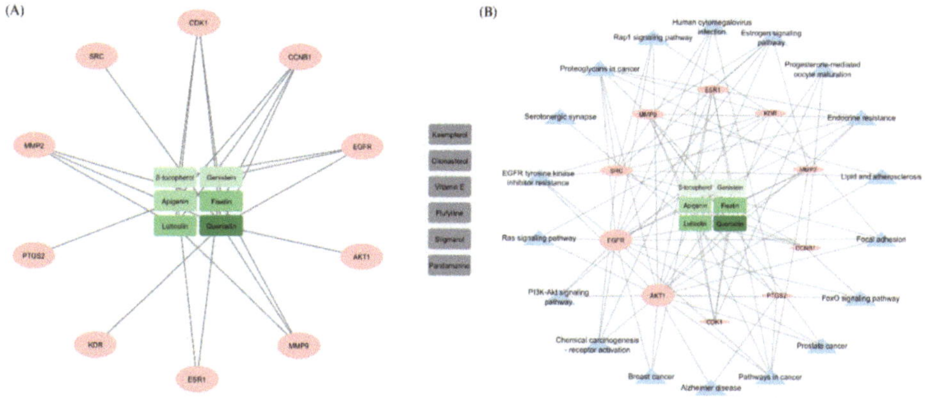

Fig. 2. Compound–target and compound–target–pathway networks for six PAL compounds and ten hub T2DM targets derived from 97 overlapping proteins. Targets were predicted by SwissTargetPrediction and intersected with T2DM. Enrichment used DAVID with $p < 0.05$. Networks visualized in Cytoscape. Panels: **a** compound–target. **b** compound–target–pathway.

The pharmacological network uses squares to represent natural compounds, circles to represent targets (Fig. 2a), and triangles to represent biochemical pathways. The size of circles reflects the number of connections between targets, related compounds and pathways (Fig. 2b).

3.5 Molecular Docking

Six hit compounds (Fig. 3) with the strongest interactions with 10 potential targets were selected for molecular docking analysis. The 3D structures of selected protein targets were downloaded from the Protein Data Bank (PDB), including AKT1 (8UW9), CCNB1 (4YC3), CDK1 (6GU6), EGFR (8F1Y), ESR1 (1UOM), KDR (2QU5), MMP2 (8H78), MMP9 (1GKC), PTGS2 (5KIR), and SRC (2H8H). Redocking between targets and their co-crystallized ligands showed RMSD values ranging from 0.8772 to 4.5454. Additionally, Metformin, a common therapeutic agent for the treatment of T2DM, was included as a positive control to compare binding affinities, providing an objective evaluation of the potential interaction of natural compounds with T2DM-related targets.

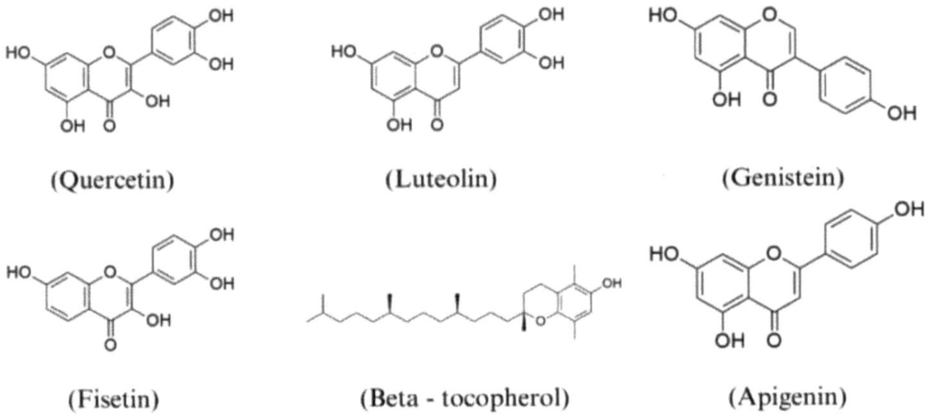

Fig. 3. 2D structures of six potential compounds in PAL identified via network analysis

Molecular docking analysis showed that the binding affinities of six compounds ranged from –10.0 to –5.3 kcal/mol. (Table 2). Notably, the binding affinity of Metformin ranged from –5.5 to –4.1 kcal/mol. These results have indicated that phytochemicals derived from PAL exhibit significant binding affinities and interactions with T2DM-related targets, some of them showed superior interactions compared to Metformin.

Table 2. Molecular docking binding affinities (kcal/mol) of six PAL compounds and metformin against hub T2DM targets (more negative = stronger predicted binding). Target key: A = AKT1, B = MMP2, C = MMP9, D = CDK1, E = CCNB1, F = ESR1, G = EGFR, H = SRC, I = PTGS2, J = KDR.

Compounds/Target protein	A	B	C	D	E	F	G	H	I	J
Luteolin	–8.5	–9.7	–9.5	–8.4	–6.4	–8.6	–8.8	–8.9	–9.8	–8.9
Apigenin	–8.1	–7.8	–9.2	–8	–6.4	–8.3	–8.8	–8.9	–9.9	–9.3
Genistein	–7.6	–7.5	–10	–8.4	–6	–9	–8.4	–9.1	–8.5	–8.2
Fisetin	–7.6	–8.2	–9.4	–8.1	–6.4	–8.4	–8.7	–9	–9.4	–8.5
β–tocopherol	–8.7	–6.6	–8.3	–7.4	–5.3	–7.8	–7.7	–9	–7.6	–8.3
Quercetin	–7.8	–8.2	–9.4	–8.2	–6.2	–8.5	–8.9	–9.1	–9.8	–7.9

The docking complexes between hit compounds and central targets (ligand–protein) formed significant interactions, including hydrogen bonds, Van der Waals forces and hydrophobic interactions. In addition, the compounds established other interactions such as π–interactions and alkyl bonds with the targets. Specific interactions between hit compounds and their respective protein was illustrated in Fig. 4, including β-tocopherol with AKT1, Quercetin with EGFR, Luteolin with ESR1, and Genistein with MMP9. These findings highlighted the potential of PAL-derived compounds in modulating key targets associated with the pathogenesis of T2DM.

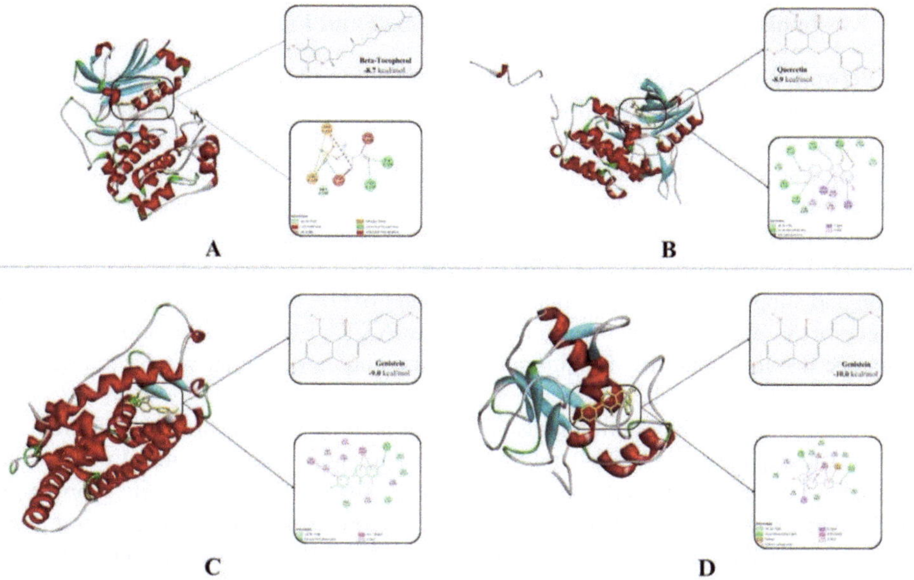

Fig. 4. Interactions between compounds and proteins. **a** ß-tocopherol—AKT1; **b** Quercetin—EGFR; **c** Genistein—ESR1; **d** Genistein—MMP9.

4 Discussion and Conclusion

4.1 Integration of Pharmacological Network and Molecular Docking

Molecular docking revealed six key compounds, five flavonoids (Luteolin, Apigenin, Quercetin, Genistein, Fisetin) and one tocopherol (ß-tocopherol), as strong binders to several potential therapeutic targets for T2DM. These compounds showed binding affinities to AKT1 ranging from –8.7 to –7.6 kcal/mol., a central protein in the PI3K/AKT/PKB signaling pathway responsible for regulating glucose use, insulin production, and lipid metabolism. Modulating AKT1 may improve insulin sensitivity and reduce cardiovascular complications in T2DM.

The compounds also bound strongly to MMP9, an enzyme involved in extracellular matrix (ECM) degradation. Elevated MMP9 in T2DM can lead to ECM disruption, contributing to atherosclerosis and metabolic imbalance. EGFR also demonstrated notable binding affinities and involved in several signaling pathways of T2DM, including PI3K/AKT, FoxO, Estrogen, Focal adhesion, and Rap1. These pathways have essential roles in regulating the survival and function of pancreatic β-cells, as well as promoting insulin secretion [8].

These findings highlighted the potential of the six hit compounds to improve insulin sensitivity and support for therapeutic strategies of T2DM through multiple biochemical pathways, thereby contributing to the limitation of complications associated with the progression of disease.

4.2 Mechanistic Insights into PI3K/AKT, FoxO and Estrogen Signaling

Our network pharmacology analysis identified the PI3K/AKT, FoxO, and estrogen signalling as the most enriched pathways associated with PAL targets, with AKT1, EGFR, ESR1, and MMP9 forming a central interaction module. These axes are closely linked to glucose homeostasis, insulin sensitivity, and β-cell survival. AKT1 is a pivotal mediator of insulin signalling and glucose uptake in insulin-responsive tissues, whereas FoxO1/3a regulate gluconeogenic enzymes, oxidative stress responses, and cell survival mechanisms [11]. Estrogen receptor α (ESR1) integrates estrogen signals to enhance insulin action and suppress hepatic gluconeogenesis via FoxO1, and also contributes to β-cell protection [8]. Thus, simultaneous modulation of AKT1, EGFR, ESR1, and MMP9 may influence insulin signalling, inflammatory injury, and extracellular matrix remodelling in T2DM [8, 11].

Although the six key compounds share overlapping targets, they are predicted to engage these pathways through partially complementary mechanisms. Luteolin and fisetin have been documented to upregulate SIRT1/3/6 and FOXO3a, while suppressing the NADPH oxidase subunit p47^phox under high-glucose conditions, reducing reactive oxygen species and improving cellular redox balance [12]. Other flavonoids in PAL, including quercetin and apigenin, have been widely reported to enhance insulin signalling and support β-cell survival through modulation of PI3K/AKT and stress-responsive pathways. In contrast, genistein, a phytoestrogenic isoflavone, may preferentially exert its effects via ESR1- and ERK-related signalling to improve insulin sensitivity and β-cell function. In parallel, β-tocopherol is expected to complement these actions by stabilising cellular membranes and reducing lipid peroxidation, thereby reducing oxidative damage in insulin-responsive tissues and pancreatic islets.

Taken together, the convergence of these compounds on AKT1, EGFR, ESR1, and MMP9 provides a plausible mechanistic explanation for the multi-target profile of PAL. By reinforcing PI3K/AKT signalling, mitigating FoxO-driven gluconeogenesis and apoptosis, and fine-tuning estrogen-related pathways, this network-level synergy may help to improve insulin sensitivity, support glucose uptake, and protect β-cells from chronic oxidative and inflammatory stress, effects that extend beyond those expected from single-target interventions [8, 11, 12].

4.3 Applications and Limitations of the Study

This study demonstrated the therapeutic potential of *Pandanus amaryllifolius* leaves (PAL) in supporting the treatment of type 2 diabetes mellitus (T2DM) through their ability to modulate multiple biological targets associated with the disease's pathogenesis. Compounds derived from PAL showed promising effects in enhancing insulin sensitivity, protecting pancreatic β-cells, and mitigating diabetes-related complications. Through molecular docking and network pharmacology analysis, flavonoid and tocopherol compounds were identified with strong binding affinities to key proteins such as AKT1 and EGFR, which play pivotal roles in the PI3K/AKT signaling pathway—crucial in T2DM therapy.

Moreover, virtual screening studies have suggested that PAL may inhibit Protein Tyrosine Phosphatase 1B (PTP1B), a potential therapeutic target for both T2DM and obesity, with up to 94.38% efficacy [9]. Additionally, *in vitro* experiments showed PAL's ability to inhibit alpha-amylase activity, thereby contributing to blood glucose reduction and antioxidant effects [10]. These findings provide a scientific basis for considering PAL as a complementary or alternative therapy for T2DM.

However, further experimental studies are required to validate the predicted biological pathways involved in PAL's antidiabetic effects. The safety and efficacy of these bioactive compounds should be explored through more in-depth *in vitro* and *in vivo* studies to comprehensively assess their translational potential. As the present insights arise from curated databases and computational inference, additional experimental work would be valuable to consolidate these hypotheses. In particular, target engagement (e.g., SPR/ITC or CETSA), pathway readouts (p-AKT, FOXO1 nuclear exclusion), and functional assays (glucose uptake, insulin secretion, GTT/ITT), alongside basic safety profiling, could further substantiate the proposed AKT–FoxO–estrogen mechanisms and strengthen the biological relevances of our findings.

5 Conclusion

Six key compounds in PAL, including Luteolin, Apigenin, Quercetin, Genistein, Fisetin, and β-tocopherol—have been demonstrated to interact with ten important proteins, including AKT1, EGFR, ESR1, MMP2, MMP9, CCNB1, CDK1, PTGS2, KDR, and SRC. These proteins play essential roles in regulating insulin sensitivity, promoting insulin secretion, protecting pancreatic β-cells, and reducing T2DM complications.

Given the rising prevalence of T2DM and its associated economic burden worldwide, the compounds from PAL present great potential as safer and more cost-effective therapeutic options. This opens new avenues for the development of plant-based treatments, offering significant benefits in the prevention and management of T2DM. However, further research is still needed to evaluate their safety and clinical efficacy, in order to fully determine the therapeutic value of these compounds and the practical application of PAL in T2DM treatment.

References

1. Rossboth, S., Lechleitner, M., Oberaigner, W.: Risk factors for diabetic foot complications in type 2 diabetes—a systematic review. Endocrinol. Diabetes Metab. **4**, e00175 (2021)
2. Omale, S., Amagon, K.I., Johnson, T.O., Bremner, S.K., Gould, G.W.: A systematic analysis of anti-diabetic medicinal plants from cells to clinical trials. PeerJ **11**, e14639 (2023)
3. Chiabchalard, A., Nooron, N.: Antihyperglycemic effects of pandanus amaryllifolius Roxb. Leaf Extract. Pharmacogn. Mag. **11**(41), 117 (2015)
4. Noor, F., Tahir Ul Qamar, M., Ashfaq, U.A., Albutti, A., Alwashmi, A.S.S., Aljasir, M.A.: Network pharmacology approach for medicinal plants: review and assessment. Pharmaceuticals, **15**(5), 572 (2022)
5. Stanzione, F., Giangreco, I., Cole, J.C.: Use of molecular docking computational tools in drug discovery. In: Witty, D.R., Cox, B. (eds.) Progress in medicinal chemistry, vol. 60, pp. 273–343. Elsevier, Amsterdam (2021)
6. Bhuyan, B., Sonowal, R.: An overview of Pandanus amaryllifolius Roxb. Ex Lindl. and its potential impact on health. Curr. Trends Pharm. Res. **8**(1), 138–157 (2021)

7. Batool, S., et al.: Network pharmacology and bioinformatics approach reveals the multi-target pharmacological mechanism of Fumaria Indica in the treatment of liver cancer. Pharmaceuticals **15**(6), 654 (2022)
8. Yan, H., et al.: Estrogen improves insulin sensitivity and suppresses gluconeogenesis via the transcription factor foxo1. Diabetes **68**(2), 291–304 (2019)
9. Hsing, H.Y., Rathnasamy, S., Dianita, R., Wahab, H.A.: Docking-based virtual screening in search for natural PTP1B inhibitors in treating type-2 diabetes mellitus and obesity. Biomed. Res. Ther. **7**(1), 3579–3592 (2020)
10. Vo, T.S., Le, P.U., Ngo, D.H.: *In Vitro* hypoglycemic and radical scavenging activities of certain medicinal plants. Exp. Appl. Biomed. Res. **23**(4), 291–298 (2022)
11. Huang, X., Liu, G., Guo, J., Su, Z.: The PI3K/AKT pathway in obesity and type 2 diabetes. Int. J. Biol. Sci. **14**(11), 1483–1496 (2018)
12. Kim, A., Lee, W., Yun, J.M.: Luteolin and fisetin suppress oxidative stress by modulating sirtuins and forkhead box O3a expression under *in vitro* diabetic conditions. Nutr. Res. Pract. **11**(5), 430–434 (2017)

Salt Tolerance Evaluation of Local Soybean Cultivars in Vietnam at the Germination Stage

Xuan-Quynh Tran[1,2], Nguyen Phuong Thao[1,2,3], and Xuan Lan Thi Hoang[1,2](✉)

[1] Applied Biotechnology for Crop Development Research Unit, School of Biotechnology, International University, VNU-HCM, Ho Chi Minh City, Vietnam
htlxuan@hcmiu.edu.vn

[2] Viet Nam National University, Ho Chi Minh City, Vietnam

[3] Research Center for Infectious Diseases, International University, Ho Chi Minh City, Vietnam

Abstract. Soybean (*Glycine max* L.) is a vital oilseed crop that contributes to global food security and enhances soil fertility through nitrogen fixation. However, its productivity is increasingly threatened by salinity stress, particularly in Vietnam, where soil salinization is intensifying due to climate change. Therefore, identifying salt-tolerant cultivars is crucial for maintaining yields in affected areas. This study assessed the salinity tolerance of twelve Vietnamese soybean cultivars with diverse genetic backgrounds at both germination and early seedling stages. Plants were exposed to salt concentrations ranging from 50 to 200 mM NaCl for examining key physiological parameters, including germination rate (GR), germination energy (GE), germination index (GI), vigor index (VI), hypocotyl and radicle lengths, number of secondary roots (NSR), and fresh and dry biomass. The findings highly demonstrate that salinity significantly impaired all traits in a concentration-dependent manner. Detailed analyses at 150 mM NaCl concentration revealed that most traits declined by more than 50% compared to controls, with VI and NSR being particularly affected. The plant salt tolerance of local soybeans and the reference genotype Williams 82 was classified using the Membership Function Value (MFV), which integrates most physiological parameters. Based on the average MFV of 10 parameters (MFVS), two cultivars (DT22 and MTD860–1) were assigned as salt-tolerant, two (DT2010 and MTD720) as highly salt-sensitive, and the remaining nine as moderately salt-tolerant. These findings provide information about early-stage salt stress responses in soybeans and the identified cultivars may serve as genetic resources for breeding programs targeting enhanced salinity resilience in soybeans.

Keywords: Germination stage · Salt stress · Soybean · Stress tolerance

1 Introduction

Soybean (*Glycine max* (L.) Merr.) is an indispensable global crop, ranking as the fourth most widely cultivated crop worldwide and serving as the largest global source of plant proteins for livestock feed, typically containing approximately 40% crude protein content in soybean seeds [1]. In addition, soybeans are extensively processed for human

L. Vu et al. (Eds.): PBC 2025, 2026.
https://doi.org/10.1007/978-3-032-29579-8_16

consumption and use in a range of commercial commodities, including soybean-based foods and drinks, pharmaceutical products, industrial biodiesel, paints, and plastics [2]. Ecologically, soybean contributes significantly to soil health through its natural ability to fix atmospheric nitrogen, thereby enhancing soil fertility [3]. However, soybean is highly sensitive to abiotic stresses such as salinity, drought, extreme temperatures, atmospheric pollutants, and soil-borne heavy metals. This vulnerability is particularly concerning with the witnesses of increasing food demand, climate change, and widespread agricultural land degradation, all of which place immense pressure on agricultural activities [1]. Currently, more than 7% of the world's land area is adversely affected by either sodicity or salinity, posing a serious challenge to agricultural practices for high yield production and food security [4].

In Vietnam, soil salinization is a recurring environmental issue, exacerbated by climate change and saline intrusion [5]. Coastal lowlands and fertile deltas—key agricultural regions of the country, prone to suffer consequences of rising sea levels and inland saltwater penetration [6]. For example, intensive saline intrusion in the Mekong River Delta can reach 30–60 km inland. This in turn causes severe impacts on economic development and livelihoods; mainly hindering cultivation of cereals and fruits as well as affecting daily lives of local communities in these regions. Therefore, the need for research, evaluation, and utilization of suitable soybean varieties with high resilience and strong yield potential becomes increasingly urgent [7]. Among the proposed measures, the identification and selection of salt-tolerant soybean varieties holds substantial ecological, financial, and social implications for the local population.

Previous research indicates that plants are most vulnerable to salt stress during the germination and seedling stages. The germination phase, which marks the beginning of a plant's life cycle and foundation for plant growth, is featured with intense metabolic activities and rapid cellular expansion, making it one of the most critical and sensitive periods under salinity stress [4]. High salt concentrations not only reduce the overall germination rate but also inhibit the radicle emergence, its elongation rate and subsequent development of lateral roots, all of which are crucial for seedling establishment [8]. The underlying mechanisms has been known to involve ion toxicity, dehydration and oxidative stress [9, 10]. Reduced water potential in the surrounding environment hampers the seed's ability to absorb water to initiate germination, leading to critical osmotic stress. On the other hand, the portion of water taken up onto plant cells also results in the endogenous accumulation of dissolve ions such as Na^+ and Cl^-. At excessive levels, such ions rapidly raise ion toxicity within seed tissues, causing severe disruption in metabolic activity required for plant growth. Additionally, under the impact of oxidative damage caused by the stress-induced production of reactive oxygen species (ROS), crucial physiological processes such as photosynthesis, respiration, and nutrient transport are strongly interfered, finally result in stunting growth [11, 12].

This research focuses on evaluating the salinity tolerance during the germination stage of 12 commonly grown soybean varieties in Vietnam. The study aims to enhance the current knowledge of salt tolerance within local soybean germplasms, supporting breeding and selection programs. Furthermore, the classification information of soybean groups based on their salinity tolerance during the early seedling stage would assist farmers to select suitable varieties for growing under saline soil conditions.

2 Materials and Methods

2.1 Plant Materials

The soybean profile used in this research comprised of 12 local soybean genotypes from DT and MTD varieties that are commonly grown in the Red River Delta and the Mekong River Delta, including DT2010, DT22, DT26, DT37, DT51, MTD176, MTD455–3, MTD720, MTD765–2, MTD860–1, MTD868–1, and MTD885–3. In addition, the research model soybean Williams 82 (W82) was also included in the experiments as a reference genotype. DT varieties were provided by the Field Crops Research Institute (Hai Duong) and the Agricultural Genetics Institute (Hanoi), MTD varieties were provided by Can Tho University (Can Tho), and W82 was provided by Thai Nguyen University of Agriculture and Forestry (Thai Nguyen). A preliminary experiment was carried out to verify the seed viability before used for experiments, confirming a germination rate approximately 100% [13].

2.2 Seed Preparation and Germination Conditions

The cultivation conditions of soybean varieties were conducted following the previously described method with minor modification [14, 15]. The seeds were sterilized with 0.5% sodium hypochlorite solution for 10 min, rinsed 3 times with sterile distilled water and soaked in water for 30 min at 37 °C. After that, the seeds with good quality were selected and incubated in Petri dishes (9 cm in diameter), between two layers of filter paper containing 10 mL of NaCl solution at concentrations of 50, 100, 150 and 200 mM. Seed batches treated with distilled water were used as controls. For seed germination assay, these plates were placed at a temperature of 25 ± 2 °C for 7 days, with the first three days kept under dark conditions. Salt solution with a determined concentration was added to the petri dish regularly to avoid moisture loss during incubation. For analyses, the seeds were considered germinated when their radicle lengths exceeded 2 mm. The experiment was repeated 3 times with 10 seeds for each replication.

2.3 Evaluation of Salt Stress on Germination and Seed Vigor

In the germination assay, the number of germinated seeds and/or their growth characters were recorded daily from the second to the seventh day following previous research [16, 17]. Germination energy (GE) and germination rate (GR) were assessed on the third and seventh days of the assay, using Eqs. (1) and (2), respectively.

$$GE = G_3 / T \times 100\% \tag{1}$$

$$GR = G_7 / T \times 100\% \tag{2}$$

In which, G_3 is the number of germinated seeds after three days, G_7 is the number of germinated seeds after seven days, and T is the total number of seeds used for the germination assay.

The germination index (GI) for each cultivar was determined based on the daily count of germinated seeds throughout the experimental period, calculated as the sum of the ratios of G_t to D_t (Eq. (3)).

$$GI = \sum (G_t/D_t) \tag{3}$$

In which, G_t is the number of seeds germinated on the day of examination, D_t is the corresponding number of days since assay initiation (DAG) for germination checking.

The vigor index (VI) was calculated by multiplying the GR value with the corresponding seedling length, which was the sum of hypocotyl length (HL) and radicle length (RL) (Eq. 4) [18].

$$VI = GR \times (HL + RL) \tag{4}$$

All measured parameters were compared against a control group, which consisted of seeds grown under non-saline (salt-free) conditions. To evaluate the impact of salt stress, the relative value of each trait was computed by dividing the measurements obtained under salt treatment by those from the control group.

The coefficient of variation (CV) was used to evaluate the relative variability of each germination assay and physiological trait across soybean cultivars under salinity conditions, calculated as the ratio of standard deviation and mean using Eq. (5) [16].

$$CV\ (\%) = (SD/X) \times 100 \tag{5}$$

In which, SD is the standard deviation of the measurements and X is the mean value of the corresponding trait.

To facilitate comparison across genotypes and salinity levels, germination and seedling traits were transformed into relative values by normalizing each stress-treated measurement to its corresponding control (Eq. 6 and 7).

$$\text{Relative value} = \text{stress/control} \times 100\,(\%) \tag{6}$$

$$\text{Relative reduction}\ (\%) = 1 - (\text{stress/control}) \times 100\,(\%) \tag{7}$$

2.4 Determination of Shoot and Root-Related Traits of Soybean Seedlings Following Germination

On the 7th DAG, hypocotyl and radicle lengths of germinated seeds were measured. The HL was measured from the base of the radicle to the position of the seed leaves and the RL was measured from the base of the radicle to the tip of the root.

Total fresh weight of radicle and hypocotyl was also recorded the 7th DAG. To measure the corresponding total dry weight, the tissues were oven-dried at 70 °C for 48 h, after which measurements were taken.

2.5 Determination of Optimal Salt Stress Concentration for Salt Tolerance Ranking

Based on the data obtained from the four tested concentrations of NaCl applied to seed germination assay, which were 50, 100, 150, and 200 mM, the salt level that caused salt-injury index (SII) of 50% of the control for each studied parameter was analyzed. Details for calculating SII were described in Sect. 2.6.

2.6 Evaluation of Salt Tolerance

The salt tolerance of local soybean varieties was evaluated using the Membership Function Value (MFV) [19]. To describe the differences in salt tolerance among soybean germplasms, the salt-tolerance index (STI) of each physiological parameter was also calculated, using Eq. (8).

$$\mathrm{STI_i} = \mathrm{V_{in}} / \mathrm{V_{ic}} \tag{8}$$

Here, $\mathrm{STI_i}$ is the STI of trait *i*, $\mathrm{V_{in}}$ and $\mathrm{V_{ic}}$ represent the values of trait *i* in the salt-stressed and control treatments, respectively.

The membership function value (MFV) for the salt tolerance of each trait at the optimal salt concentration was calculated according to Eq. (9) [17]:

$$\mathrm{X_i} = (\mathrm{X} - \mathrm{X_{min}}) / (\mathrm{X_{max}} - \mathrm{X_{min}}) \tag{9}$$

In which, $\mathrm{X_i}$ is the MFV of the STI of a trait of a soybean variety, X is the STI of a trait of a variety, $\mathrm{X_{max}}$ and $\mathrm{X_{min}}$ are the largest and smallest STI values of a trait recorded in all varieties. The salt tolerance of each variety was evaluated based on the average MFV value of the traits [17, 19]. The MFV value ranges from 0 to 1. For each local soybean variety, the average MFV of salt tolerance (MFVS) is the average value of the MFVs of the evaluated traits.

According to previous method of salt tolerance classification using MFV, the salt tolerance levels of soybean varieties were separated into five grades based on mean value (X_{mean}) and standard deviation (SD) of MFVS, with Xi as MFVS value of particular variety: (1) $Xi \geq X_{mean} + 1.64SD$, highly salt-tolerant (HST); (2) $X_{mean} + 1.64SD > Xi \geq X_{mean} + SD$, salt-tolerant (ST); (3) $X_{mean} + SD > Xi \geq X_{mean}—SD$, moderately salt-tolerant (MST); (4) $X_{mean}\text{-}SD > Xi \geq X_{mean}-1.64SD$, salt-sensitive (SS); (5) $X_{mean}-1.64SD > Xi$, highly-salt sensitive (HSS) [20]. The normality of MFVS was verified through the Shapiro–Wilk test, complemented by visual inspection of the normal Q–Q plot produced in GraphPad Prism.

In addition, the SII was also obtained to determine the salinity concentration causing 50% damage to the germination indices of soybeans [17]. For each trait, the avrage SII value was measured based on average STI value of that trait across 13 soybean varieties in our study, according to Eq. (10). Next, mean SII of each trait was plotted against the corresponding NaCl concentrations to construct a linear equation in the form of $y = ax + b$. from which the NaCl concentration with SII at 0.5 ($\mathrm{SII_{50}}$) was calculated according to Eq. (11):

$$\mathrm{SII} = 1 - \mathrm{STI} \tag{10}$$

$$X = \frac{(0.5 - b)}{a} \tag{11}$$

In which, X is corresponding NaCl concentration at SII of 0.5 (50% of control), b and a were identified from previous linear equation with mean SII at 4 NaCl concentrations as input. For each model, the linear regression analysis reported the slope, coefficient of determination (R^2), and 95% confidence intervals, and incorporated residual diagnostics to assess compliance with the proportional-injury assumption of the linear model.

2.7 Statistical Analysis

Data from all experiments were sorted and processed using Microsoft Excel software. SPSS 26 (IBM Corp., Armonk, NY, USA) was employed to perform analysis of variance (ANOVA), Tukey's honest test was used for determining significant differences between local soybean varieties under different salinity levels (p-value = 0.05).

To avoid multicollinearity and prevent overfitting, the initial 10-predictor model was replaced with a Principal Component Regression (PCR) approach. Two principal components were extracted using Principal Component Analysis (PCA), and their scores were used as independent variables in a final multiple linear regression model predicting the MFVS values. The final relative contribution of each original trait to MFVS (*ie.,* the trait weights) was calculated by multiplying the product of the trait's component loading and the standardized regression coefficient of the corresponding principal component, as described in Eq. 12 [16].

$$W = (L_1 \times \beta_1) + (L_2 \times \beta_2) \tag{12}$$

Hierarchical cluster analysis, implemented in SPSS using Ward's method and squared Euclidean distance, was conducted to evaluate the clustering of soybean cultivars according to their salt tolerance characters, as defined by the scores of principal components PC1 and PC2 [16].

3 Results

3.1 Effects of Different Salinity Concentrations on Seed Germination and Early Growth

Following treatments with different salt levels (0, 50, 100, 150, and 200 mM NaCl), germination parameters for thirteen soybean genotypes had been recorded from day 2nd to day 7th of the seed germination assay. As shown in Table 1 and **Supplementary Table S1**, negative effects caused by salt were observed in comparison with the control (0 mM NaCl treatment), and as salinity levels increased, there was a progressive decline in all measured parameters. With 50 mM NaCl treatment, there were minor reductions in average relative GR, GE, and GI by 5%-10% compared to the control. The decreases became more pronounced at 100 mM NaCl, ranging from 8% –23% respectively, and were significant at 150 mM, with declines of 14% - 42%. The largest reduction was observed for relative GI, which decreased from 21.96% in the control to 12.79% at 150 mM NaCl (*i.e.* reduction by 42%). Under 200 mM NaCl conditions, the relative reduction of GR, GE, GI reached 30%, 52%, and 64% respectively. Similarly, following the increasing salinity concentrations, the relative VI remarkably decreased by 32.48% (50 mM NaCl treatment), up to 85%-93% (NaCl concentrations of 150 mM and higher). In agreement with these indices, inhibition degrees of hypocotyl and radicle elongation were significantly reduced under high salt conditions, whereby the average relative HL and RL were shorter by 84–91% and 79–87%, respectively. The relative number of secondary roots (NSR) showed the sharpest decline, ranging from 40% (with low salt treatment) to 89% (with modest salt treatment). Meanwhile, with high NaCl concentrations such as 150 and 200 mM, there was almost no secondary roots appearing.

Fresh and dry biomass parameters including hypocotyl fresh weight (HFW), radicle fresh weight (RFW), hypocotyl dried weight (HDW), and radicle dried weight (RDW) also showed the drastic reductions, with a higher salt adverse effects on shoot growth following treatments with NaCl 100 mM or higher concentration (Table 1). Average relative HFW and HDW of 150 mM NaCl-treated seeds displayed a reduction up to 81% and 75%, respectively, whereas their means for relative RFW and RDW were reduced 76% and 67%, respectively. With 200 mM NaCl application, germination, growth, lateral root formation, and biomass accumulation were significantly inhibited by the salt stress. Among the examined parameters, in overall, GR, GE, and GI are least influenced by salt stress at both low and high concentrations whereas VI and NSR were shown to be the most salt-sensitive indicators.

Salt stress exerts differential impacts on soybean seed germination, with each germination trait exhibiting varying levels of sensitivity. The coefficients of variation for the evaluated traits differed across salt concentrations, generally following the trend 200 mM > 150 mM > 100 mM > 50 mM > control, with the exception of RL at 50–100 mM, RFW and RDW at 100 mM NaCl. Under all four salt stress conditions, the coefficients of variation for the remaining germination indices exceeded 10%, suggesting that salinity stress amplified the phenotypic variation among genotypes. This increased variability facilitates the assessment and differentiation of salt tolerance levels among soybean cultivars.

Table 1. Comparison of indices under salt stress at the soybean germination stage. This experimental data is based on the results of 13 soybean varieties under different NaCl treatments. The values are Max, Min, Average, SE, and CV of all varieties (three replicates) at each concentration. CV: coefficient of variation; SE: standard error; GR: germination rate; GE: germination energy; GI: germination index; VI: vigor index; HL: hypocotyl length; RL: radicle length; NSR: number of secondary roots; HFW: hypocotyl fresh weight; RFW: radicle fresh weight; HDW: hypocotyl dried weight; RDW: radicle dried weight. Different lowercase letters indicate significant difference in mean of the same measured trait among different NaCl conditions.

NaCl treatment	Index	GR (%)	GE (%)	GI	VI	HL (cm)	RL (cm)	NSR	HFW (g)	RFW (g)	HDW (g)	RDW (g)
Control	Max	100.00	100.00	25.00	1904.13	10.26	10.31	30.43	0.2842	0.1794	0.0194	0.0120
	Min	86.67	83.33	13.45	882.20	5.41	4.42	7.17	0.1535	0.0851	0.0095	0.0053
	Average	96.15^{c}	95.38^{c}	21.96^{d}	1500.46^{d}	7.77^{d}	7.68^{d}	18.39^{c}	0.2182^{d}	0.1287^{d}	0.0154^{d}	0.0083^{c}
	SE	3.03	3.65	1.78	197.94	0.27	0.36	1.07	0.0066	0.0051	0.0006	0.0003
	CV	5.45	6.62	14.01	22.85	18.89	25.81	31.81	16.4478	21.6767	20.0798	20.6456
50 mM	Max	100.00	100.00	23.90	1258.33	7.03	7.95	16.70	0.1982	0.1331	0.0176	0.0095
	Min	73.33	60.00	9.58	437.27	2.38	2.76	0.73	0.0989	0.0420	0.0057	0.0033
	Average	91.28^{bc}	86.67^{c}	19.77^{cd}	1013.09^{c}	4.88^{c}	6.20^{c}	10.98^{b}	0.1634^{c}	0.0973^{c}	0.0127^{c}	0.0072^{c}
	SE	4.62	8.01	2.33	155.90	0.25	0.29	0.83	0.0057	0.0049	0.0006	0.0004
	CV	8.76	16.01	20.45	26.65	28.13	25.70	41.36	19.0525	27.6468	27.4831	28.0531
100 mM	Max	100.00	96.67	22.73	852.20	3.30	4.80	5.07	0.0992	0.0755	0.0104	0.0057
	Min	73.33	56.67	8.88	300.13	1.28	1.88	0.00	0.0474	0.0330	0.0042	0.0025
	Average	88.46^{bc}	80.51^{bc}	16.82^{c}	612.60^{b}	2.11^{b}	3.28^{b}	2.11^{a}	0.0739^{b}	0.0533^{b}	0.0069^{b}	0.0041^{b}
	SE	4.95	8.01	2.10	97.92	0.11	0.15	0.32	0.0033	0.0021	0.0004	0.0002
	CV	9.68	17.22	21.58	27.68	28.80	24.62	82.14	24.7654	21.3608	29.9813	21.8930
150 mM	Max	96.67	86.67	17.48	343.13	1.71	2.70	0.70	0.0656	0.0477	0.0068	0.0038
	Min	63.33	33.33	7.26	117.73	0.55	0.90	0.00	0.0189	0.0191	0.0020	0.0020
	Average	82.56^{b}	67.69^{b}	12.79^{b}	227.53^{a}	1.26^{ab}	1.58^{a}	0.06^{a}	0.0421^{a}	0.0308^{a}	0.0039^{a}	0.0027^{ab}
	SE	6.63	11.04	2.07	45.64	0.07	0.10	0.04	0.0026	0.0016	0.0003	0.0001
	CV	13.90	28.24	27.99	34.74	31.97	33.83	343.21	33.9907	28.9464	35.1844	21.5780
200 mM	Max	93.33	80.00	13.62	182.13	1.26	2.28	0.00	0.0373	0.0306	0.0040	0.0028
	Min	36.67	13.33	3.23	34.23	0.06	0.48	0.00	0.0013	0.0112	0.0002	0.0017
	Average	66.92^{a}	45.64^{a}	7.83^{a}	101.66^{a}	0.69^{a}	0.99^{a}	0.00^{a}	0.0200^{a}	0.0211^{a}	0.0021^{a}	0.0022^{a}

(continued)

Table 1. *(continued)*

NaCl treatment	Index	GR (%)	GE (%)	GI	VI	HL (cm)	RL (cm)	NSR	HFW (g)	RFW (g)	HDW (g)	RDW (g)
	SE	8.55	11.90	1.70	28.28	0.08	0.10	0.00	0.0022	0.0013	0.0002	0.0001
	CV	22.13	45.16	37.72	48.18	60.43	53.30	-	59.8762	32.9793	57.3101	15.7075

3.2 Analyses of Genotype-Based Salt Effects on Germination Indices

Salinity stress exerted a concentration-dependent inhibitory effect on all germination parameters. In comparison to control, at 50 mM NaCl, all cultivars had minor decrease in GR, GE, GI with no significance, as shown in Table 2. However, DT37, MTD720, and MTD868–1 showed a decrease in VI values by nearly half in comparison with the non-stressed groups. The fold decrease was calculated by dividing the value under control conditions by the corresponding value under each NaCl treatment, allowing assessment of the reduction across NaCl concentrations. At 100 mM NaCl, DT2010 exhibited significant reduction with RGE, RGI, and RVI by 1.58, 1.69, 2.94 times respectively, while MTD720 showed the highest reduction in GR (decrease 27%) and VI (by 70%) among the examined soybean varieties. Following 150 mM NaCl treatment, significant reductions were observed, with the extent of decline varying across cultivars. At this salt concentration, MTD720 continued to show the highest decrease in relative RGR, RGE, and RGI by 2–3 times, and RVI up to 14.45 times. Fold-change analysis revealed that GE and VI were the most affected traits, with the highest decrease of VI by over 10-fold in susceptible varieties.

3.3 Analyses of Genotype-Based Salt Effects on Shoot and Root-Related Traits at the Germination Stage

According to Table 3, shoot and root traits were highly affected by salt stress and in a genotype-independent manner. Under 50 mM NaCl conditions, DT37 showed the highest reduction in almost examined parameters, including 2.27 times in relative RHL, 9.77 times in RNSR, and 1.67 times in RHDW, followed by MTD455–3 with 1.7-fold decrease in RRL and MTD868–1 with decreases in RHFW, RRFW, and RRDW by 1.8–2.2 times. At salt concentration of 100 mM, MTD176 and DT2010 showed significantly declined growth in comparison with the corresponding controls. Under this condition, average HL and HFW of MTD176 declined most whereas DT2010 showed substantial reduction in all parameters, especially RL, RFW, HDW, and RDW, and continued to exhibit sever reduction under higher salt conditions. In addition, 100 mM NaCl-treated DT51 unsuccessfully formed lateral roots. It has been noted that between elongation and biomass accumulation of shoot and root tissues, the former parameter was more severely impacted by salt stress, which led to growth inhibition.

3.4 Determination of Optimal Salt Stress Concentration

To identify the relevant salt stress concentration for further salt tolerance analysis, all indicators examined in this study were utilized, including NSR at 150mM and 200mM salt conditions. The average SII caused by each indicator over all 13 soybean genotypes included W82 was calculated based on the data of average STI of studied soybeans and analyzed using linear regression analysis in order to identify at what concentration of NaCl would result in a reduction by 50% in the value of examined trait/parameter of the control. Across all germination and seedling traits, R^2 values ranged from 0.7449 to 0.9864, indicating that NaCl concentrations explained a substantial proportion of the variation in SII. The 95% CI intervals for the slope parameters did not include zero for any significant relationships, confirming a positive linear trend. In addition, residual diagnostics further supported the assumptions of the linear regression model. For all evaluated parameters, the residuals were randomly distributed around zero without systematic patterns, curvature, or heteroscedasticity (**Supplementary Fig. S1, Table S2**). No influential outliers were detected. Collectively, these results confirm that the linear model is appropriate for describing the increase in SII across the 50–200 mM NaCl range.

It has been observed that GR and GE were least affected by salt stress, showing SII values decreasing to 50% of control (SII_{50}) when the plants are subjected to 339.24 mM and 205.02 mM NaCl treatments, respectively (Fig. 1a and b). By contrast, other germination indices including GI and VI had SII_{50} values at 139.08 mM and 84.05 mM, respectively (Fig. 1c and b). For other physiological parameters (HL, RL, NSR, HFW, RFW, HDW, RDW), the NaCl concentrations at which these SII_{50} values decreased to 50% were 62.54, 105.43, 37.68, 88.90, 97.95, 105.86, 122.91 mM, respectively (Fig. 1e–k). After calculation, the average salt concentration corresponding to SIIs of 11 parameters at 50% was 126.24 mM. Although most seeds germinated when exposed to 100–200 mM NaCl, the growth of seedlings was significantly inhibited. At 150 mM NaCl concentration, all physiological indicators were more sensitive to salinity relative to seed germination, showing severe growth inhibition by salt stress. To ensure that seedling growth and biomass accumulation would contribute to the measurements in salt tolerance analyses, 150 mM NaCl was selected as salt concentration to next classify soybean varieties for their salt tolerance capacity.

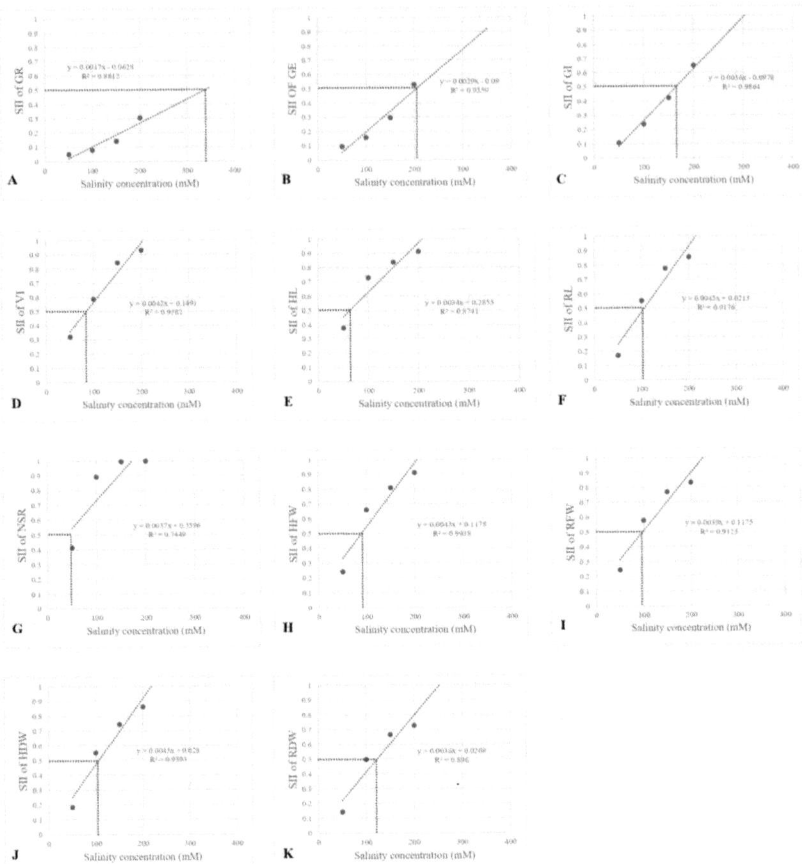

Fig. 1. Profile of NaCl concentrations that cause negative effects on 11 indicators used in this study, so-called salt-injury index (SII). Based on this, the corresponding NaCl concentration for causing SII by 50% (0.5) was extrapolated for each indicator, including germination rate (GR) **a**, germination energy (GE) **b,** germination index (GI) **c**, vigor index (VI) **d**, hypocotyl length (HL) **e**, radicle length (RL **f**, number of secondary roots (NSR) **g**, hypocotyl fresh weight (HFW) **h**, radicle fresh weight (RFW) **i**, hypocotyl dried weight (HDW) **j**, and radicle dried weight (RDW) **(K)**. Data in the figure were graphed from the SII means of 13 soybean varieties for each indicator following NaCl treatments of 50, 100, 150 and 200 mM.

Table 2. Germination indicators of soybean varieties under salinity stress induced by NaCl solution treatment. Different lowercase letters present significant differences among treatment conditions of particular soybean varieties, as analyzed by one-way ANOVA. GR: germination rate; GE: germination energy; GI: germination index; VI: vigor index.

Soybean varieties and salt treatments		GR (%)	GE (%)	GI	VI
DT22	0 mM	$93.33^{a}\pm3.33$	$100.00^{c}\pm0.00$	$25.00^{c}\pm0.41$	$1130.80^{c}\pm135.29$
	50 mM	$96.67^{a}\pm3.33$	$96.67^{c}\pm3.33$	$23.90^{bc}\pm0.52$	$1227.97^{c}\pm151.97$
	100 mM	$93.33^{a}\pm3.33$	$90.00^{bc}\pm5.78$	$20.09^{b}\pm0.42$	$717.47^{bc}\pm151.07$
	150 mM	$83.33^{a}\pm6.67$	$76.67^{ab}\pm3.33$	$14.72^{a}\pm1.04$	$311.00^{ab}\pm22.87$
	200 mM	$83.33^{a}\pm3.33$	$60.00^{a}\pm5.78$	$10.91^{a}\pm1.63$	$165.07^{a}\pm22.16$
DT26	0 mM	$100.00^{b}\pm0.00$	$100.00^{d}\pm0.00$	$22.43^{c}\pm0.50$	$1300.67^{d}\pm191.19$
	50 mM	$100.00^{b}\pm0.00$	$100.00^{d}\pm0.00$	$22.60^{c}\pm0.17$	$1197.00^{cd}\pm156.47$
	100 mM	$100.00^{b}\pm0.00$	$83.33^{c}\pm3.33$	$17.71^{bc}\pm0.99$	$673.33^{bc}\pm136.92$
	150 mM	$96.67^{a}\pm3.33$	$66.67^{b}\pm3.33$	$12.31^{b}\pm2.01$	$186.83^{ab}\pm61.21$
	200 mM	$73.33^{a}\pm12.02$	$53.33^{a}\pm3.33$	$8.43^{a}\pm0.98$	$56.97^{a}\pm16.21$
DT37	0 mM	$90.00^{b}\pm5.78$	$83.33^{c}\pm8.82$	$13.45^{c}\pm0.37$	$882.20^{c}\pm112.61$
	50 mM	$83.33^{b}\pm8.82$	$60.00^{bc}\pm5.78$	$9.58^{bc}\pm0.67$	$437.27^{b}\pm92.92$
	100 mM	$73.33^{b}\pm3.33$	$56.67^{bc}\pm6.68$	$8.88^{b}\pm0.90$	$300.13^{ab}\pm32.44$
	150 mM	$66.67^{b}\pm3.33$	$33.33^{ab}\pm6.68$	$7.26^{b}\pm1.46$	$120.00^{ab}\pm31.68$
	200 mM	$36.67^{a}\pm3.33$	$13.33^{a}\pm8.82$	$3.23^{a}\pm0.10$	$34.23^{a}\pm5.44$
DT51	0 mM	$100.00^{b}\pm0.00$	$100.00^{b}\pm0.00$	$24.10^{b}\pm0.44$	$1447.67^{c}\pm243.64$
	50 mM	$96.67^{b}\pm3.33$	$96.67^{b}\pm3.33$	$22.73^{b}\pm0.75$	$902.37^{bc}\pm77.24$
	100 mM	$96.67^{b}\pm3.33$	$96.67^{b}\pm3.33$	$22.73^{b}\pm0.75$	$505.17^{ab}\pm87.17$
	150 mM	$90.00^{b}\pm0.00$	$86.67^{b}\pm3.33$	$16.10^{b}\pm1.01$	$208.20^{a}\pm19.39$
	200 mM	$70.00^{a}\pm5.78$	$23.33^{a}\pm3.33$	$5.68^{a}\pm0.30$	$38.97^{a}\pm5.46$
DT2010	0 mM	$100.00^{b}\pm0.00$	$100.00^{c}\pm0.00$	$22.60^{c}\pm0.33$	$1629.00^{c}\pm218.57$
	50 mM	$96.67^{ab}\pm3.33$	$80.00^{bc}\pm5.78$	$19.43^{c}\pm1.28$	$1076.70^{bc}\pm140.57$
	100 mM	$90.00^{ab}\pm5.78$	$63.33^{ab}\pm3.33$	$13.36^{b}\pm0.68$	$553.43^{ab}\pm205.44$
	150 mM	$93.33^{ab}\pm3.33$	$53.33^{a}\pm3.33$	$9.62^{a}\pm0.84$	$131.77^{a}\pm49.90$
	200 mM	$80.00^{a}\pm5.78$	$43.33^{a}\pm8.82$	$7.10^{a}\pm0.37$	$42.73^{a}\pm5.70$
MTD176	0 mM	$96.67^{b}\pm3.33$	$96.67^{b}\pm3.33$	$23.73^{c}\pm0.20$	$1904.13^{c}\pm251.79$
	50 mM	$90.00^{b}\pm5.78$	$90.00^{b}\pm5.78$	$21.67^{c}\pm0.65$	$1215.83^{bc}\pm104.44$
	100 mM	$93.33^{b}\pm3.33$	$93.33^{b}\pm3.33$	$16.20^{b}\pm0.14$	$711.53^{ab}\pm201.65$
	150 mM	$80.00^{ab}\pm5.78$	$76.67^{b}\pm3.33$	$15.30^{b}\pm1.39$	$229.60^{a}\pm13.81$
	200 mM	$60.00^{a}\pm5.78$	$50.00^{a}\pm5.78$	$8.22^{a}\pm0.68$	$99.03^{a}\pm15.28$
MTD455-3	0 mM	$100.00^{a}\pm0.00$	$100.00^{b}\pm0.00$	$22.93^{c}\pm0.58$	$1897.33^{d}\pm202.28$
	50 mM	$96.67^{a}\pm3.33$	$96.67^{ab}\pm3.33$	$19.40^{b}\pm0.29$	$1129.10^{c}\pm67.94$
	100 mM	$93.33^{a}\pm6.67$	$93.33^{ab}\pm6.67$	$18.37^{b}\pm0.34$	$721.67^{bc}\pm153.39$
	150 mM	$93.33^{a}\pm6.67$	$86.67^{ab}\pm6.67$	$17.48^{b}\pm1.07$	$319.20^{ab}\pm31.64$
	200 mM	$76.67^{a}\pm8.82$	$73.33^{a}\pm6.67$	$13.62^{a}\pm1.09$	$155.63^{a}\pm28.76$
MTD720	0 mM	$100.00^{a}\pm0.00$	$90.00^{c}\pm0.00$	$20.26^{c}\pm1.09$	$1701.00^{c}\pm253.27$
	50 mM	$73.33^{a}\pm8.82$	$60.00^{bc}\pm5.78$	$14.57^{b}\pm1.29$	$846.47^{b}\pm194.34$

(continued)

Table 2. *(continued)*

Soybean varieties and salt treatments		GR (%)	GE (%)	GI	VI
	100 mM	$73.33^{a}\pm6.67$	$60.00^{bc}\pm5.78$	$13.27^{b}\pm0.47$	$515.93^{ab}\pm98.85$
	150 mM	$63.33^{a}\pm12.02$	$33.33^{ab}\pm8.82$	$8.09^{a}\pm0.98$	$117.73^{a}\pm8.56$
	200 mM	$60.00^{a}\pm11.55$	$13.33^{a}\pm8.82$	$4.43^{a}\pm1.29$	$107.47^{a}\pm47.57$
MTD765-2	0 mM	$96.67^{b}\pm3.33$	$96.67^{b}\pm3.33$	$21.90^{c}\pm1.06$	$1576.23^{d}\pm113.36$
	50 mM	$96.67^{b}\pm3.33$	$96.67^{b}\pm3.33$	$20.23^{c}\pm0.20$	$1154.67^{c}\pm59.99$
	100 mM	$93.33^{b}\pm6.67$	$86.67^{b}\pm13.33$	$18.31^{bc}\pm2.12$	$808.93^{b}\pm33.13$
	150 mM	$80.00^{b}\pm5.78$	$80.00^{ab}\pm5.78$	$14.58^{b}\pm0.78$	$259.20^{a}\pm20.97$
	200 mM	$53.33^{a}\pm6.67$	$46.67^{a}\pm8.82$	$6.94^{a}\pm0.96$	$108.27^{a}\pm14.18$
MTD860-1	0 mM	$86.67^{b}\pm3.33$	$86.67^{b}\pm3.33$	$19.30^{b}\pm0.52$	$1405.57^{b}\pm435.96$
	50 mM	$83.33^{ab}\pm3.33$	$83.33^{b}\pm3.33$	$18.44^{b}\pm0.96$	$998.70^{ab}\pm121.70$
	100 mM	$83.33^{ab}\pm6.67$	$83.33^{b}\pm6.67$	$16.61^{b}\pm0.93$	$638.43^{ab}\pm96.64$
	150 mM	$76.67^{ab}\pm3.33$	$60.00^{ab}\pm11.55$	$9.57^{a}\pm1.14$	$343.13^{a}\pm64.49$
	200 mM	$66.67^{a}\pm3.33$	$33.33^{a}\pm3.33$	$6.44^{a}\pm0.27$	$136.33^{a}\pm36.76$
MTD868-1	0 mM	$86.67^{b}\pm6.67$	$86.67^{b}\pm6.67$	$21.14^{c}\pm1.92$	$996.33^{c}\pm170.50$
	50 mM	$83.33^{b}\pm3.33$	$76.67^{ab}\pm3.33$	$18.22^{bc}\pm0.26$	$517.03^{b}\pm83.93$
	100 mM	$80.00^{ab}\pm5.78$	$70.00^{ab}\pm5.78$	$15.24^{b}\pm1.54$	$314.80^{ab}\pm12.10$
	150 mM	$66.67^{ab}\pm6.67$	$56.67^{ab}\pm8.82$	$9.20^{a}\pm1.13$	$164.60^{ab}\pm5.83$
	200 mM	$56.67^{a}\pm3.33$	$46.67^{a}\pm8.82$	$7.36^{a}\pm0.56$	$104.59^{a}\pm3.27$
MTD885-3	0 mM	$100.00^{a}\pm0.00$	$100.00^{a}\pm0.00$	$24.93^{c}\pm0.00$	$1847.33^{d}\pm133.71$
	50 mM	$93.33^{a}\pm6.67$	$93.33^{a}\pm6.67$	$22.53^{c}\pm1.51$	$1258.33^{cd}\pm134.12$
	100 mM	$93.33^{a}\pm6.67$	$93.33^{a}\pm6.67$	$20.87^{bc}\pm2.06$	$852.20^{bc}\pm225.01$
	150 mM	$93.33^{a}\pm6.67$	$86.67^{a}\pm3.33$	$15.64^{ab}\pm0.99$	$294.47^{ab}\pm10.06$
	200 mM	$93.33^{a}\pm3.33$	$80.00^{a}\pm0.00$	$12.28^{a}\pm0.15$	$182.13^{a}\pm11.56$
W82	0 mM	$100.00^{b}\pm0.00$	$100.00^{c}\pm0.00$	$23.76^{c}\pm0.93$	$1787.67^{c}\pm162.61$
	50 mM	$96.67^{b}\pm3.33$	$96.67^{bc}\pm3.33$	$23.73^{c}\pm0.71$	$1208.80^{bc}\pm187.32$
	100 mM	$86.67^{b}\pm8.82$	$76.67^{bc}\pm6.67$	$16.97^{b}\pm1.23$	$650.77^{ab}\pm124.50$
	150 mM	$90.00^{b}\pm5.78$	$83.33^{ab}\pm6.67$	$16.45^{b}\pm1.02$	$272.10^{a}\pm44.31$
	200 mM	$60.00^{a}\pm5.78$	$56.67^{a}\pm3.33$	$7.11^{a}\pm0.20$	$90.10^{a}\pm24.22$

Table 3. Shoot and root-associated parameters of soybean varieties with and without salinity stress induced by NaCl solution treatment. Different lowercase letters present significant differences among treatment conditions of particular soybean varieties, as analyzed by one-way ANOVA. HL: hypocotyl length; RL: radicle length; NSR: number of secondary roots; HFW: hypocotyl fresh weight; RFW: radicle fresh weight; HDW: hypocotyl dried weight; RDW: radicle dried weight.

Soybean varieties and salt treatments		HL (cm)	RL (cm)	NSR	HFW (g)	RFW (g)	HDW (g)	RDW (g)
DT22	0 mM	$6.81^{d}\pm0.31$	$5.23^{d}\pm0.29$	$20.23^{d}\pm1.05$	$0.1945^{e}\pm0.0077$	$0.1020^{c}\pm0.0044$	$0.0138^{d}\pm0.0007$	$0.0069^{d}\pm0.0003$
	50 mM	$5.31^{c}\pm0.20$	$7.35^{c}\pm0.56$	$16.70^{c}\pm1.57$	$0.1683^{d}\pm0.0059$	$0.1086^{c}\pm0.0047$	$0.0135^{d}\pm0.0006$	$0.0073^{d}\pm0.0003$
	100 mM	$2.10^{b}\pm0.05$	$3.80^{b}\pm0.22$	$4.43^{b}\pm0.55$	$0.0821^{c}\pm0.0023$	$0.0609^{b}\pm0.0020$	$0.0076^{c}\pm0.0002$	$0.0047^{c}\pm0.0002$
	150 mM	$1.47^{ab}\pm0.05$	$2.27^{a}\pm0.13$	$0.00^{a}\pm0.00$	$0.0545^{b}\pm0.0020$	$0.0425^{a}\pm0.0015$	$0.0048^{b}\pm0.0002$	$0.0033^{b}\pm0.0001$
	200 mM	$0.82^{a}\pm0.05$	$2.28^{a}\pm0.08$	$0.00^{a}\pm0.00$	$0.0262^{a}\pm0.0023$	$0.0306^{a}\pm0.0018$	$0.0023^{a}\pm0.0002$	$0.0023^{a}\pm0.0001$
DT26	0 mM	$7.38^{e}\pm0.24$	$5.53^{c}\pm0.49$	$11.10^{c}\pm1.36$	$0.1914^{d}\pm0.0097$	$0.1084^{d}\pm0.0066$	$0.0120^{c}\pm0.0008$	$0.0064^{c}\pm0.0005$
	50 mM	$5.64^{d}\pm0.20$	$6.33^{c}\pm0.53$	$7.33^{b}\pm1.02$	$0.1686^{d}\pm0.0077$	$0.0871^{c}\pm0.0063$	$0.0119^{c}\pm0.0008$	$0.0056^{c}\pm0.0005$
	100 mM	$2.14^{c}\pm0.10$	$2.44^{b}\pm0.21$	$0.90^{a}\pm0.23$	$0.0803^{c}\pm0.0047$	$0.0434^{b}\pm0.0024$	$0.0064^{b}\pm0.0004$	$0.0033^{b}\pm0.0002$
	150 mM	$0.86^{b}\pm0.11$	$1.04^{a}\pm0.08$	$0.00^{a}\pm0.00$	$0.0293^{b}\pm0.0036$	$0.0253^{a}\pm0.0016$	$0.0024^{a}\pm0.0003$	$0.0020^{a}\pm0.0001$
	200 mM	$0.16^{a}\pm0.06$	$0.60^{a}\pm0.02$	$0.00^{a}\pm0.00$	$0.0034^{a}\pm0.0014$	$0.0145^{a}\pm0.0009$	$0.0003^{a}\pm0.0001$	$0.0017^{a}\pm0.0001$
DT37	0 mM	$5.41^{e}\pm0.23$	$4.42^{d}\pm0.36$	$7.17^{b}\pm0.90$	$0.1535^{d}\pm0.0078$	$0.0851^{d}\pm0.0043$	$0.0095^{d}\pm0.0005$	$0.0053^{c}\pm0.0003$
	50 mM	$2.38^{d}\pm0.11$	$2.76^{c}\pm0.21$	$0.73^{a}\pm0.27$	$0.0788^{c}\pm0.0041$	$0.0492^{c}\pm0.0022$	$0.0057^{c}\pm0.0003$	$0.0033^{b}\pm0.0002$
	100 mM	$1.48^{c}\pm0.08$	$1.88^{b}\pm0.18$	$0.10^{a}\pm0.07$	$0.0474^{b}\pm0.0032$	$0.0330^{b}\pm0.0022$	$0.0042^{b}\pm0.0002$	$0.0025^{a}\pm0.0002$
	150 mM	$0.80^{b}\pm0.07$	$0.97^{a}\pm0.08$	$0.00^{a}\pm0.00$	$0.0258^{a}\pm0.0025$	$0.0191^{a}\pm0.0011$	$0.0029^{ab}\pm0.0003$	$0.0020^{a}\pm0.0001$
	200 mM	$0.30^{a}\pm0.06$	$0.65^{a}\pm0.04$	$0.00^{a}\pm0.00$	$0.0117^{a}\pm0.0024$	$0.0135^{a}\pm0.0010$	$0.0018^{a}\pm0.0004$	$0.0022^{a}\pm0.0001$
DT51	0 mM	$6.15^{d}\pm0.37$	$8.33^{d}\pm0.63$	$13.77^{c}\pm2.05$	$0.2209^{d}\pm0.0125$	$0.1395^{d}\pm0.0075$	$0.0165^{e}\pm0.0007$	$0.0085^{c}\pm0.0006$
	50 mM	$3.70^{c}\pm0.18$	$5.70^{c}\pm0.47$	$7.60^{b}\pm0.85$	$0.1582^{c}\pm0.0071$	$0.1118^{c}\pm0.0047$	$0.0140^{d}\pm0.0006$	$0.0083^{c}\pm0.0003$
	100 mM	$1.34^{b}\pm0.04$	$2.82^{b}\pm0.15$	$0.00^{a}\pm0.00$	$0.0613^{b}\pm0.0021$	$0.0518^{b}\pm0.0018$	$0.0060^{c}\pm0.0002$	$0.0040^{b}\pm0.0002$
	150 mM	$0.94^{b}\pm0.05$	$1.37^{a}\pm0.08$	$0.00^{a}\pm0.00$	$0.0431^{b}\pm0.0092$	$0.0249^{a}\pm0.0010$	$0.0033^{b}\pm0.0002$	$0.0024^{a}\pm0.0001$
	200 mM	$0.06^{a}\pm0.03$	$0.50^{a}\pm0.02$	$0.00^{a}\pm0.00$	$0.0016^{a}\pm0.0010$	$0.0114^{a}\pm0.0004$	$0.0004^{a}\pm0.0002$	$0.0024^{a}\pm0.0001$
DT2010	0 mM	$6.88^{d}\pm0.31$	$9.41^{d}\pm0.79$	$19.87^{c}\pm1.93$	$0.2059^{d}\pm0.0090$	$0.1274^{d}\pm0.0067$	$0.0166^{d}\pm0.0019$	$0.0082^{d}\pm0.0004$
	50 mM	$4.02^{c}\pm0.20$	$7.07^{c}\pm0.44$	$10.13^{b}\pm1.19$	$0.1303^{c}\pm0.0060$	$0.0894^{c}\pm0.0070$	$0.0116^{c}\pm0.0004$	$0.0063^{c}\pm0.0003$
	100 mM	$1.28^{b}\pm0.05$	$2.46^{b}\pm0.23$	$0.13^{a}\pm0.09$	$0.0508^{b}\pm0.0024$	$0.0375^{b}\pm0.0046$	$0.0046^{b}\pm0.0003$	$0.0029^{b}\pm0.0002$
	150 mM	$0.55^{a}\pm0.09$	$0.90^{ab}\pm0.08$	$0.00^{a}\pm0.00$	$0.0189^{a}\pm0.0030$	$0.0191^{a}\pm0.0017$	$0.0020^{ab}\pm0.0003$	$0.0020^{ab}\pm0.0001$

(continued)

Table 3. *(continued)*

Soybean varieties and salt treatments		HL (cm)	RL (cm)	NSR	HFW (g)	RFW (g)	HDW (g)	RDW (g)
	200 mM	$0.06^{a}\pm0.04$	$0.48^{a}\pm0.02$	$0.00^{a}\pm0.00$	$0.0013^{a}\pm0.0007$	$0.0112^{a}\pm0.0012$	$0.0002^{a}\pm0.0001$	$0.0017^{a}\pm0.0001$
MTD176	0 mM	$10.26^{d}\pm0.29$	$9.33^{c}\pm0.87$	$25.20^{c}\pm2.26$	$0.2842^{d}\pm0.0108$	$0.1576^{d}\pm0.0082$	$0.0187^{d}\pm0.0007$	$0.0090^{d}\pm0.0003$
	50 mM	$5.66^{c}\pm0.27$	$7.95^{c}\pm0.63$	$15.53^{b}\pm1.22$	$0.1935^{c}\pm0.0080$	$0.1237^{c}\pm0.0053$	$0.0154^{c}\pm0.0006$	$0.0082^{d}\pm0.0002$
	100 mM	$1.65^{b}\pm0.03$	$3.65^{b}\pm0.35$	$2.67^{a}\pm0.52$	$0.0630^{b}\pm0.0017$	$0.0562^{b}\pm0.0015$	$0.0057^{b}\pm0.0002$	$0.0041^{c}\pm0.0002$
	150 mM	$1.19^{ab}\pm0.03$	$1.69^{ab}\pm0.14$	$0.00^{a}\pm0.00$	$0.0393^{ab}\pm0.0021$	$0.0323^{a}\pm0.0016$	$0.0034^{a}\pm0.0002$	$0.0028^{b}\pm0.0002$
	200 mM	$0.69^{a}\pm0.05$	$0.99^{a}\pm0.07$	$0.00^{a}\pm0.00$	$0.0225^{a}\pm0.0021$	$0.0231^{a}\pm0.0014$	$0.0023^{a}\pm0.0002$	$0.0018^{a}\pm0.0001$
MTD455-3	0 mM	$8.67^{e}\pm0.24$	$10.31^{d}\pm0.83$	$30.43^{c}\pm1.99$	$0.2522^{e}\pm0.0075$	$0.1794^{d}\pm0.0091$	$0.0194^{e}\pm0.0005$	$0.0120^{d}\pm0.0005$
	50 mM	$5.70^{d}\pm0.18$	$6.06^{c}\pm0.50$	$14.53^{b}\pm1.06$	$0.1982^{d}\pm0.0094$	$0.1049^{c}\pm0.0040$	$0.0169^{d}\pm0.0004$	$0.0091^{c}\pm0.0003$
	100 mM	$2.58^{c}\pm0.09$	$3.28^{b}\pm0.26$	$3.10^{a}\pm0.57$	$0.0992^{c}\pm0.0031$	$0.0592^{b}\pm0.0023$	$0.0098^{c}\pm0.0003$	$0.0051^{b}\pm0.0002$
	150 mM	$1.61^{b}\pm0.04$	$1.81^{ab}\pm0.12$	$0.03^{a}\pm0.03$	$0.0656^{b}\pm0.0014$	$0.0399^{a}\pm0.0018$	$0.0068^{b}\pm0.0002$	$0.0033^{a}\pm0.0001$
	200 mM	$1.01^{a}\pm0.03$	$0.99^{a}\pm0.07$	$0.00^{a}\pm0.00$	$0.0323^{a}\pm0.0009$	$0.0304^{a}\pm0.0010$	$0.0038^{a}\pm0.0001$	$0.0025^{a}\pm0.0001$
MTD720	0 mM	$7.49^{d}\pm0.17$	$9.52^{c}\pm0.83$	$18.47^{b}\pm1.74$	$0.1945^{d}\pm0.0078$	$0.1204^{c}\pm0.0074$	$0.0136^{d}\pm0.0009$	$0.0084^{c}\pm0.0006$
	50 mM	$3.43^{c}\pm0.15$	$7.92^{c}\pm0.74$	$14.53^{b}\pm1.95$	$0.1126^{c}\pm0.0069$	$0.1096^{c}\pm0.0087$	$0.0101^{c}\pm0.0008$	$0.0083^{c}\pm0.0007$
	100 mM	$1.69^{b}\pm0.05$	$3.55^{b}\pm0.52$	$3.10^{a}\pm0.88$	$0.0569^{b}\pm0.0036$	$0.0529^{b}\pm0.0049$	$0.0055^{b}\pm0.0004$	$0.0041^{b}\pm0.0004$
	150 mM	$0.88^{a}\pm0.05$	$1.12^{a}\pm0.10$	$0.00^{a}\pm0.00$	$0.0270^{a}\pm0.0022$	$0.0223^{a}\pm0.0017$	$0.0029^{a}\pm0.0002$	$0.0022^{a}\pm0.0001$
	200 mM	$0.63^{a}\pm0.08$	$0.97^{a}\pm0.08$	$0.00^{a}\pm0.00$	$0.0217^{a}\pm0.0030$	$0.0205^{a}\pm0.0015$	$0.0020\ ^{a}\pm0.0003$	$0.0020^{a}\pm0.0001$
MTD765-2	0 mM	$8.60^{d}\pm0.22$	$7.67^{c}\pm0.54$	$18.63^{d}\pm1.53$	$0.2362^{d}\pm0.0081$	$0.1394^{c}\pm0.0065$	$0.0170^{d}\pm0.0005$	$0.0089^{c}\pm0.0004$
	50 mM	$4.90^{c}\pm0.16$	$7.12^{c}\pm0.56$	$10.57^{c}\pm1.08$	$0.1660^{c}\pm0.0068$	$0.1331^{c}\pm0.0081$	$0.0130^{c}\pm0.0006$	$0.0089^{c}\pm0.0006$
	100 mM	$2.59^{b}\pm0.10$	$4.80^{b}\pm0.48$	$5.07^{b}\pm0.88$	$0.0925^{b}\pm0.0037$	$0.0755^{b}\pm0.0042$	$0.0081^{b}\pm0.0003$	$0.0057^{b}\pm0.0003$
	150 mM	$1.59^{a}\pm0.45$	$1.64^{a}\pm0.12$	$0.00^{a}\pm0.00$	$0.0518^{a}\pm0.0020$	$0.0325^{a}\pm0.0016$	$0.0051^{a}\pm0.0002$	$0.0033^{a}\pm0.0002$
	200 mM	$1.14^{a}\pm0.37$	$0.90^{a}\pm0.05$	$0.00^{a}\pm0.00$	$0.0373^{a}\pm0.0017$	$0.0229^{a}\pm0.0008$	$0.0040^{a}\pm0.0002$	$0.0022^{a}\pm0.0001$
MTD860-1	0 mM	$8.53^{d}\pm0.54$	$7.34^{c}\pm0.91$	$15.50^{b}\pm2.43$	$0.2284^{d}\pm0.0163$	$0.1444^{d}\pm0.0096$	$0.0161^{d}\pm0.0012$	$0.0096^{c}\pm0.0007$
	50 mM	$5.16^{c}\pm0.27$	$6.96^{c}\pm0.69$	$15.53^{b}\pm1.75$	$0.1764^{c}\pm0.0101$	$0.1177^{c}\pm0.0080$	$0.0135^{c}\pm0.0007$	$0.0095^{c}\pm0.0007$
	100 mM	$2.38^{b}\pm0.08$	$4.29^{b}\pm0.36$	$3.87^{a}\pm0.75$	$0.0868^{b}\pm0.0035$	$0.0660^{b}\pm0.0032$	$0.0069^{b}\pm0.0003$	$0.0050^{b}\pm0.0003$
	150 mM	$1.71^{ab}\pm0.08$	$2.70^{ab}\pm0.28$	$0.70^{a}\pm0.32$	$0.0598^{ab}\pm0.0033$	$0.0477^{ab}\pm0.0038$	$0.0051^{ab}\pm0.0003$	$0.0038^{ab}\pm0.0002$
	200 mM	$0.88^{a}\pm0.06$	$1.13^{a}\pm0.11$	$0.00^{a}\pm0.00$	$0.0307^{a}\pm0.0025$	$0.0273^{a}\pm0.0025$	$0.0029^{a}\pm0.0002$	$0.0028^{a}\pm0.0001$
MTD868-1	0 mM	$8.79^{d}\pm0.30$	$7.63^{c}\pm0.71$	$19.23^{c}\pm1.66$	$0.2003^{d}\pm0.0094$	$0.1078^{d}\pm0.0049$	$0.0122^{d}\pm0.0007$	$0.0065^{c}\pm0.0003$

(continued)

Table 3. *(continued)*

	50 mM	$2.99^{c}\pm0.21$	$3.57^{b}\pm0.39$	$8.07^{b}\pm1.11$	$0.0789^{c}\pm0.0063$	$0.0450^{c}\pm0.0038$	$0.0061^{c}\pm0.0005$	$0.0036^{b}\pm0.0003$
	100 mM	$1.86^{b}\pm0.15$	$1.99^{a}\pm0.32$	$1.77^{a}\pm0.62$	$0.0473^{b}\pm0.0033$	$0.0303^{b}\pm0.0022$	$0.0040^{b}\pm0.0003$	$0.0024^{a}\pm0.0002$
	150 mM	$1.19^{ab}\pm0.13$	$1.26^{a}\pm0.11$	$0.00^{a}\pm0.00$	$0.0239^{a}\pm0.0027$	$0.0220^{ab}\pm0.0013$	$0.0023^{a}\pm0.0002$	$0.0021^{a}\pm0.0002$
	200 mM	$0.50^{a}\pm0.11$	$0.96^{a}\pm0.08$	$0.00^{a}\pm0.00$	$0.0108^{a}\pm0.0023$	$0.0161^{a}\pm0.0010$	$0.0011^{a}\pm0.0002$	$0.0018^{a}\pm0.0001$
MTD885-3	0 mM	$9.25^{e}\pm0.16$	$9.22^{d}\pm0.54$	$19.33^{c}\pm1.49$	$0.2373^{e}\pm0.0063$	$0.1593^{d}\pm0.0165$	$0.0175^{d}\pm0.0004$	$0.0094^{d}\pm0.0004$
	50 mM	$7.03^{d}\pm0.17$	$6.39^{c}\pm0.47$	$12.87^{b}\pm0.82$	$0.1833^{d}\pm0.0056$	$0.1034^{c}\pm0.0040$	$0.0176^{d}\pm0.0005$	$0.0082^{c}\pm0.0003$
	100 mM	$3.30^{c}\pm0.18$	$3.23^{b}\pm0.20$	$0.67^{a}\pm0.27$	$0.0964^{c}\pm0.0046$	$0.0570^{b}\pm0.0023$	$0.0097^{c}\pm0.0004$	$0.0047^{b}\pm0.0002$
	150 mM	$1.62^{b}\pm0.06$	$1.55^{a}\pm0.09$	$0.00^{a}\pm0.00$	$0.0481^{b}\pm0.0010$	$0.0358^{ab}\pm0.0011$	$0.0046^{b}\pm0.0001$	$0.0029^{a}\pm0.0001$
	200 mM	$1.05^{a}\pm0.03$	$0.91^{a}\pm0.05$	$0.00^{a}\pm0.00$	$0.0281^{a}\pm0.0014$	$0.0262^{a}\pm0.0011$	$0.0028^{a}\pm0.0001$	$0.0021^{a}\pm0.0001$
W82	0 mM	$9.43^{e}\pm0.24$	$8.44^{d}\pm0.69$	$20.20^{c}\pm2.13$	$0.2573^{e}\pm0.0097$	$0.1172^{d}\pm0.0063$	$0.0184^{d}\pm0.0005$	$0.0086^{d}\pm0.0004$
	50 mM	$6.86^{d}\pm0.27$	$5.77^{c}\pm0.52$	$8.63^{b}\pm1.06$	$0.1916^{d}\pm0.0099$	$0.0844^{c}\pm0.0052$	$0.0151^{cd}\pm0.0009$	$0.0068^{c}\pm0.0005$
	100 mM	$2.73^{c}\pm0.12$	$3.71^{b}\pm0.27$	$1.40^{a}\pm0.36$	$0.0873^{c}\pm0.0037$	$0.0493^{b}\pm0.0038$	$0.0104^{bc}\pm0.0033$	$0.0039^{b}\pm0.0003$
	150 mM	$1.59^{b}\pm0.10$	$1.47^{a}\pm0.17$	$0.00^{a}\pm0.00$	$0.0494^{b}\pm0.0032$	$0.0295^{a}\pm0.0026$	$0.0048^{ab}\pm0.0004$	$0.0027^{ab}\pm0.0002$
	200 mM	$0.88^{a}\pm0.08$	$0.59^{a}\pm0.02$	$0.00^{a}\pm0.00$	$0.0220^{a}\pm0.0017$	$0.0171^{a}\pm0.0007$	$0.0024^{a}\pm0.0002$	$0.0020^{a}\pm0.0002$

3.5 Salt Tolerance Evaluation

The MFV values of each indicator following 150 mM NaCl treatment were calculated and classified for salt tolerance of each soybean genotype. Because of complete inhibition in NSR under 150 mM NaCl concentration, NSR was not utilized in MFVS analysis. The MFVS values of the 13 genotypes had a mean of 0.507 and a standard deviation (SD) of 0.213. Based on the salinity responses and overall salt tolerance inferred from the MFVS values, no soybean varieties felled into the range of HST ($X_i \geq X_{mean} + 1.64SD$, mean MFVS > 0.856) and SS ($X_{mean} - SD > X_i \geq X_{mean} - 1.64SD$, 0.294 > mean MFVS ≥ 0.157). Thirteen soybean varieties were divided into three grades: (1) 2 genotypes were salt-tolerant (ST) ($X^- + 1.64SD > X_i \geq X^- + 1SD$, 0.856 > mean MFVS ≥ 0.720); (2) 9 genotypes were moderately salt-tolerant (MST) ($X^- + 1SD > X_i \geq X^- - 1SD$, 0.720 > mean MFVS ≥ 0.294); and (3) 2 were highly salt-sensitive (HSS) ($X^- - 1.64SD > X_i$, 0.157 > mean MFVS) (Table 4). To verify the suitability of the five-class MFVS classification, the distribution of MFVS values was analyzed using Shapiro–Wilk test and Normal Q-Q plot. The The Shapiro–Wilk test indicated no significant deviation from normality ($p = 0.5554$), supporting the assumption of an approximately Gaussian distribution. Additionally, the Normal Q–Q plot (**Supplementary Fig. S2**) shows that MFVS values align closely with the theoretical normal line. Together, these results confirm that the MFVS distribution is sufficiently normal to justify the use of equal-interval, five-class MFV grouping (HST to HSS). In detail, DT22 and MTD860 were ST varieties, DT2010 and MTD720 were HSS varieties while other cultivars classified as MST varieties including the reference soybean W82.

3.6 Correlation Analysis of Parameter STI

To study the possible relationship between indicators of 13 soybean cultivars under 150 mM NaCl conditions, the STI for germination rate (STI_GR), germination energy (STI_GE), germination index (STI_GI), vigor index (STI_VI), hypocotyl length (STI_HL), radicle length (STI_RL), hypocotyl fresh weight (STI_SFW), radicle fresh weight (STI_RFW), hypocotyl dried weight (STI_HDW), and radicle dried weight (STI_RDW) were calculated and analyzed by Pearson Correlation (Table 5). At the *p*-levels of 0.01 and 0.05, there were significant strong correlations between STIs of VI and any of HL, RL, HFW, RFW, HDW, and RDW. There were also positive correlations between all physiological indicators (HL, RL, HFW, RFW, HDW, RDW), and the highest correlation coefficient (0.931) was between STI_RL and STI_RDW. From the Correlation table, VI under stress is strongly associated with growth traits like length, fresh and dry biomass of hypocotyl and radicle, meanwhile the growth traits themselves displayed interrelated correlation under salinity stress conditions.

3.7 Principal Component Regression and Relative Trait Weights Contribution to Salinity Tolerance (MFVS)

To avoid multicollinearity and prevent overfitting, PCA was performed on the relative values of the ten indicators (RGR, RGE, RGI, RVI, RHL, RRL, RHFW, RRFW, RHDW, and RRDW) under 150 mM NaCl. Two principal components with eigenvalues greater than 1.0 were extracted, labeled PC1 and PC2. According to Table 6, these two components explained 81.361% of the total variance in the original ten traits, indicating that most of the data's complexity could be effectively reduced to two independent dimensions. The eigenvalue of the first principal component (PC1) was 5.699, which accounted for 56.994% of the variance. This component was associated with traits related to plant biomass and vigor (Table 6), with the most influential traits including RVI (component loading of 0.945), RHFW (component loading of 0.922), and RHL (component loading of 0.913). The second principal component (PC2) had an eigenvalue of 2.437, explaining an additional 24.37% of the variance. This component was primarily defined by traits associated with seedling germination metrics, specifically RGI (component loading of 0.857), RGE (component loading of 0.819), and RGR (component loading of 0.756). According to **Supplementary Table S3**, there was no correlation between PC1 and PC2, confirming their independence.

The subsequent multiple linear regression model (Principal Component Regression) using the two component scores to predict MFVS yielded a R^2 value of 0.998 and a highly significant F-statistic ($F(2, 10) = 2613.12$, $p < 0.001$), with PC1 as the primary contributor ($\beta = 0.971$), indicating that variation in seedling vigor explains the majority of MFVS differences among genotypes (Tables 7 and 8). In addition, PC2 served as secondary contributor, with smaller regression coefficient ($\beta = 0.234$). According to Table 8, relative hypocotyl fresh weight (RHFW) showed the greatest relative trait weight (0.943), followed by relative vigor index (RVI, 0.934). These results confirmed that RHFW and RVI are the most influential traits determining salinity tolerance, whereas relative germination rate (RGR) served as the least influential trait with relative trait weight of 0.348.

A hierarchical cluster analysis based on the Ward method using squared Euclidean distance of 10 was conducted to examine the grouping of soybean cultivars based on their unique salt tolerance strategies, as defined by the scores of the two Principal Components PC1 and PC2 (Fig. 2). Based on the mean PC scores, 3 distinct groups were identified with high positive mean PC1 scores, high positive mean PC2 scores, and negative mean scores for both PC1 and PC2. The first group exhibited strongly positive PC1 scores, indicating superior performance in traits related to seedling vigor and biomass accumulation. The second group showed high positive PC2 scores, reflecting strong germination-associated responses under salinity. In contrast, the third group displayed negative mean values for both PC1 and PC2, representing cultivars highly sensitive to salt stress across both germination and seedling growth traits.

According to the Cluster Membership results in Table 9, DT22 and MTD860–1 were classified into Cluster 1, characterized by strong salt tolerance associated with seedling vigor and biomass accumulation. In contrast, DT37, MTD868–1, and MTD720 were grouped into the salt-sensitive cluster (Cluster 3). The remaining cultivars, DT26, DT37, DT51, MTD176, MTD455–3, MTD765–2, MTD885–3, and W82, were assigned to Cluster 2, which represents strong salt tolerance driven primarily by germination-related traits.

Table 4. MFV values of each studied indicator under 150 mM NaCl conditions. Parameters including STI of each trait: GR, germination rate; GE, germination energy; GI, germination index; VI, vigor index; HL, hypocotyl length; RL, radicle length; HFW, hypocotyl fresh weight; RFW, radicle fresh weight; HDW, hypocotyl dried weight; RDW, radicle dried weight. Salt tolerance classification: ST, salt tolerant; MST, moderately salt tolerance; HSS, highly salt sensitive.

Varieties	GR	GE	GI	VI	HL	RL	HFW	RFW	HDW	RDW	MFVS	Salt tolerance classification	Ranking
DT22	0.7786	0.7985	0.5220	1.0000	0.8068	1.0000	1.0000	0.5439	0.9883	1.0000	0.8438	ST	1
DT26	1.0000	0.5970	0.4128	0.3616	0.2216	0.2728	0.3253	0.4627	0.3457	0.2927	0.4292	MST	9
DT37	0.3222	0.0597	0.3872	0.3246	0.4005	0.3680	0.4049	0.4131	0.8033	0.5695	0.4053	MST	10
DT51	0.8000	1.0000	0.7411	0.3625	0.4334	0.2059	0.5484	0.1584	0.3457	0.1641	0.4759	MST	8
DT2010	0.9000	0.3284	0.0728	0.0567	0.0000	0.0000	0.0000	0.0000	0.0000	0.0000	0.1358	HSS	12
MTD176	0.5828	0.8518	0.6761	0.2496	0.2133	0.2539	0.2467	0.3050	0.2666	0.2868	0.3933	MST	11
MTD455–3	0.9000	1.0000	1.0000	0.4811	0.6293	0.2368	0.8933	0.4018	1.0000	0.1327	0.6675	MST	3
MTD720	0.0000	0.0000	0.0000	0.0000	0.2217	0.0651	0.2496	0.1956	0.4032	0.0768	0.1212	HSS	13
MTD765–2	0.5828	0.9213	0.7339	0.4627	0.6241	0.3524	0.6768	0.4613	0.7804	0.5414	0.6137	MST	5
MTD860–1	0.7538	0.6487	0.2664	0.8499	0.7165	0.8076	0.9024	1.0000	0.8533	0.6483	0.7447	ST	2
MTD868–1	0.4077	0.5712	0.0996	0.4664	1.0000	0.9104	0.5402	0.9012	0.5919	0.7128	0.6201	MST	4
MTD885–3	0.9000	1.0000	0.6290	0.4382	0.5665	0.2154	0.5886	0.4147	0.6189	0.2757	0.5647	MST	7
W82	0.8000	0.9328	0.8069	0.4033	0.5235	0.2333	0.5318	0.5642	0.6103	0.2989	0.5705	MST	6

Table 5. Correlation analysis between salt-tolerance indices of germination rate (STI_GR), germination energy (STI_GE), germination index (STI_GI), vigor index (STI_VI), hypocotyl length (STI_HL), radicle length (STI_RL), hypocotyl fresh weight (STI_SFW), radicle fresh weight (STI_RFW), hypocotyl dried weight (STI_HDW), and radicle dried weight (STI_RDW) of 13 soybean varieties treated with 150 mM NaCl solution.

	STI_GR	STI_GE	STI_GI	STI_VI	STI_HL	STI_RL	STI_HFW	STI_RFW	STI_HDW	STI_RDW
STI_GR	1									
STI_GE	0.660*	1								
STI_GI	0.482	0.812**	1							
STI_VI	0.343	0.450	0.264	1						
STI_HL	–0.012	0.388	0.178	0.757**	1					
STI_RL	–0.039	0.110	–0.148	0.835**	0.816**	1				
STI_HFW	0.235	0.528	0.456	0.892**	0.816**	0.661*	1			
STI_RFW	–0.009	0.160	–0.070	0.695**	0.791**	0.806**	0.609*	1		
STI_HDW	–0.016	0.253	0.383	0.749**	0.758**	0.590*	0.895**	0.591*	1	
STI_RDW	–0.103	0.066	–0.068	0.809**	0.740**	0.931**	0.619*	0.712**	0.639*	1

*. Correlation is significant at the 0.05 level (2-tailed).
**. Correlation is significant at the 0.01 level (2-tailed)

Table 6. Principal component (PC) analysis method with two extracting components. Relative value of parameters as model input: germination rate (RGR), germination energy (RGE), germination index (RGI), vigor index (RVI), hypocotyl length (RHL), radicle length (RRL), hypocotyl fresh weight (RHFW), radicle fresh weight (RRFW), hypocotyl dried weight (RHDW), radicle dried weight (RRDW).

Determination of indicators	150 mM	
	PC1	PC2
RGR	0.176	0.756
RGE	0.449	0.819
RGI	0.290	0.857
RVI	0.945	0.072
RHL	0.913	–0.113
RRL	0.865	–0.394
RHFW	0.922	0.205
RRFW	0.805	–0.305
RHDW	0.844	0.009
RRDW	0.834	–0.389
Eigenvalue	5.699	2.437
Variance contribution rate %	56.994	24.367
Cumulative contribution rate %	56.994	81.361

Table 7. Principal component regression with MFVS as the dependent variable and the 2 Component Scores (PC1 Score and PC2 Score) as the independent variables. Relative value of parameters as model input: germination rate (RGR), germination energy (RGE), germination index (RGI), vigor index (RVI), hypocotyl length (RHL), radicle length (RRL), hypocotyl fresh weight (RHFW), radicle fresh weight (RRFW), hypocotyl dried weight (RHDW), radicle dried weight (RRDW).

Model	R	R square	Adjusted R square	Std. error of the estimate	Change statistics				
					R square change	F change	df1	df2	Sig. F change
1	.999[a]	0.998	0.998	0.01019	0.998	2613.124	2	10	0.000

a. Predictors: (Constant), REGR factor score 2 for analysis 1, REGR factor score 1 for analysis 1.

Table 8. Relative trait weights. Relative value of parameters as model input: germination rate (RGR), germination energy (RGE), germination index (RGI), vigor index (RVI), hypocotyl length (RHL), radicle length (RRL), hypocotyl fresh weight (RHFW), radicle fresh weight (RRFW), hypocotyl dried weight (RHDW), radicle dried weight (RRDW).

	Component loading		Regression coefficients		
Relative traits	PC 1	PC 2	PC 1	PC 2	Relative Trait Weight
β			0.971	0.234	
RGR	0.176	0.756			0.3478
RGE	0.449	0.819			0.6276
RGI	0.290	0.857			0.4821
RVI	0.945	0.072			0.9344
RHL	0.913	−0.113			0.8601
RRL	0.865	−0.394			0.7477
RHFW	0.922	0.205			**0.9432**
RRFW	0.805	−0.305			0.7103
RHDW	0.844	0.009			0.8216
RRDW	0.834	−0.389			0.7188

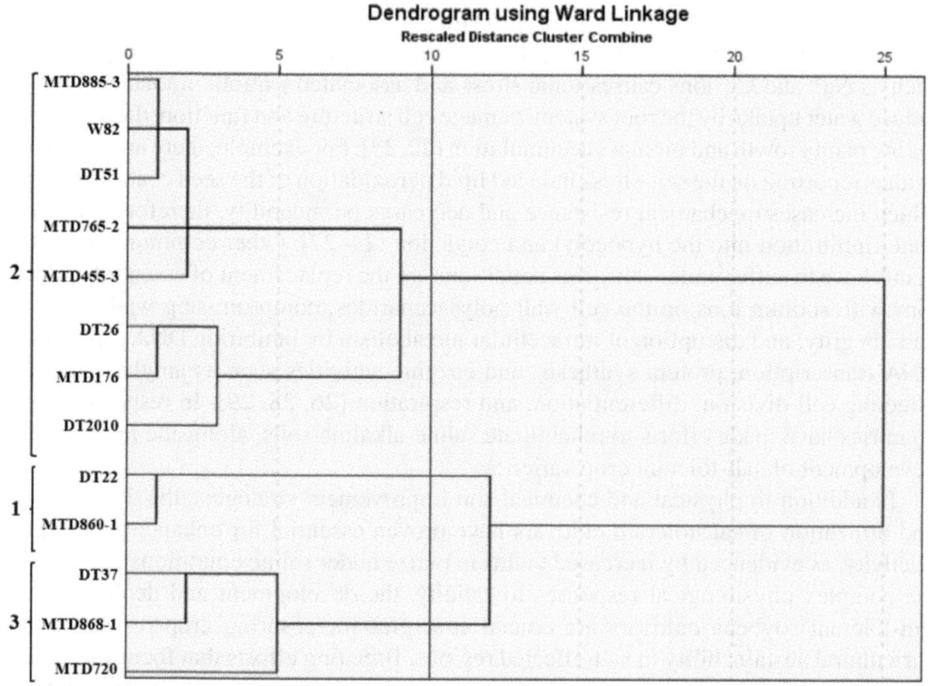

Fig. 2. Hierarchical cluster analysis based on the Furthest Neighbor to evaluate the salt tolerance of 13 soybean genotypes with squared Euclidean distance at 10 (red line). Three clusters were identified according to Cluster Membership.

Table 9. Cluster Membership analysis following Hierarchical cluster analysis with distinct means of 3 clusters.

Ward method		REGR factor score 1 for analysis 1	REGR factor score 2 for analysis 1
1	Mean	1.5105	–0.4717
	Std. deviation	0.2672	0.2879
2	Mean	–0.2877	0.6572
	Std. deviation	0.7525	0.4961
3	Mean	–0.2397	–1.4380
	Std. deviation	1.1330	0.2134

4 Discussion

Soil salinity is an escalating global threat, projected to affect nearly 50% of arable land by 2050 [21]. In general, salt stress disrupts physiological and biochemical processes, ultimately reducing crop yields and threatening food security. Under severe conditions, the stress could critically impair plant growth, particularly during early developmental stages such as seed germination [21]. Excessive uptake of certain ions onto plant cells such as Na^+ and Cl^- ions causes ionic stress and associated osmotic imbalance, which reduce water uptake by the root system, damage cell structure and function, therefore hindering plant growth and biomass accumulation [22, 23]. For example, there are numerous studies reporting on the salt-stress induced lipid peroxidation of the seed coat membrane, which increases mechanical resistance and decreases permeability, therefore hindering water infiltration into the hypocotyl and cotyledon [24–27]. Other common issues that plants have to suffer under salt stress conditions are the replacement of essential calcium ions with sodium ions on the cell wall polysaccharides, compromising wall elasticity and integrity; and disruption of intracellular metabolism by inhibiting DNA replication, RNA transcription, protein synthesis, and enzyme activities such as amylase, thereby affecting cell division, differentiation, and respiration [26, 28, 29]. In response, many countries have made efforts to rehabilitate saline-alkaline soils, alongside research and development of salt-tolerant crop varieties.

In addition to physical and chemical soil improvement strategies, the development and utilization of salt-tolerant cultivars have proven essential for enhancing crop productivity, as evidenced by increased yields in maize under saline conditions [30]. Given the complex physiological responses to salinity, the development and deployment of salt-tolerant soybean cultivars are crucial strategies for ensuring crop resilience and agricultural sustainability in salt-affected regions. Breeding efforts that focus on screening and evaluating germplasm under salt stress conditions offer long-term solutions to mitigate yield losses while enhancing ecological and economic stability. Previous studies have identified several reliable indicators for assessing salt tolerance during germination stage, including germination rate, germination energy, vigor index, seed germination index, hypocotyl length, radicle length, total fresh weight, and total dry weight [4, 16, 31]. The analytical methods for comparison for salt tolerance among various germplasms have been also well established such as salt-tolerance index, analysis of variance (ANOVA), correlation analysis of individual traits, and the calculation of mean functioning value (MFV) for salt-tolerance classification [4, 16, 17, 31].

In our study, it is also observed that the adverse effects of salinity became increasingly severe with rising NaCl concentrations. Notably, the vigor index and number of secondary roots were drastically reduced, with complete inhibition (100% reduction) following 200 mM NaCl treatment, while other parameters exhibited reductions ranging from 30.4% to 91% (Table 1). Across 13 soybean varieties examined, germination parameters including germination rate, germination energy, and germination index showed significant deviations from the corresponding control counterparts when they were treated with NaCl at concentrations of 150 mM or higher. At high salt concentrations, the reduction in germination rate, germination energy, and germination index indicates delayed and impaired germination. This suggests that salinity stress adversely affects seed viability, energy, and the overall speed of germination [32, 33]. Under low salinity concentration (50 mM), most varieties displayed no effect to slight reduction in germination indices. Notably, cultivar DT22 showed improved performance, with increased germination rate (GR) and vigor index (VI), suggesting a stimulatory effect of low salt level. This record aligns with previous findings, which has been explained by enhancement of water absorption and activation of seed metabolism to promote seed germination by mild salinity [34]. However, at higher salt levels, all varieties experienced significant reductions in germination potential and seed vigor. In particular, MTD720 emerged as the most salt-sensitive cultivar, indicating a lower capacity for salt tolerance compared to other varieties.

Regarding evaluation on shoot and root-associated traits, plant growth inhibition was clearly observed even at low and moderate salinity conditions such as 50 mM and 100 mM NaCl. Higher salt concentrations led to substantial reductions in shoot and root elongation, as well as their corresponding biomass. The extent of salt-induced growth inhibition varied among the genotypes. For instance, DT37 and MTD720 exhibited pronounced sensitivity to 50 mM NaCl, with hypocotyl lengths reduced by more than twofold. At 100 mM NaCl, several cultivars showed distinct patterns of growth inhibition: MTD176 experienced the most severe reduction in hypocotyl growth, DT2010 showed the greatest decrease in root growth, and DT51 failed to develop lateral root formation (Table 3). These findings emphasize that salt stress has a more profound inhibitory effect on vegetative growth traits compared to germination parameters, particularly at higher salt concentrations. Notably, at 150 mM NaCl, DT2010 demonstrated significant reductions across all measured parameters, ranging from 4- to 12.58-fold compared with the control counterpart.

After screening eleven parameters for the salt-injury index at 50% of the control by establishing linear regression with the equation $y = ax + b$ for SII value of each trait at 0.5 (50%), the recommended salt concentrations were identified. In average, the salt concentration that could cause SII_{50} for most of the indicators was 126.24 mM. Based on this, 150 mM NaCl was considered as an appropriate threshold for subsequent salt tolerance analyses, as it effectively differentiates salt sensitivity among soybean genotypes without causing complete inhibition of growth. This is particularly relevant given that vegetative growth traits tend to be more susceptible to salinity stress than germination parameters.

Elevated salt concentrations impede the ability of soybean seeds to absorb water, consequently limiting various aspects of seedling growth [35]. Importantly, plant salt tolerance is not determined by a single factor but rather results from a complex interplay of biochemical and molecular processes [36]. As such, a comprehensive evaluation of salinity tolerance should integrate multiple indicators and their interrelationships rather than relying on a single parameter. In this study, we employed ANOVA and correlation analysis to assess various salt tolerance indicators at the identified optimal salt concentration of 150 mM NaCl. The results revealed strong correlations between seed vigor index and physiological growth traits as well as interrelationships among growth traits, particularly between radicle elongation and radicle dried biomass.

By calculating the MFV of each parameter and comparing the average MFV (MFVS) under stringent salinity condition (150 mM NaCl) across different soybean cultivars, a comprehensive evaluation of their salt tolerance could be achieved. Each soybean variety was assigned an MFVS index, where a higher MFVS indicates greater salt tolerance. Based on the analyses, DT22 and MTD860–3 were assigned as salt-tolerant varieties, DT2010 and MTD720 were assigned as highly salt-sensitive varieties while other cultivars were classified as moderately salt-tolerant varieties, including W82. Connecting our findings with previous reports, W82 consistently demonstrated moderate salt tolerance across both germination and vegetative growth stages [37].

To further strengthen the evaluation of salinity tolerance and eliminate multicollinearity among indicators, a Principal Component Regression (PCR) model was applied using PCA-derived component scores. The two extracted components explained 81.36% of the total variance, with PC1 primarily representing seedling vigor and biomass traits and PC2 reflecting germination-related traits. The PCR model showed excellent predictive accuracy for MFVS ($R^2 = 0.998$), indicating high reliability of this approach. Trait weight analysis further revealed RHFW and RVI as the most influential contributors to salinity tolerance, whereas RGR had the smallest contribution. These results demonstrate that the PCA–PCR framework provides a robust and biologically meaningful method for assessing soybean salinity tolerance, although validation across a broader set of genotypes would further enhance its general applicability.

In addition to MFVS-based ranking, hierarchical clustering of the PC1 and PC2 scores revealed grouping patterns that did not fully match the MFVS rank order, highlighting the value of a multivariate perspective. While MFVS summarizes the overall level of salt tolerance, the clustering analysis differentiated cultivars according to their underlying tolerance strategies. Cultivars with high positive PC1 scores formed a vigor-dominant tolerant group, maintaining superior seedling growth under salinity, whereas those with positive PC2 showed germination-tolerant group. In contrast, those with negative scores on both components represented a sensitivity profile marked by weak post-germination performance. By distinguishing tolerance mechanisms rather than solely ranking overall performance, this strategy-based classification provides a more informative framework for breeding programs.

5 Conclusion

This study evaluated twelve Vietnamese soybean cultivars (DT2010, DT22, DT26, DT37, DT51, MTD176, MTD455–3, MTD720, MTD765–2, MTD860–1, MTD868–1, and MTD885–3) for their responses to salinity stress during both germination and post-germinative growth. The results indicated that seedling development was adversely affected by higher salt concentrations, particularly beyond 150 mM NaCl. Based on membership function value (MFV) analysis, DT22 was identified to have the highest salt tolerance capacity, followed closely by MTD860, as evidenced by strong germination and growth performance under stress. Several cultivars, including DT37, DT51, MTD455–3, MTD765–2, MTD860–1, MTD868–1 and MTD885–3, were classified as moderately salt-tolerant. In contrast, DT2010 and MTD720 were identified as highly salt-sensitive. These findings provide valuable information about the salt-tolerance capacity of local soybeans, which can be used as a reference by farmers in selecting suitable varieties grown in saline-prone environments, and by researchers in selecting relevant genetic resources used in breeding programs or molecular studies focused on developing resilient soybean varieties.

Acknowledgements. This research is funded by International University, VNU-HCM under grant number SV2023–04-BT.

References

1. Msarie, M.W., et al.: Enhancing soybean salt tolerance with gsno and silicon: a comprehensive physiological, biochemical, and genetic study. Int. J. Mol. Sci. **26**, 609 (2025). https://doi.org/10.3390/ijms26020609
2. Liu, K.: Soybean: Overview. In: Reference module in food science. Elsevier (2016). https://doi.org/10.1016/B978-0-08-100596-5.00028-7
3. Fraanje, W., Garnett, T., Breewood, H.: Soy: Food, feed, and land use change. University of Oxford, Oxford, UK (2020). https://doi.org/10.56661/47e58c32
4. Han, L., et al.: A comprehensive evaluation of soybean germplasm resources for salt tolerance during germination. Plants. **14**, 791 (2025). https://doi.org/10.3390/plants14050791
5. Phu, N.V.: The impacts of climate change on agricultural production and sustainable agriculture of smallholder farmers in Vietnam. Vietnam J. Agric. Sci. **4**, 1100–1107 (2021). https://doi.org/10.31817/vjas.2021.4.2.10
6. Veettil, B.K., Acharki, S., Puri, V., Teck, V.: Coastal environmental changes In Ninh Thuan Province. South-Central Vietnam. PLOS ONE. **20**, e0313382 (2025). https://doi.org/10.1371/journal.pone.0313382
7. Soil salinity prediction using satellite-based variables and machine learning: case study in Tra Vinh Province, Mekong Delta, Vietnam. Vietnam. J. Earth Sci. https://vjs.ac.vn/jse/article/view/22438. Last accessed 2025/05/23
8. Guan, R.-X., et al.: Salt tolerance in soybeans: focus on screening methods and genetics. Plants. **13**, 97 (2023). https://doi.org/10.3390/plants13010097
9. Hasanuzzaman, M., Raihan, M.R.H., Nowroz, F., Fujita, M.: Insight into the mechanism of salt-induced oxidative stress tolerance in soybean by the application of Bacillus subtilis: Coordinated actions of osmoregulation, ion homeostasis, antioxidant defense, and methylglyoxal detoxification. Antioxidants. **11**, 1856 (2022). https://doi.org/10.3390/antiox11101856

10. Jaramillo Roman, V., Den Toom, L.A., Castro Gamiz, C., Van Der Pijl, N., Visser, R.G.F., Van Loo, E.N., Van Der Linden, C.G.: Differential responses to salt stress in ion dynamics, growth and seed yield of European Quinoa varieties. Environ. Exp. Bot. **177**, 104146 (2020). https://doi.org/10.1016/j.envexpbot.2020.104146
11. Luo, Q., Yu, B., Liu, Y.: Differential sensitivity to chloride and sodium ions in seedlings of *Glycine Max* And *G. Soja* under nacl stress. J. Plant Physiol. **162**, 1003–1012 (2005). https://doi.org/10.1016/j.jplph.2004.11.008
12. Hou, Q.-Z., Sun, K., Zhang, H., Su, X., Fan, B.-Q., Feng, H.-Q.: The responses of photosystem II and intracellular ATP production of *Arabidopsis* leaves to salt stress are affected by extracellular ATP. J. Plant Res. **131**, 331–339 (2018). https://doi.org/10.1007/s10265-017-0990-9
13. Li, W., Zhang, H., Zeng, Y., Xiang, L., Lei, Z., Huang, Q., Li, T., Shen, F., Cheng, Q.: A salt tolerance evaluation method for sunflower (*Helianthus annuus* L.) at the seed germination stage. Sci. Rep. **10**, 10626 (2020). https://doi.org/10.1038/s41598-020-67210-3
14. Hoang, X.L.T., et al.: The drought-mediated soybean GmNAC085 functions as a positive regulator of plant response to salinity. Int. J. Mol. Sci. **22**, 8986 (2021). https://doi.org/10.3390/ijms22168986
15. Shelke, D.B., Pandey, M., Nikalje, G.C., Zaware, B.N., Suprasanna, P., Nikam, T.D.: Salt responsive physiological, photosynthetic and biochemical attributes at early seedling stage for screening soybean genotypes. Plant Physiol. Biochem. PPB. **118**, 519–528 (2017). https://doi.org/10.1016/j.plaphy.2017.07.013
16. Zhou, X., Tian, Y., Qu, Z., Wang, J., Han, D., Dong, S.: Comparing the salt tolerance of different spring soybean varieties at the germination stage. Plants. **12**, 2789 (2023). https://doi.org/10.3390/plants12152789
17. Wu, H., Guo, J., Wang, C., Li, K., Zhang, X., Yang, Z., Li, M., Wang, B.: An effective screening method and a reliable screening trait for salt tolerance of Brassica napus at the Germination Stage. Front. Plant Sci. **10**, (2019). https://doi.org/10.3389/fpls.2019.00530
18. Abdul-Baki (1973) Vigor determination in soybean seed by multiple criteria1. Crop Science—Wiley Online Library. https://acsess.onlinelibrary.wiley.com/doi/abs/10.2135/cropsci1973.0011183x001300060013x. Last accessed 2025/05/28
19. Screening diverse soybean genotypes for drought tolerance by membership function value based on multiple traits and drought-tolerant coefficient of yield. BMC Plant Biology. https://bmcplantbiol.biomedcentral.com/articles/10.1186/s12870-020-02519-9. Last accessed 2025/05/28
20. Chen, X., Min, D., Yasir, T.A., Hu, Y.-G.: Evaluation of 14 morphological, yield-related and physiological traits as indicators of drought tolerance in Chinese winter bread wheat revealed by analysis of the membership function value of drought tolerance (MFVD). Field Crops Res. **137**, 195–201 (2012). https://doi.org/10.1016/j.fcr.2012.09.008
21. Soil Salinity: A threat to global food security. Agronomy Journal. https://acsess.onlinelibrary.wiley.com/doi/abs/10.2134/agronj2016.06.0368. Last accessed 2025/05/31
22. Giambalvo, D., et al.: Nitrogen availability drives mycorrhizal effects on wheat growth, nitrogen uptake and recovery under salt stress. Agronomy **12**, 2823 (2022). https://doi.org/10.3390/agronomy12112823
23. Modulation in growth, biochemical attributes and proteome profile of rice cultivars under salt stress—PubMed. https://pubmed.ncbi.nlm.nih.gov/31733605/. Last accessed 2025/05/29
24. Kesh, H., Devi, S., Kumar, N., Kumar, A., Kumar, A., Dhansu, P., Sheoran, P., Mann, A., Kesh, H., Devi, S., Kumar, N., Kumar, A., Kumar, A., Dhansu, P., Sheoran, P., Mann, A.: Insights into physiological, biochemical and molecular responses in wheat under salt stress. In: Wheat—recent advances. IntechOpen (2022). https://doi.org/10.5772/intechopen.102740
25. Uçarlı, C.: Effects of salinity on seed germination and early seedling stage. In: Abiotic stress in plants. IntechOpen (2020). https://doi.org/10.5772/intechopen.93647

26. Dabravolski, S.A., Isayenkov, S.V.: The regulation of plant cell wall organisation under salt stress. Front. Plant Sci. **14**, (2023). https://doi.org/10.3389/fpls.2023.1118313
27. Xiao, F., Zhou, H.: Plant salt response: Perception, signaling, and tolerance. Front. Plant Sci. **13**, (2023). https://doi.org/10.3389/fpls.2022.1053699
28. Liu, J., et al.: Cell wall components and extensibility regulate root growth in Suaeda salsa and Spinacia oleracea under Salinity. Plants. **11**, 900 (2022). https://doi.org/10.3390/plants11070900
29. Sunita, K., Mishra, I., Mishra, J., Prakash, J., Arora, N.K.: Secondary metabolites from halotolerant plant growth promoting rhizobacteria for ameliorating salinity stress in plants. Front. Microbiol. **11**, (2020). https://doi.org/10.3389/fmicb.2020.567768
30. Ye, X., et al.: Transcriptome profiling of *Puccinellia Tenuiflora* during seed germination under a long-term saline-alkali stress. BMC Genomics **20**, 589 (2019). https://doi.org/10.1186/s12864-019-5860-5
31. A salt tolerance evaluation method for sunflower (*Helianthus annuus* L.) at the seed germination stage. Scientific Reports. https://www.nature.com/articles/s41598-020-67210-3. Last accessed 2025/05/25
32. Xu, N., et al.: Effects of salt stress on seed germination and respiratory metabolism in different *Flueggea Suffruticosa* Genotypes. PeerJ **11**, e15668 (2023). https://doi.org/10.7717/peerj.15668
33. Shumye, G., Gedamu, Y.: Effect of salinity on seed germination of some tomato (*Lycopersicon Esculentum* Mill.) Varieties. J. Arid. Agric. **7**, 76–82 (2021). https://doi.org/10.25081/jaa.2021.v7.6588
34. Xu, Y.: The effect of salt concentration on the germination of soybeans. Adv. Eng. Technol. Res. **8**, 624–624 (2023). https://doi.org/10.56028/aetr.8.1.624.2023
35. Atta, K., Mondal, S., Gorai, S., Singh, A.P., Kumari, A., Ghosh, T., Roy, A., Hembram, S., Gaikwad, D.J., Mondal, S., Bhattacharya, S., Jha, U.C., Jespersen, D.: Impacts of salinity stress on crop plants: improving salt tolerance through genetic and molecular dissection. Front. Plant Sci. **14**, (2023). https://doi.org/10.3389/fpls.2023.1241736
36. Iyengar, E., Reddy, M.: Handbook of photosynthesis, USA (1996)
37. Do, T.D., et al.: Mapping and confirmation of loci for salt tolerance in a novel Soybean Germplasm. Fiskeby III. Theor. Appl. Genet. **131**, 513–524 (2018). https://doi.org/10.1007/s00122-017-3015-0

AKAP8L is a Novel Potential Prognostic Biomarker in Clear Cell Renal Carcinoma Patients with COVID-19

Anh P. Hoang Le[1,2], Duc Phuc Ngo[2], and Minh Nam Nguyen[2,3,4](✉)

[1] School of Biomedical Engineering, International University, Vietnam National University Ho Chi Minh City, Ho Chi Minh, Vietnam

[2] Vietnam National University Ho Chi Minh City, Ho Chi Minh, Vietnam
nmnam@uhsvnu.edu.vn

[3] Department of Biomedical Engineering, Faculty of Medicine, University of Health Sciences, Vietnam National University Ho Chi Minh City, Ho Chi Minh, Vietnam

[4] Research Center for Genetics and Reproductive Health, University of Health Sciences, Vietnam National University Ho Chi Minh, Ho Chi Minh, Vietnam

Abstract.
Introduction The recent COVID-19 pandemic threatened the success rate of clear cell renal cell carcinoma (ccRCC) treatment due to its direct effects on kidney tissues. COVID infection was eventually incorporated into the ccRCC treatment paradigm. Finding the optimal therapeutic approach according to patients' specific conditions is required. This study investigated *AKAP8L* and validated its prognostic significance for both ccRCC and patients with COVID-19 comorbidity, and compared its predictive accuracy with that of an existing clinical biomarker for ccRCC prognosis. **Materials and methods:** GTEX, TCGA, and GSE datasets were retrieved online. RNA sequencing data (TPM) of normal subjects ($n = 85$), ccRCC patients ($n = 533$), COVID-infected groups, and control samples ($n = 3$ per group) were evaluated. The *AKAP8L* level in each dataset was used to identify the association between the expression of the gene with the risk of poor prognosis in ccRCC and COVID-19 patients. A t-test, ANOVA, and multivariate analysis were performed to detect differences among the groups and compare the association between the expression level of *AKAP8L* and the risk of poor outcomes. **Results:** *AKAP8L* was elevated 30 times in renal tissues responding to ccRCC ($p < 0.001$) and increased 30 to 70 TPM with each stage progression. *AKAP8L* also showed an increase of 36 TPM to 799.6 TPM in response to COVID infection. High transcription of *AKAP8L* also exceeded sex, age, metastasis, tumor stages, and pathology associated with poor survival outcomes. **Conclusion:** *AKAP8L* is a promising biomarker for ccRCC prognosis in COVID-19 circumstances.

Keywords: *AKAP8L* · Biomarker · Prognosis · Clear cell renal cell carcinoma · COVID-19

L. Vu et al. (Eds.): PBC 2025, 2026.
https://doi.org/10.1007/978-3-032-29579-8_17

1 Introduction

Renal carcinoma was listed among the top ten most common cancers worldwide [1–3]. It is defined as a carcinoma originating from cells within the renal cortex or renal tubular epithelium. Among renal cancers, ccRCC is the most common subtype, along with papillary RCC and chromophobe RCC, which account for 85% of the total cases [4]. The majority of ccRCCs stem from anomalies in the proximal convoluted tubule [5]. In the Asian region, the risk of getting this cancer has increased in recent years [5]. The cause for this shift is hypothesized to be the insufficiency in availability and adaptability of early detection, access to healthcare, and treatment [6, 7]. In the midst of COVID-19 pandemic persistence, treatment for ccRCC must also include COVID within the decision-making paradigm [8]. COVID-19 appears to act as a catalyst that accelerates the progression of renal diseases, as the infection can target renal tissues, increasing acute kidney injury and diminishing the success rate of ccRCC treatment [9]. It was commonly hypothesized that SARS-CoV-2, the culprit of COVID-19, accesses targets through ACE2-receptors, which by chance are highly expressed in renal tissues, especially in the cortex [10]. This might explain the hypothesis behind the focal destruction of the kidney. Conversely, cancer treatments and pathology can expose COVID-19 patients to more severe impacts, including immunosuppression from either treatments or their specific conditions and egregious immune responses such as septic shock [11]. Precision medicine is then considered to minimize the risk of treatment failures and increase the prognostic efficiency within such conditions.

Precision medicine, or personalized medicine, aims to use molecular data in addition to more traditional approaches like investigating medical history and symptoms. The goal is to provide the most benefit while minimizing risk by finding the medical approach that is the best fit for the patient's specific condition. The application of precision medicine is anticipated to improve all areas of medicine, including predicting an individual's risk of carrying a disease, disease prognosis, and risks of side effects, as well as predicting the efficiency of the treatment [12]. Since the discovery of the Von Hippel-Lindau gene 25 years ago, academics have shifted their focus to studying the underlying molecular behavior of ccRCC, which promises prospective therapeutic potential in terms of prognostic and treatment strategies for this cancer [13]. Even though multiple attempts in this field have been made and many molecular markers, including *CAIX, PTEN, and CXCR4*, have shown great potential in treatment decisions in mRCC, none of them has significantly improved the current prognostic systems [14]. Additionally, none of the existing biomarkers has been validated for COVID-19 infection prognosis accordingly, which is crucial in the current COVID-19 epidemic situation. Therefore, this study attempted to investigate a novel target, adding an option to the list of prominent prognostic biomarkers for both ccRCC and COVID-19.

The A-kinase anchoring proteins (AKAPs) belong to a family of structural proteins involved in the binding of cAMP-dependent protein kinase A (PKA) and anchoring of the regulatory subunit of PKA, adjacent to the substrate [15, 16]. Recently, an increasing number of AKAP family members have been identified as different cancer biomarkers. For instance, *AKAP4* plays a pivotal role in tumorigenesis by promoting cellular growth, migration, and invasion, as noted in cervical, colorectal, and esophageal cancers [17–19]; *AKAP9* is implicated in cancer development and metastasis by modulating different interacting proteins [20–24]; and the target of the present study, *AKAP8L*, a nuclear A-kinase anchor protein that regulates transcription and alternative splicing, plays an essential role in tumorigenesis by facilitating cancer cell growth and suppressing apoptosis [25, 26]. Notably, the N-terminal region of *AKAP8L* interacts with *mTORC1* to regulate translation, cell growth, and cell proliferation [27, 28]. The mTOR signaling pathway has been shown to be primarily responsible for cancer growth, implying that *AKAP8L* expression changes would be an indicator of carcinoma progression [29]. The protein may also regulate gene methylation and the cell cycle, particularly mitotic DNA condensation [25, 30]. A recent study revealed that *AKAP8L* overexpression was directly correlated with poor survival outcomes in esophageal carcinoma [31]. Surprisingly, the protein-encoding gene was significantly increased in COVID-19-infected samples [32], implying that *AKAP8L* may have a synergistic effect on ccRCC patients who are also infected with COVID-19. However, the underlying molecular relationship between *AKAP8L* and COVID-19 is unknown, despite the fact that its mediator, DHX9, has been linked to human immunity and its responses to SARS-CoV-2 infection [33]. The relationship between *DHX-9* and *AKAP8L* will be further investigated in the present study. Therefore, studying *AKAP8L* along with COVID-19 might resolve the mystery of ccRCC underlying molecular mechanisms and open a new door for ccRCC prognosis with COVID infection.

Due to the lack of publicly available dataset contains ccRCC tumor samples from COVID-19–positive patients, *AKAP8L* will be investigated as an independent biomarker in renal carcinoma, and COVID-19 samples will be used to test how well it might be able to predict outcomes.

2 Materials and Methods

2.1 Tissue and Blood Databases

Expression profiles of *AKAP8L* in human bodies, specifically renal tissues, were downloaded from the online Human Protein Atlas database (open access: https://www.proteinatlas.org/). The study used four distinct types of datasets: (1) transcription levels in normal tissues (GTEX-dataset, n = 85); (2) translation levels in different tissues; (3) cellular expression levels; and (4) expression in diverse cancers.

2.2 Gene Interaction Profile

Data about the genetic interaction of *AKAP8L* with neighboring molecules were retrieved from an open-access online database: http://string-db.org/ (STRING database) [34]. However, only co-expression genes were discussed, while co-mentioned genes were omitted. Their co-expression score will be considered to predict the strength of the interaction [35].

2.3 MRNA Dataset

Kidney/ccRCC/TCGA-dataset (Firehose Legacy) (n = 534; 64.4% male; 35.6% female; 269/534 (50.4%) Stage I samples, 57/534 (10.7%) Stage II samples, 125/534 (23.4%)

Stage III samples and 83/534 (15.5%) Stage IV samples) was retrieved from cBioPortal (http://cbioportal.org), hosted by the Center for Molecular Oncology at Memorial Sloan Kettering Cancer Center (MSK) [36]. This platform carries patients' clinical data, focusing on Cancer Genomics, which was originally collected at MSK. Each dataset includes patients' clinical data, their mRNA, and protein expression (median values and Z-score). As this study focuses on interpreting mRNA expression, only the mRNA sequencing profile was analyzed. A correlation between the RNA expression level and the survival outcomes of patients with ccRCC, as well as association with different demographic factors (sex, age, cancer stages, tumor classification), were investigated in this dataset to determine the prognostic values of the selected mRNA in ccRCC.

GSE152586, GSE162113, and GSE161881 were retrieved from the NCBI-GEO Accession Portal to assess the association between the expression of the mRNA marker of interest in renal tissue and COVID infection. In brief, GSE152586 was collected from alveolar type II cells in human primary lung tissue (*Homo sapiens*) in two cohorts (n = 6: control = 3; SARS-CoV-2-48hpi = 3) [37]. The analysis was conducted to inspect the diagnostic values of the selected mRNA by determining the level of differential significance between its expression level in infected samples and the control cohort. GSE162113 includes RNA-sequencing data from mouse (*Mus musculus*) vital organ biopsy in 13 models (n = 13: control = 7; SARS-CoV-2-infected = 6), collected at two distinct time points: three- and seven-days post-infection (dpi) [32]. This data was investigated to detect the changes in gene expression throughout the period of infection in different tissues to evaluate the potential of the selected mRNA as a prognostic biomarker for COVID-19. Finally, the GSE161881 dataset was built on six Vero-E6 cell samples (originating from an epithelial layer isolated from the kidney of an African green monkey, *Chlorocebus sabaeus*) [38]. Within this batch, three samples were flow-through infected with SARS-CoV-2 while three samples were infected with a concentrated viral culture. The cells were subjected to RNA-sequencing eight hours post-infection (hpi). This dataset is expected to clarify the correlation between the viral load and gene expression, emphasizing the prognostic values of the selected mRNA.

Any missing or replicated data points will be removed during the data cleaning. A resulted total of 533 subjects in the ccRCC-TCGA dataset were subjected to further analysis. *AKAP8L* mRNA expressions were then categorized into two sub-groups based on the recorded expression level: High (expression level > median value) and Low (expression level $\leq$ median value). The number of subjects in GSE datasets remained unchanged after data screening.

Regarding normalization and batch correction, the public RNA sequencing data has been preprocessed and normalized using either FPKM (fragments per kilobase of transcript per million fragments mapped) and TPM (transcripts per million) normalization methods—which indicated their resulted expression units as well. To facilitate the comparison amongst different datasets, we convert all units into TPM using the equations below, followed by batch correction using ComBat package in R [39]:

$$FPKM = \frac{rawfragmentscounts}{featurelengthofgene \times totalreadcounts} \times 10^9 \tag{1}$$

$$TPM = \frac{\frac{rawfragmentcounts}{featurelengthofgene}}{\Sigma\frac{rawcounts}{featurelength}} \times 10^6 = \frac{FPKM}{\Sigma FPKM} \times 10^6 \quad (2)$$

2.4 Statistical Analysis

Processed RNA sequencing data was the primary input for five major analyses. Firstly, R-studio was implemented to build the Kaplan-Meier survival curve to demonstrate the ccRCC survival outcome according to the expression level of *AKAP8L*. The prodlim function was used to build the survival curve. The Chi-square test was used to examine the categorical differences between two levels of expression (low vs high) with survival outcomes of ccRCC patients, applied for two investigated genes ($\alpha = 0.05$). A one-way ANOVA was then used to detect differences between the quantitative expression levels of *AKAP8L* among four stages of ccRCC. The forest plots were used to determine their hazardous ratios for poor prognosis in cancer based on *AKAP8L* and other factors such as cancer stages, gender (Male, Female), metastasis (M0, M1, MX), and tumor pathology (1, 1a,1b, 2, 2a, 2b, 3, 3a, 3b, 3c, 4). Pearson correlation was used to detect any possible association between the transcriptomic counts and the progression of ccRCC (classified in four stages: I, II, III, IV). The 'ggforest ()' function was used to perform risk comparison tests amongst diverse indicators. Regarding SARS-CoV-2 datasets (GSE152586, GSE162113, and GSE161881), two-way ANOVA was applied to detect significant differences between *AKAP8L* expression levels in different cohorts, mainly control and SARS-CoV-2 infected.

All statistical analyses, except for survival curve and hazard ratio analyses, were conducted on the GraphPad Prism v.8.1.0 platform. The conducted analytical approach was designed based on the experimental design concept that a shared factor, A, can be used to examine both B and C, which has been applied successfully in previous studies with similar settings [40, 41]. In details, COVID-19–associated gene expression data (GSE152586, GSE162113, and GSE161881) was compared with the Kidney/ccRCC TCGA dataset (Firehose Legacy) transcriptomes to assess the potential molecular overlap between viral response pathways and ccRCC biology. Beyond these standard statistical and bioinformatic analyses, which were conducted in the R environment, no computational simulations or in silico experiments were performed. Datasets were compared according to APA-style: $0.01 < p \leq 0.05$ (*), $0.001 < p \leq 0.01$ (**), $p \leq 0.001$ (***), adjusted using Benjamini-Hochberg standards. The results were expressed as mean $\pm$ standard deviation (std). The analysis flow was briefly demonstrated in Fig. 1.

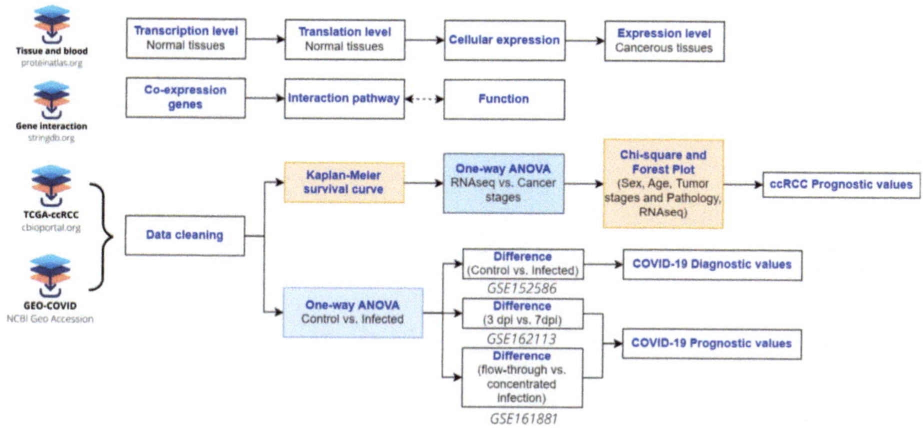

Fig. 1. Analysis workflow diagram for this study. Orange-colored boxes represent tests conducted on the R-studio platform. Blue-colored boxes represent tests conducted on the GraphPad Prism platform.

3 Results

3.1 Expression of AKAP8L in Tissues

As described in Fig. 2, *AKAP8L* showed low tissue specificity. The expression of *AKAP8L* at the transcriptional level in renal tissues is moderately low compared to other tissues, especially when considering bone marrow, lymphatic glands, and the brain (Fig. 2B–D). The GTEX-dataset showed RNA expression level of *AKAP8L* in normal renal tissues is 29.27 ± 9.1 nTPM for renal cortex ($n = 85$) and 22.95 ± 5.393 nTPM for renal medulla ($n = 4$) ($p = 0.174$, Fig. 3). On the other hand, the translational level of *AKAP8L* is high, which occurs in multiple other organs, not only the urinary system but also pancreatic, endocrine, muscle tissues, and others (Fig. 2B and G), indicating the highly regulated translation of *AKAP8L* takes place without tissue specificity. Taking a closer look, immunohistochemistry results indicated that this protein is located predominantly in the nucleus (Fig. 2E). Referring to the TCGA dataset, *AKAP8L* RNA expression in different tumors varies without significant differences (Fig. 2F). Its low specificity for tissues and cancer types was thus confirmed. The TCGA-mutation datasets showed that *AKAP8L* has a low rate of mutation (0.7%), implying that the significant elevation of these genes' transcripts in ccRCC patients was not caused by either direct genetic mutation or hereditary but possibly due to gene interactions.

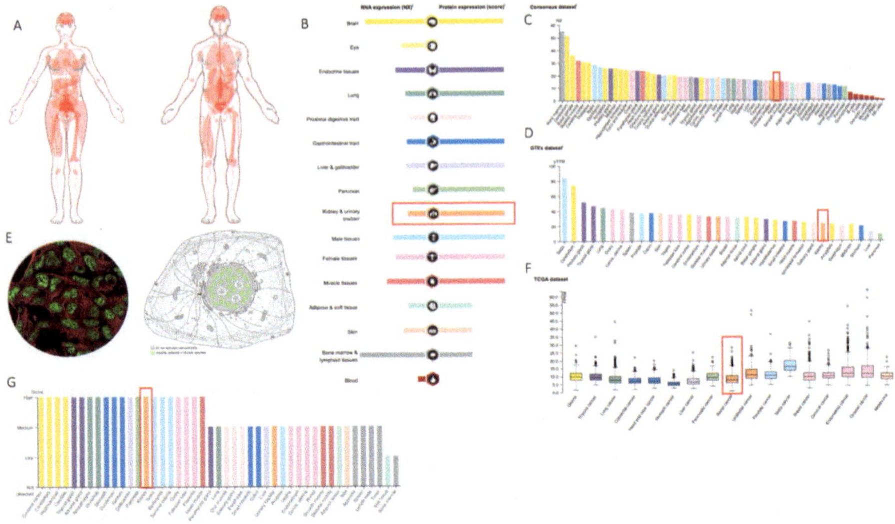

Fig. 2. Expression of AKAP8L in tissues. (**A**) distribution of AKAP8L protein in the human body. (**B**) comparison between transcription and translation levels in different tissues. Transcriptional level in different tissues according to Consensus dataset (**C**) and GTEX dataset (**D**). Distribution of AKAP8L protein at the cellular level (**E**) and transcriptional level (**F**) according to different cancers in TCGA-dataset. (**G**) translational level in different tissues according to score.

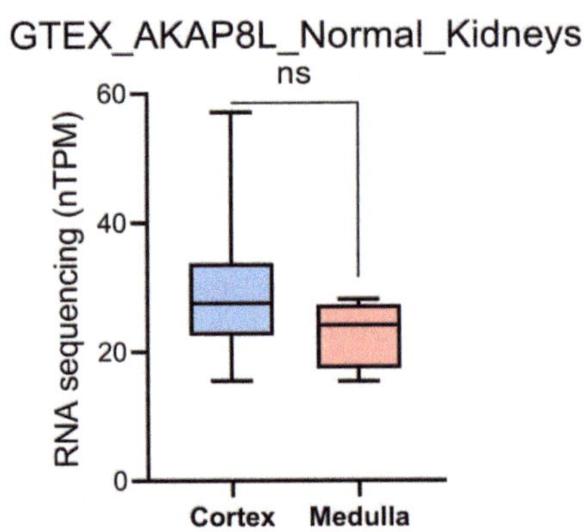

Fig. 3. RNA expression levels of *AKAP8L* are relatively uniform in normal renal tissues. RNA expression levels (nTPM) of *AKAP8L* in renal cortex (n = 85) and renal medulla (n = 4) according to GTEX-dataset *(Independent t-test; $p > 0.05$ (ns), $0.01 < p \leq 0.05$ (*), $0.001 < p \leq 0.01$ (**), $p \leq 0.001$ (***).*

3.2 Prognostic Value of *AKAP8L* for CcRCC

In order to determine the relationship between *AKAP8L* mRNA expression and the progression of ccRCC, the level of expression was evaluated in four cancer stages, along with tumor grades, metastasis status, tumor pathology and its effects on the patient's survival outcome, using data provided in Kidney/ccRCC/TCGA-dataset (Firehose Legacy) dataset. Overall, high expressions of *AKAP8L* were found to compromise poor survival outcomes of the patients ($p = 5.16 \times 10$–7, Fig. 4A). Regarding the correlation between RNA expression and cancer progression, *AKAP8L* showed a highly significant difference between late-stage patients (Stage III: $p = 0.043$; Stage IV: $p < 0.001$) and patients in stage-I (Fig. 4B). Notably, *AKAP8L* renal transcription in Stage-I patients was 713.6 ± 16.71 nTPM (Fig. 4F), approximately 30-times higher than in normal renal tissues (Fig. 3), and each progression stage added around 30 to 70 nTPM to RNA counts (Fig. 4F). There was also a significant ascending linear trend in *AKAP8L* RNA expression amongst four stages of ccRCC progression (slope = 51.49; $p < 0.0001$, alerting $R^2 = 0.9904$, Fig. 4G).

Regarding tumor pathology, there were significant differences between Grade 2 - 3 tumors and Grade 4 tumors, with Grade 4 exceeding in RNA counts ($p < 0.001$ and $p = 0.032$, respectively, Fig. 4C). Grade 4 tumors also exhibited a notable increase of 80.8 ± 90.5 nTPM compared to Grade 1 tumors. These results indicated that Grade 4 – undifferentiated tumors with the most abnormal cells possibly yielded a higher expression level of *AKAP8L*, implying a relationship with tumor cell morphology. The analyzed results also showed a significantly higher count of *AKAP8L* in metastasized tumors compared to non-metastasized tumors (Mean difference (MD) = 140.4 ± 40.30 nTPM), suggesting a relationship between the concerned gene and the metastasis status ($p < 0.001$, Fig. 4D). Considering tumor physiology, there were no significant differences amongst different classes of tumor physiology (including tumor invasiveness and size, Fig. 4E). The relationship between *AKAP8L* and the mentioned parameters will be further discussed in a later session based on interpreted results.

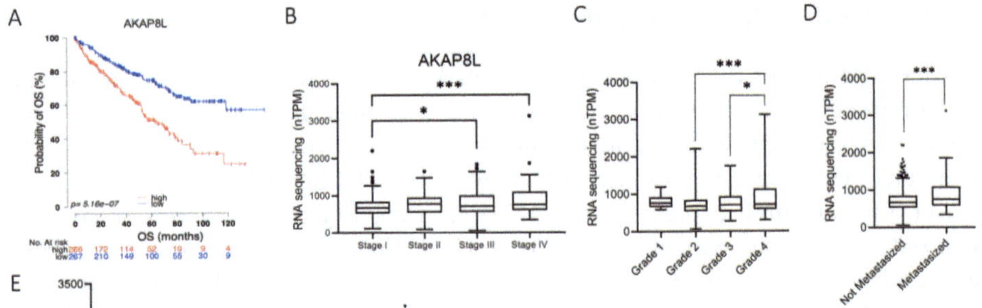

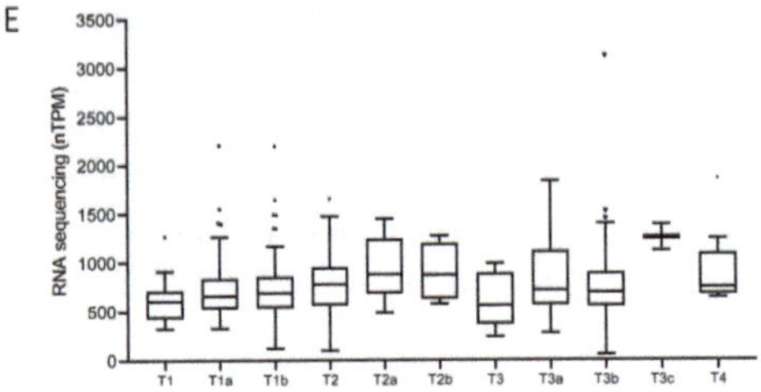

F

Stages	I	II	III	IV
AKAP8L	713.6 ± 16.71	780.4 ± 43.49	810.8 ± 33.28	872.5 ± 46.78

G

Variables	p-value	Slope	R^2 (alerting)	R^2 (effect size)
AKAP8L	< 0.0001	51.49	0.9904	0.03222

Fig. 4. The expression profile of *AKAP8L* in ccRCC associated with the disease pathology and progression. (**A**) Kaplan-Meier survival curve of *AKAP8L (observed survival (OS) rate (%) vs observed survival (OS) time (months); mean OS time: 42.18 months)*. (**B**) Turkey box-and-whisker plots showing alterations in RNA counts (nTPM) of *AKAP8L* according to four stages of ccRCC (Stage I = 267; II = 57; III = 126; IV = 86). (**C**) differences in RNA expression level (nTPM) of *AKAP8L* according to four tumor grades (Grade 1 = 14; Grade 2 = 229; Grade 3 = 207; Grade 4 = 76). (**D**) differences in RNA expression level (nTPM) according to Metastasis status (Metastasized = 79, Not metastasized = 422). (**E**) differences in RNA expression level (nTPM) according to tumor pathology. (**F**) Table showing the mean ± std values in four stages of ccRCC. (**G**) Table showing four parameters in linear trend-fitting between the expression level according to four stages of ccRCC. *(One-way ANOVA, $0.01 < p \leq 0.05$ (*), $0.001 < p \leq 0.01$ (**), $p \leq 0.001$ (***))*

Multivariate analysis was then conducted to evaluate the effects of *AKAP8L* on the risks of poor outcomes in ccRCC patients, compared to other factors, including sex, age, metastasis status, tumor stages, and tumor pathology. In Fig. 5, a high expression level of *AKAP8L* had significant fatal risks compared to a low expression level (HR high level = 1.66 h low level, p = 0.0031). The high expression level of *AKAP8L* transparently compromised higher mortality rate compared to the low level. On the other hand, this marker also showed a comparable level of significance compared to the effects of advanced tumor-stages III/IV on the outcome of the disease (p = 0.0031 to p < 0.001, respectively). In detail, stages III/IV showed 5–20 times increased mortality compared to Stages I/II. Age groups higher than 50-year-old also expressed a higher hazard rate compared to ≤ 50-year-old group (p = 0.0136, approximately twice the risk rate of the reference). Regarding metastasis status, there was no distinction among non-metastasized (M0), metastasized (M1, p = 0.4478) and unmeasurable (MX, p = 0.3287). This conclusion is similar to tumor pathology. Nevertheless, it can be deduced that smaller tumors carry higher risks compared to those of larger sizes, as T1 patients showed insignificantly higher risks compared to T2, T3, and T4 patients. In summary, except for tumors at advanced stages, high levels of *AKAP8L* can be a promising indicator for high fatality in ccRCC patients.

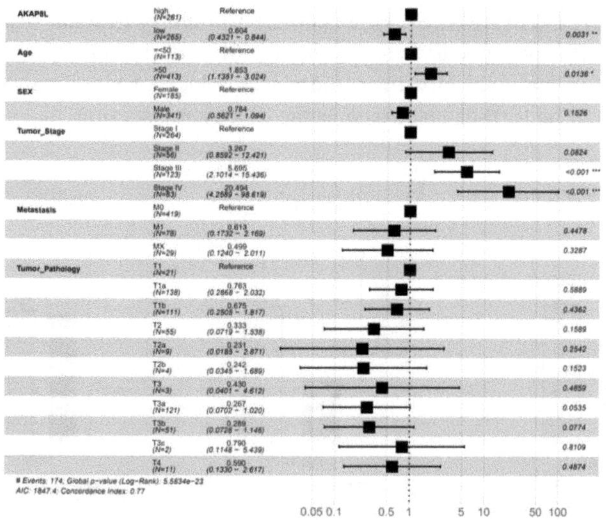

Fig. 5. Multivariate Cox regression analysis of overall survival in patients with ccRCC. Forest plot showing multivariate test results comparing hazard ratio of different factors in survival outcomes of ccRCC patients. Hazard ratios (HRs) were estimated using a multivariate Cox proportional hazards model adjusted for major clinical covariates. Black squares denote HR values, and horizontal error bars represent 95% confidence intervals *(written values indicated the significance of the association between studied factors and the disease pathology, horizontal axis indicated Log-rank of each factor when compared to others; 0.01 < p ≤ 0.05 (*), 0.001 < p ≤ 0.01 (**), p ≤ 0.001 (***)).*

3.3 Prognostic Value of *AKAP8L* for COVID-19 Infection

Prognostic values of *AKAP8L* for COVID-19 were investigated using three datasets: GSE152586, GSE162113, and GSE161881. In Fig. 6B, *AKAP8L* RNA counts showed a significant difference between two cohorts (MD = 29.45 ± 2.62 FPKM = 68.35 ± 4.78 TPM, p < 0.001). Specifically, RNA counts increased from 60.2 to 128.7 TPM in infected alveolar cells at 48 hpi. This bolstered the significance of COVID-diagnostic capability of *AKAP8L* in alveolar cells. Regarding the association between the expression of investigated genes and the viral load, *AKAP8L* showed an insignificant difference (p = 0.066) (Fig. 6A). However, it is notable that concentrated infection yielded higher RNA counts of 557.0 TPM compared to 402.7 TPM. Conclusively, the RNA count was directly proportional to the viral load (concentrated viral infection versus flow-through infection) (MD = 154.3 ± 57.93TPM). This implied the promising value of *AKAP8L* for specifically predicting the viral load. Next, analysis was conducted on GSE161881-dataset to investigate the changes in *AKAP8L* RNA expression throughout the period of infection (at two-time points: 3 dpi and 7 dpi). In Fig. 6C, *AKAP8L* evidently had capability of differentiating the infected group from the control group, as noted at 7 dpi (MD = 799.6 TPM, p < 0.001) compared to 3 dpi (MD = 36 TPM, p = 0.570). Additionally, the control group showed a significant decrease (MD = 611.6, p < 0.001) between 3 dpi and 7 dpi, while the infected group showed a minor increase by comparison (MD = 152 TPM; p = 0.995). This indicated that the persistence of COVID infection resulted in an increase in *AKAP8L* expression, particularly in renal tissues. This result was consistent with the primary results in Fig. 6B that infection with SARS-CoV-2 induced *AKAP8L* expresion. However, there was a lack of significance in the Vero-6 cohorts, which will be further discussed in the discussion section.

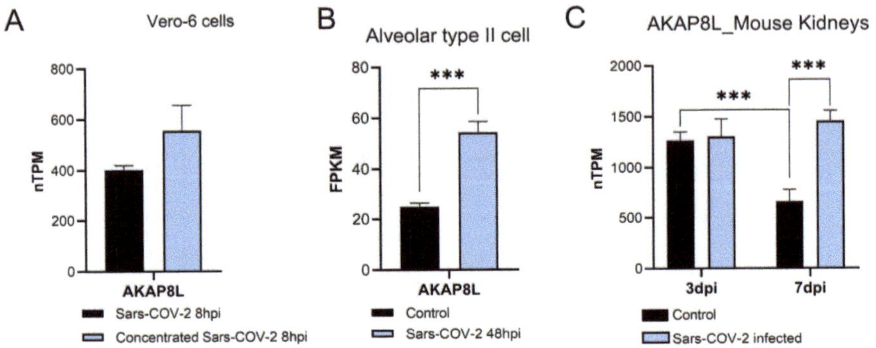

Fig. 6. Expression of *AKAP8L* under COVID-infection. (**A**) Boxplot comparing AKAP8L RNA counts during COVID infection (nTPM) in Vero E6 cells between concentrated and flow-through infection of SARS-CoV-2 (flow-through = 3; concentrated = 3). (**B**) Boxplot comparing AKAP8L RNA counts during COVID infection (nTPM) in Alveolar type II cells between control and SARS-CoV-2 48 hpi (control = 3; SARS-CoV-2 48 hpi = 3). (**C**) Boxplot comparing AKAP8L RNA counts during COVID infection (nTPM) in mouse kidneys between 3 dpi and 7 dpi (control 3 dpi = 3; SARS-CoV-2 3 dpi = 3; control 7 dpi = 4; SARS-CoV-2 7 dpi = 3) *(A-B: Independent t-test, C: two-way ANOVA;* $0.01 < p \le 0.05$ *(*),* $0.001 < p \le 0.01$ *(**),* $p \le 0.001$ *(***))*

3.4 Protein-Protein Interaction of *AKAP8L*

Based on the String database, AKAP8L showed co-expression with proteins involved in nucleic acid processing and immune responses. Particularly, in Fig. 7, *AKAP8L* co-expresses with *SFPQ* (Co-expression Score (CS) = 0.626), *DHX9* (CS = 0.931), and *PRPF40A* (CS = 0.874). According to their CS, *DHX9* was the closest companion of *AKAP8L*, and could be considered as a factor that bolsters its molecular alterations according to both ccRCC and COVID-19. Their relationship will be further discussed in the sequential section.

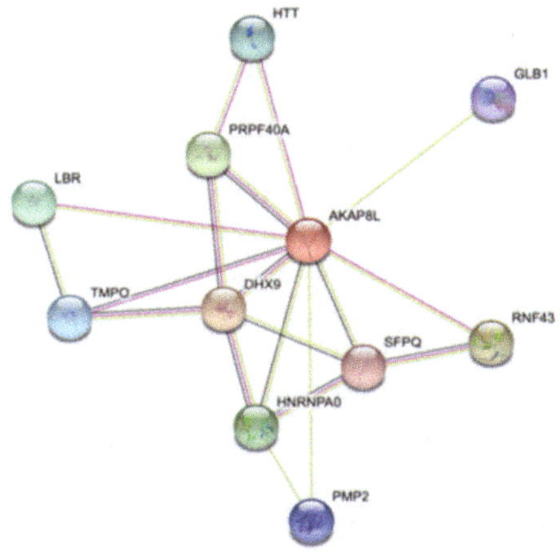

Fig. 7. Protein interaction of *AKAP8L*. Black lines represent co-expression while other colors represent co-mentioning or co-appearance in an experiment.

4 Discussion

Clear-cell renal cell carcinoma is the most common histological subtype of renal cancer in adults and accounts for 75–85% of all cases [42]. Although localized ccRCC can be treated by surgical resection or ablation of the tumour, advanced ccRCC remains a highly lethal disease [43]. With the recent outbreak of COVID-19, the clinical outcomes of ccRCC patients are being seriously affected, predominantly by the popular SARS-CoV-2, which causes the target's destruction of the renal tissues [9, 10]. Hence, a unique novel kidney biomarker urgently demands the prognosis of ccRCC patients during this COVID-19 pandemic.

AKAP8L is a scaffolding protein that binds to the regulatory subunit of cyclic adenosine monophosphate (cAMP)-dependent protein kinase (PKA) and mediates cell-cycle regulation and mitotic DNA condensation [16, 44]. A recent study demonstrated that *AKAP8L* interacts with mTORC1 to regulate cell growth and proliferation [27]. The altered expression of AKAP8L is associated with a poor prognosis for esophageal and colon cancer [31, 45]. Moreover, *AKAP8L* is expressed in kidney tissues [46], and SARS-CoV-2 binds to ACE2 receptors in the renal cortex [10]. However, the role of *AKAP8L* in ccRCC patients with COVID-19 and the underlying molecular mechanism remain unclear. Thus, the current study aims to explore the expression patterns and prognostic values of *AKAP8L* in ccRCC and COVID-19.

In this study, high expression of *AKAP8L* was associated with advanced tumour grades, metastasis, and poor survival outcomes in ccRCC. *AKAP8L* also showed distinctive elevation in stage III-IV compared to stage I-II and demonstrated significant linear trends. These results suggested that *AKAP8L* plays a critical role in cancer cell proliferation, cell survival, and invasion. The findings are consistent with previous studies on AKAPs in various cancers, including esophageal, colorectal, and lung cancer [31, 45, 47–51]. In multivariate analysis, as expected, we found that age greater than 50 and advanced tumour stage were significantly correlated with poor survival outcomes for ccRCC. Notably, higher transcriptional levels of *AKAP8L* were associated with a poor prognosis for ccRCC after adjusting the confounding variables. These results indicate that *AKAP8L* would be an independent prognostic predictor in ccRCC patients.

To explore the effect of *AKAP8L* expression pattern on COVID-19 diagnosis and prognosis, the present study found that *AKAP8L* RNA counts showed a significant difference between the two cohorts, and the RNA counts increased from 60.2 to 128.7 TPM in infected alveolar cells at 48 hpi. This result implies that *AKAP8L* could be a COVID-19 diagnostic marker in alveolar cells. Recently, Denise et al. identified some laboratory biomarkers as COVID-19 diagnostic and prognostic markers [52], and the present study is in line with that. Interestingly, even after 7 dpi, *AKAP8L* showed a significant change in transcriptional level in response to COVID infection, increasing by 152 TPM compared to 3 dpi. This showed the long-lasting response of *AKAP8L* according to COVID-19 infection, which might benefit the long-term prognosis of this disease. However, to investigate the effect of viral load on *AKAP8L* expression, the study showed an insignificant association between the expression of the investigated gene and the viral load. This result is consistent with the recent studies on viral load and COVID-19 prognosis [42].

The shift in *AKAP8L* transcription level according to two diseases can be explained based on its genetic interaction: *AKAP8L* and *SFPQ* are involved in nuclear activities, namely RNA-processing (hnRNPM-splicing) and DNA repair [53–55]. This can happen in a variety of cancer progressions, including renal carcinoma [56, 57]. However, the comparatively low CS indicated that their association remains elusive. *DHX9* is a nuclear protein that plays a pivotal role in maintaining genomic stability by unwinding R-loops in impaired nucleic acids [56]. It has been claimed as an indicator and a therapeutic target for cancer. It was proven to interact with multiple cancer indicators such as *Lin28*, *SOX4*, and *p65* in their transcription and translation pathways [59, 60]. *AKAP8L* was found to interact with *DHX9* during its function and interaction with *mTORC1*, with unknown mechanisms [27]. *PRPF40A* is present in nucleic acid processing, namely pre-mRNA processing and taking part in spliceosomes, and seems related to tumour hypoxia, yet the underlying mechanisms remain unknown [61]. Regardless, *PRPF40A* is still among the primary targets for therapeutic and diagnostic cancer. In summary, although the link between *AKAP8L* and renal carcinoma is obscure, its companions have supported the association as they all participate in pathways related to cancer progression, and *AKAP8L* interacts with *mTORC1*, which involves tumour growth.

Regarding the association between *AKAP8L* and COVID-19, *DHX9*—its highly related companion, plays a direct role in the antiviral response of the immune system. *DHX9* recognizes viral nucleic acids and initiates the immune response cascade, as well as enriching cytokine production in macrophages through RNA polymerase II recruitment [33]. Their interaction might explain the elevation in *AKAP8L* under the effects of COVID-19 infection.

In summary, *AKAP8L* competitively holds value in both ccRCC and COVID-19 prognosis. According to an analysis of data from diverse aspects, it has been noted that upregulation of *AKAP8L* at transcriptional level is a response to both ccRCC and COVID-19 prognoses. However, it responds differently to these two diseases: while the occurrence of ccRCC in patients' systems could stimulate *AKAP8L* to transcript 30-times higher than its original level, the infection of COVID only increased the level from 36 TPM to 799.6 TPM, still less than thrice the original value. To explain the response of *AKAP8L* to these diseases, this study has gone through its interaction with neighbouring genes as, according to TCGA-dataset, this gene has a low mutation rate in ccRCC patients (0.7%), implying that it might not be the direct cause. *AKAP8L* was proven to interact with mTORC1—a well-known protein involved in tumour growth. Its other interactors, *SFPQ*, *DHX9*, and *PRPF40A*, are involved in nucleic acid processing related to tumour generation and progression [53–55, 58, 61]. The interaction with carcinoma-related factors suggested that *AKAP8L* might involve itself in this pathway, explaining the highly correlated response to the progression of ccRCC. Notably, its significant elevation in metastasized samples suggests participation in the metastasis pathway. These hypotheses will need further experiments and investigation to be confirmed. Surprisingly, *AKAP8L*'s strong interactor—*DHX9* (CS = 0.931), also recognizes viral nucleic acids and initiates the immune response cascade, which might be the explanation behind *AKAP8L*'s significant response to COVID-19 infection [33]. However, similar to its association with ccRCC, *AKAP8L*'s response to COVID infection will need further investigation to clarify. Overall, it can be concluded that *AKAP8L* is a promising biomarker for ccRCC prognostic, especially during the COVID-19 pandemic.

5 Conclusion

Overall, this study provided evidence that *AKAP8L* is a promising prognostic biomarker for clear cell renal cell carcinoma (ccRCC) and COVID-19. The study demonstrated an association of *AKAP8L* with tumor-related proteins, namely *mTORC1*, and was noted to have low tissue specificity; this suggested prognosis applications of the gene in other cancer types.

Acknowledgments. We would like to thank International University and School of Medicine, Vietnam National University HCMC for giving us collaborating opportunities. We appreciate Dr. Huong Ha and Dr. Khon Huynh for their contribution and our friend, Mr. Quang Nguyen, for his support throughout the process.

Author Contributions. MNN conceptualized the study. APHL, MNN designed the study and planed the experiments. MNN administrated the project. MNN supervised the project. APHL collected and cleaned the data. APHL and DPN performed the analysis. APHL prepared all figures. APHL, DPN wrote original draft. MNN revised manuscript. All authors have read the journal's authorship agreement and the manuscript has been reviewed by and approved by all named authors. All authors have read and approved the final manuscript.

Data Availability Statement. Only publicly available data were used in this study, and data sources and handling of these data are described in the Materials. Further information is available from the corresponding author upon request.

Conflict of Interest Statement. All authors have read the journal's policy on disclosure of potential conflicts of interest and declare no potential conflicts of interest with respect to the research, authorship, and/or publication of this article.

References

1. Benjamin B. Health and Vital Statistics: Health and Vital Statistics (2022). https://doi.org/10.4324/9781003281900/HEALTH-VITAL-STATISTICS
2. Ferlay, J., Soerjomataram, I., Dikshit, R., Eser, S., Mathers, C., Rebelo, M., et al.: Cancer incidence and mortality worldwide: sources, methods and major patterns in GLOBOCAN 2012. Int. J. Cancer **136**, E359–E386 (2015). https://doi.org/10.1002/IJC.29210
3. Ferlay, J., Steliarova-Foucher, E., Lortet-Tieulent, J., Rosso, S., Coebergh, J.W.W., Comber, H., et al.: Cancer incidence and mortality patterns in Europe: estimates for 40 countries in 2012. Eur. J. Cancer **49**, 1374–1403 (2013). https://doi.org/10.1016/J.EJCA.2012.12.027
4. Hsieh, J.J., Purdue, M.P., Signoretti, S., Swanton, C., Albiges, L., Schmidinger, M., et al.: Renal cell carcinoma. Nat. Rev. Dis. Primers. **3** (2017) https://doi.org/10.1038/NRDP.2017.9
5. Chow, W.H., Dong, L.M., Devesa, S.S.: Epidemiology and risk factors for kidney cancer. Nat. Rev. Urol. **7**, 245–257 (2010). https://doi.org/10.1038/NRUROL.2010.46
6. Hébert, J.R., Daguise, V.G., Hurley, D.M., Wilkerson, R.C., Mosley, C.M., Adams, S.A., et al.: Mapping cancer mortality-to-incidence ratios to illustrate racial and sex disparities in a high-risk population. Cancer **115**, 2539 (2009). https://doi.org/10.1002/CNCR.24270
7. Lipworth, L., Tarone, R.E., Lund, L., McLaughlin, J.K.: Epidemiologic characteristics and risk factors for renal cell cancer. Clin. Epidemiol. **1**, 33 (2009). https://doi.org/10.2147/CLEP.S4759
8. Aeppli, S., Eboulet, E.I., Eisen, T., Escudier, B., Fischer, S., Larkin, J., et al.: Impact of COVID-19 pandemic on treatment patterns in metastatic clear cell renal cell carcinoma. ESMO Open 2020;5. https://doi.org/10.1136/ESMOOPEN-2020-000852
9. Mihalopoulos, M., Dogra, N., Mohamed, N., Badani, K., Kyprianou, N.: COVID-19 and kidney disease: molecular determinants and clinical implications in renal cancer. Eur. Urol. Focus **6**, 1086 (2020). https://doi.org/10.1016/J.EUF.2020.06.002
10. Fan, C., Lu, W., Li, K., Ding, Y., Wang, J.: ACE2 Expression in kidney and testis may cause kidney and testis infection in COVID-19 Patients. Front Med (Lausanne) **7**, 563893 (2021). https://doi.org/10.3389/FMED.2020.563893/FULL
11. Al-Quteimat, O.M., Amer, A.M.: The Impact of the COVID-19 Pandemic on Cancer Patients. Am. J. Clin. Oncol. **43** (2020). https://doi.org/10.1097/COC.0000000000000712
12. Vargas, A.J., Harris, C.C.: Biomarker development in the precision medicine era: lung cancer as a case study. Nat. Rev. Cancer **16**, 525–537 (2016). https://doi.org/10.1038/NRC.2016.56
13. McGuire, B.B., Fitzpatrick, J.M.: Biomarkers in renal cell carcinoma. Curr. Opin. Urol. **19**, 441–446 (2009). https://doi.org/10.1097/MOU.0B013E32832F0C68
14. Ljungberg, B., Albiges, L., Abu-Ghanem, Y., Bedke, J., Capitanio, U., Dabestani, S., et al.: European association of urology guidelines on renal cell carcinoma: the 2022 update. Eur. Urol. **82**, 399–410 (2022). https://doi.org/10.1016/J.EURURO.2022.03.006
15. Wong, W., Scott, J.D.: AKAP signalling complexes: focal points in space and time. Nat. Rev. Mol. Cell Biol. **5**:12 5:959–70 (2004). https://doi.org/10.1038/nrm1527
16. Taskén, K., Aandahl, E.M.: Localized effects of cAMP mediated by distinct routes of protein kinase A. Physiol. Rev. **84**, 137–167 (2004). https://doi.org/10.1152/PHYSREV.00021.2003
17. Saini, S., Jagadish, N., Gupta, A., Bhatnagar, A., Suri, A.: A novel cancer testis antigen, A-Kinase anchor protein 4 (AKAP4) is a potential biomarker for breast cancer. PLoS One **8** (2013). https://doi.org/10.1371/JOURNAL.PONE.0057095
18. Li, S., Qin, X., Li, Y., Guo, A., Ma, L., Jiao, F., et al.: AKAP4 mediated tumor malignancy in esophageal cancer. Am J Transl Res **8**, 597 (2016)
19. Jagadish, N., Parashar, D., Gupta, N., Agarwal, S., Purohit, S., Kumar, V., et al.: A-kinase anchor protein 4 (AKAP4) a promising therapeutic target of colorectal cancer. J. Exp. Clin. Cancer. Res. **34** (2015). https://doi.org/10.1186/S13046-015-0258-Y

20. Lee, J.H., Lee, E.S., Kim, Y.S., Won, N.H., Chae, Y.S.: BRAF mutation and AKAP9 expression in sporadic papillary thyroid carcinomas. Pathology **38**, 201–204 (2006). https://doi.org/10.1080/00313020600696264
21. Kabbarah, O., Nogueira, C., Feng, B., Nazarian, R.M., Bosenberg, M., Wu, M., et al.: Integrative genome comparison of primary and metastatic melanomas. PLoS One **5** (2010). https://doi.org/10.1371/JOURNAL.PONE.0010770
22. Truong, T., Sauter, W., McKay, J.D., Hosgood, H.D., Gallagher, C., Amos, C.I., et al.: International lung cancer consortium: coordinated association study of 10 potential lung cancer susceptibility variants. Carcinogenesis **31**, 625–633 (2010). https://doi.org/10.1093/CARCIN/BGQ001
23. Frank, B., Wiestler, M., Kropp, S., Hemminki, K., Spurdle, A.B., Sutter, C., et al.: Association of a common AKAP9 variant with breast cancer risk: a collaborative analysis. J. Natl. Cancer Inst. **100**, 437–442 (2008). https://doi.org/10.1093/JNCI/DJN037
24. Hu, Z.Y., Liu, Y.P., Xie, L.Y., Wang, X.Y., Yang, F., Chen, S.Y., et al.: AKAP-9 promotes colorectal cancer development by regulating Cdc42 interacting protein 4. Biochim. Biophys. Acta **1862**, 1172 (2016). https://doi.org/10.1016/J.BBADIS.2016.03.012
25. Li, Y., Kao, G.D., Garcia, B.A., Shabanowitz, J., Hunt, D.F., Qin, J., et al.: A novel histone deacetylase pathway regulates mitosis by modulating Aurora B kinase activity. Genes Dev. **20**, 2566 (2006). https://doi.org/10.1101/GAD.1455006
26. Qi, F., Yuan, Y., Zhi, X., Huang, Q., Chen, Y., Zhuang, W., et al.: Synergistic effects of AKAP95, Cyclin D1, Cyclin E1, and Cx43 in the development of rectal cancer. Int. J. Clin. Exp. Pathol. **8**, 1666 (2015)
27. Melick, C.H., Meng, D., Jewell, J.L.: A-kinase anchoring protein 8L interacts with mTORC1 and promotes cell growth. J. Biol. Chem. **295**, 8096–8105 (2020). https://doi.org/10.1074/JBC.AC120.012595
28. Sharma, A., Hoeffer, C.A., Takayasu, Y., Miyawaki, T., McBride, S.M., Klann, E., et al.: Dysregulation of mTOR signaling in fragile X syndrome. J. Neurosci. **30**, 694–702 (2010). https://doi.org/10.1523/JNEUROSCI.3696-09.2010
29. Zou, Z., Tao, T., Li, H., Zhu, X.: mTOR signaling pathway and mTOR inhibitors in cancer: progress and challenges. Cell Biosci. **10** (2020). https://doi.org/10.1186/S13578-020-00396-1
30. Bieluszewska, A., Weglewska, M., Bieluszewski, T., Lesniewicz, K., Poreba, E.: PKA-binding domain of AKAP8 is essential for direct interaction with DPY30 protein. FEBS J. **285**, 947–964 (2018). https://doi.org/10.1111/FEBS.14378
31. Luo, Q. yun, Di, T., Qiu, M.Z., Xia, Z. fei., Du, Y., Lin, R. duan, et al.: High AKAP8L expression predicts poor prognosis in esophageal squamous cell carcinoma. Cancer Cell Int. **22**:1–10 (2022). https://doi.org/10.1186/S12935-022-02492-3/FIGURES/5
32. Li, S., Ma, F., Yokota, T., Garcia, G., Palermo, A., Wang, Y., et al.: Metabolic reprogramming and epigenetic changes of vital organs in SARS-CoV-2–induced systemic toxicity. JCI Insight **6** (2021). https://doi.org/10.1172/JCI.INSIGHT.145027
33. Jiao, A., Sun, C., Wang, X., Lei, L., Liu, H., Li, W., et al.: DExD/H-box helicase 9 intrinsically controls CD8+ T cell–mediated antiviral response through noncanonical mechanisms. Sci. Adv. **8** (2022). https://doi.org/10.1126/SCIADV.ABK2691/SUPPL_FILE/SCIADV.ABK2691_TABLES_S1_TO_S5.ZIP
34. Szklarczyk, D., Gable, A.L., Nastou, K.C., Lyon, D., Kirsch, R., Pyysalo, S., et al.: The STRING database in 2021: customizable protein-protein networks, and functional characterization of user-uploaded gene/measurement sets. Nucleic Acids Res. **49**, D605–D612 (2021). https://doi.org/10.1093/NAR/GKAA1074
35. von Mering, C., Jensen, L.J., Snel, B., Hooper, S.D., Krupp, M., Foglierini, M., et al.: STRING: known and predicted protein-protein associations, integrated and transferred across organisms. Nucleic Acids Res. **33** (2005). https://doi.org/10.1093/NAR/GKI005

36. Giordano, T.J.: The cancer genome atlas research network: a sight to behold. Endocrine Pathology. **4** 25:362–5 (2014). https://doi.org/10.1007/S12022-014-9345-4
37. Katsura, H., Sontake, V., Tata, A., Kobayashi, Y., Edwards, C.E., Heaton, B.E., et al.: Human lung stem cell-based alveolospheres provide insights into SARS-CoV-2-Mediated interferon responses and pneumocyte dysfunction. Cell Stem Cell **27**, 890-904.e8 (2020). https://doi.org/10.1016/J.STEM.2020.10.005
38. Zhang, Y., Guo, R., Kim, S.H., Shah, H., Zhang, S., Liang, J.H., et al.: SARS-CoV-2 hijacks folate and one-carbon metabolism for viral replication. Nat Commun **12** (2021). https://doi.org/10.1038/S41467-021-21903-Z
39. Zhao, Y., Li, M.C, Konaté, M.M., Chen, L., Das, B., Karlovich, C., et al.: TPM, FPKM, or normalized counts? a comparative study of quantification measures for the analysis of RNA-seq Data from the NCI Patient-Derived Models Repository. J. Transl. Med. **19** (2021). https://doi.org/10.1186/S12967-021-02936-W
40. Rothman, K.J., Greenland, S., Lash, T.L.: Modern epidemiology (3rd ed.). Lippincott Williams & Wilkins (2008)
41. Nie, J., Yang, H., Liu, X., Deng, W., Fu, B.: Identification and validation of shared gene signature of kidney renal clear cell carcinoma and COVID-19. PeerJ **12**, e16927 (2024)
42. Moch, H., Cubilla, A.L., Humphrey, P.A., Reuter, V.E., Ulbright, T.M.: The 2016 WHO classification of tumours of the urinary system and male genital organs-part a: renal, penile, and testicular tumours. Eur. Urol. **70**, 93–105 (2016). https://doi.org/10.1016/J.EURURO.2016.02.029
43. Alchahin, A.M., Mei, S., Tsea, I., Hirz, T., Kfoury, Y., Dahl, D., et al.: A transcriptional metastatic signature predicts survival in clear cell renal cell carcinoma. Nat. Commun. **13** (2022). https://doi.org/10.1038/S41467-022-33375-W
44. Wong, W., Scott, J.D.: AKAP signalling complexes: focal points in space and time. Nat. Rev. Mol. Cell Biol. **5**, 959–970 (2004). https://doi.org/10.1038/NRM1527
45. Luo, Q.Y., Di, T., Chen, Z.G., Peng, J.H., Sun, J., Xia, Z.F., et al.: Novel prognostic model predicts overall survival in colon cancer based on RNA splicing regulation gene expression. Cancer Sci. **113**, 3330–3346 (2022). https://doi.org/10.1111/CAS.15480
46. Fagerberg, L., Hallstrom, B.M., Oksvold, P., Kampf, C., Djureinovic, D., Odeberg, J., et al.: Analysis of the human tissue-specific expression by genome-wide integration of transcriptomics and antibody-based proteomics. Mol. Cell. Proteomics **13**, 397–406 (2014). https://doi.org/10.1074/MCP.M113.035600
47. Diviani, D., Raimondi, F., del Vescovo, C.D., Dreyer, E., Reggi, E., Osman, H., et al.: Small-molecule protein-protein interaction inhibitor of oncogenic Rho signaling. Cell Chem. Biol. **23**, 1135–1146 (2016). https://doi.org/10.1016/J.CHEMBIOL.2016.07.015
48. Hu, S. xian, Kong, X. yu, Yuan, Y. yang, Teng, B. gang, Zhi, X. hong, Zhuang, W. xin, et al.: Relationship between AKAP95, cyclin E1, cyclin D1, and clinicopathological parameters in lung cancer tissue. Zhonghua Lao Dong Wei Sheng Zhi Ye Bing Za Zhi **31** 890–4 (2013)
49. Huang, P., Sun, Q., Zhuang, W., Peng, K., Wang, D., Yao, Y., et al.: Epac1, PDE4, and PKC protein expression and their association with AKAP95, Cx43, and cyclinD2/E1 in breast cancer tissues. Thorac Cancer **8**, 495–500 (2017). https://doi.org/10.1111/1759-7714.12475
50. Wang, S., Wang, K., Deng, Z., Jiang, Z., Wang, D., Yao, Y., et al.: Correlation between the protein expression levels of A-kinase anchor protein95, p-retinoblastoma (Ser780), cyclin D2/3, and cyclin E2 in esophageal cancer tissues. Asia Pac. J. Clin. Oncol. **15**, e162–e166 (2019). https://doi.org/10.1111/ajco.13146
51. Reggi, E., Diviani, D.: The role of A-kinase anchoring proteins in cancer development. Cell. Signal. **40**, 143–155 (2017). https://doi.org/10.1016/J.CELLSIG.2017.09.011
52. Battaglini, D., Lopes-Pacheco, M., Castro-Faria-Neto, H.C., Pelosi, P., Rocco, P.R.M.: Laboratory biomarkers for diagnosis and prognosis in COVID-19. Front Immunol **13** (2022). https://doi.org/10.3389/FIMMU.2022.857573

53. Bottini, S., Hamouda-Tekaya, N., Mategot, R., Zaragosi, L.E., Audebert, S., Pisano, S., et al.: Post-transcriptional gene silencing mediated by microRNAs is controlled by nucleoplasmic Sfpq. Nat. Commun. **8** 1 8:1–16 (2017). https://doi.org/10.1038/s41467-017-01126-x
54. Hu, X., Harvey, S.E., Zheng, R., Lyu, J., Grzeskowiak, C.L., Powell. E., et al.: The RNA-binding protein AKAP8 suppresses tumor metastasis by antagonizing EMT-associated alternative splicing. Nat Commun **11** (2020). https://doi.org/10.1038/S41467-020-14304-1
55. Jaafar, L., Li, Z., Li, S., Dynan, W.S.: SFPQ•NONO and XLF function separately and together to promote DNA double-strand break repair via canonical nonhomologous end joining. Nucleic Acids Res. **45**, 1848 (2017). https://doi.org/10.1093/NAR/GKW1209
56. Gao, Z., Chen, M., Tian, X., Chen, L., Chen, L., Zheng, X., et al.: A novel human lncRNA SANT1 cis-regulates the expression of SLC47A2 by altering SFPQ/E2F1/HDAC1 binding to the promoter region in renal cell carcinoma. RNA Biol. **16**, 940 (2019). https://doi.org/10.1080/15476286.2019.1602436
57. Yang, L., Yang, J., Jacobson, B., Gilbertsen, A., Smith, K., Higgins, L.A., et al.: SFPQ Promotes Lung Cancer Malignancy via Regulation of CD44 v6 Expression. Front Oncol **12** (2022). https://doi.org/10.3389/FONC.2022.862250/FULL
58. Gulliver, C., Hoffmann, R., Baillie, G.S.: The enigmatic helicase DHX9 and its association with the hallmarks of cancer. Future Sci OA **7** (2021). https://doi.org/10.2144/FSOA-2020-0140
59. Chellini, L., Pieraccioli, M., Sette, C., Paronetto, M.P.: The DNA/RNA helicase DHX9 contributes to the transcriptional program of the androgen receptor in prostate cancer. J. Exp. Clin. Cancer Res. **41**, 1–15 (2022). https://doi.org/10.1186/S13046-022-02384-4/FIGURES/6
60. Lee, T., Pelletier, J.: The biology of DHX9 and its potential as a therapeutic target. Oncotarget **7** 42716 (2016). https://doi.org/10.18632/ONCOTARGET.8446
61. Huo, Z., Zhai, S., Weng, Y., Qian, H., Tang, X., Shi, Y., et al.: PRPF40A as a potential diagnostic and prognostic marker is upregulated in pancreatic cancer tissues and cell lines: an integrated bioinformatics data analysis. Onco. Targets. Ther. **12**, 5037–5051 (2019). https://doi.org/10.2147/OTT.S206039

Current and Emerging Environmental Health Issues: Insights from the Junior Board Advisory Committee of the Pacific Basin Consortium for Environment and Health

Dwan Vilcins[1(✉)], Siti Nurshahida Nazli[1,2], Rahel M. Ketema[3], Injeong Kim[4], Khadija Al Nabhani[5], Avinash Kumar[6,7], and Ángel Mérida-Ortega[8]

[1] The University of Queensland, Child Health Research Centre, Children's Health and Environment Program, South Brisbane, QLD, Australia
d.vilcins@uq.edu.au

[2] Department of Environmental Health, Faculty of Health Sciences, Universiti Teknologi MARA Cawangan Pulau Pinang, Pulau Pinang, Malaysia

[3] Center for Environmental and Health Sciences, Hokkaido University, Sapporo, Japan

[4] Korea Institute of Industrial Technology, Research Institute of Sustainable Development Technology, Low-Carbon Transition R&D Department, Seobuk-Gu, Cheonan-Si, Chungcheongnam-Do, Republic of Korea

[5] Department of Chemical Engineering, Texas A&M University at Qatar, Education City, Doha, Qatar

[6] Department of Biological Sciences, Louisiana State University, Baton Rouge, LA, USA

[7] Pennington Biomedical Research Center, Baton Rouge, LA, USA

[8] Dirección de Salud Ambiental, Centro de Investigación en Salud Poblacional, Instituto Nacional de Salud Pública, Cuernavaca, Morelos, México

Abstract. The Pacific Basin, home to half of the world's population, face diverse and unevenly distributed environmental health challenges. The aim of this manuscript is to present insights into current and emerging environmental health issues from the regions or countries represented by the Junior Board members of the Pacific Basin Consortium for Environment and Health, a scientific platform for collaborative research, training, and the development of solutions to address environmental health concerns. We highlight some key environmental health issues in the Western Pacific Region, Australia, United States, Korea, Malaysia, Qatar and Mexico. Our research shows that these diverse regions share several common environmental health challenges, including the impacts of climate change, air pollution (both indoor and ambient), widespread exposure to emerging chemicals, and legacy pollutants. However, the magnitude of these environmental treads may vary among the evaluated regions due to their differences in geography, lifestyle, population density, and socio-economic development. We argue that, while the Pacific Basin might present a unique environmental contamination profile, collective action is essential to improving environmental health not only within the region, but also at the global level.

Keywords: Pacific Basin · environmental health · Western Pacific Region · Australia · United States · Korea · Malaysia · Mexico

L. Vu et al. (Eds.): PBC 2025, 2026.
https://doi.org/10.1007/978-3-032-29579-8_18

Abbreviations

PBC The Pacific Basin Consortium for Environment and Health
PFAS Polyfluoroalkyl substances
PM Particular matter
POPs Persistent Organic Pollutants

1 Introduction

Across the 20th century, huge gains were made in the field of environmental health, with dramatic decreases in water- and foodborne diseases globally. Additionally, the importance of environmental hazards in occupational settings and the effect of persistent chemical pollutants were recognised as key issues in environmental health (Groopman 2019). These discoveries lead to international treaties, such as the Stockholm Convention on Persistent Organic Pollutants (POPs) (Scheringer et al. 2022), to reduce human exposure, as well as improvements in workplace health and safety. Despite these gains, environmental health has many legacy issues to address, and new concerns are emerging. Environmental health has changed over recent decades to address the global rise of chronic diseases (Groopman 2019). Many chronic health conditions have environmental exposures as clear aetiological concern. A mapping exercise was held in 2009 by the International Life Sciences Institute (ILSI) and Environmental Sciences Institute which identified a range of environmental health exposures of concern, which included the contamination of food by chemicals such as plasticisers, additives in food, issues around regulation of chemicals, nanomaterials and nanotechnology, and the risk of co-exposures and mixture exposures (Bonnefoi et al. 2010).

Countries directly bordering the Pacific Basin, home to half of the world's population, face diverse and unevenly distributed environmental health issues, due to vast differences in geography, lifestyle, population density and socio-economic development, among others (Jagals et al. 2022). In response to the growing environmental health challenges, the Pacific Basin Consortium for Environment and Health (PBC) was established in 1986 to provide a forum for individuals and organizations with diverse professional expertise and geographic backgrounds to come together and discuss the most pressing environment and health issues of our time, engage in cooperative research, and develop and disseminate innovative strategies for addressing these issues and creating sustainable, affordable alternatives. To facilitate a network among early career researchers the PBC has a Junior Board Advisory Committee, which supports the PBC through reflecting the issues of their regions. The aim of this manuscript was to present the insights from some PBC Junior Board members on the current and emerging environmental health issues from their home region or countries located within the Pacific Basin.

1.1 Western Pacific Region

Environmental challenges in the Western Pacific region are rising due to increasing industrialization, urbanization, and poor chemical management, leading to widespread exposure to harmful substances (Suk et al. 2003; United Nations Environment Programme

(UNEP) and National Institute for Environmental Studies (NIES) 2012). Populations are increasingly exposed to pesticides, heavy metals such as lead, and plastic-related pollutants such as microplastics, phthalates, and bisphenols, particularly in areas near industrial and mining activities (Sly et al. 2024). Informal e-waste processing or mismanagement of waste, for example, in Vietnam, the Philippines, and Indonesia, is also a significant source of hazardous chemical exposure (Poudel et al. 2024; Sly et al. 2024). Such exposures often occur as complex chemical mixtures and potentially increasing the risk of adverse health effects.

The most vulnerable groups, children, pregnant women, the elderly, low-income communities, and small island populations, are disproportionately affected by chemical exposures and their associated health consequences. In small island nations, vulnerability is further intensified by geographic isolation, limited infrastructure, and dependence on imported goods, which can increase exposure to hazardous chemicals (Kishi et al. 2025; World Health Organization Regional Office for the Western Pacific 2017). For example, in Fiji, microplastics, known carriers of POPs, have been found on the beaches (Al Nabhani et al. 2022), in fish (Drova et al. 2025) and in the atmosphere (Al Nabhani et al., in press). This further indicates widespread contamination of aquatic and marine environments by microplastics, highlighting their potential risks to the ecosystem. These are of particular concern for communities with strong reliance on marine and coastal resources for food security and livelihoods.

Additionally, many countries in the region face climate-related hazards such as air pollution, including both ambient and indoor air pollution, which compromise respiratory health. These environmental threats, combined with extreme weather events such as floods, rising sea levels, and heat waves, amplify environmental exposures and intensify public health risks (World Health Organization 2024).

1.2 Australia

While Australia sits within the Western Pacific region it faces a different set of challenges to many of its regional neighbours. Climate change presents as one of the greatest challenges to environmental health across Australia (Xu et al. 2023). This country is prone to natural disasters, and major disasters likes floods, drought, bushfires and severe storms are predicted to increase. A 2024 Senate Select Committee found these events bring both risk to human health from the event itself, but also broader/longer term impacts, such as mould growth from rainfall events, bushfire smoke affecting distant communities and mental health issues (Select Committee on the Impact of Climate Risk on Insurance Premiums and Availability 2024). Work is underway to understand the impact of bushfire smoke, with the goal of understanding the biological mechanism between smoke exposure and health effects (Vilcins et al. 2024a, 2024b). Protecting human health from climate change will require an understanding of potential health intervention points. Intertwined with climate change is accelerating biodiversity loss. Over the last 200 years since colonisation, Australia has experienced the largest decline in biodiversity of any continent, with losses predicted to accelerate over the coming decade (Barraclough et al. 2023). With a loss of biodiversity comes a loss in critical ecosystems services, including those that support good health.

Australia continues to deal with the consequences of legacy pollutants, such as lead and asbestos (Brown et al. 2023; National Health and Medical Research Council 2015), as well as significant challenges with per- and polyfluoroalkyl substances (PFAS). While several large scale PFAS events, typically related to defence force activities (Hagen et al. 2024), have been researched in Australia, biological monitoring shows that serum PFAS levels are consistent across the country (Nilsson et al. 2025), meaning this exposure is ubiquitous and not related to single case events. Lastly, indoor air quality and indoor exposures continue to be a significant source of pollutants for the Australian population. Our recent work in this area found that 100% of homes contained oxidant by-products of air pollution in household dust, and that indoor air quality is typically worse than outdoor levels (Vilcins et al. 2025). As the Australian population spends more time indoors, including during extreme weather events, the importance of indoor environmental quality becomes increasingly critical for public health and well-being.

1.3 United States

The United States faces a wide array of environmental health challenges, many of which are exacerbated by industrial activities, urbanization, and climate change. Air pollution remains a major concern, particularly in densely populated urban areas where traffic emissions contribute to elevated levels of particulate matter (PM) ($PM_{2.5}$), ozone, and nitrogen dioxide, posing serious risks for respiratory and cardiovascular health (Goldsborough et al. 2023; Hou et al. 2015; Huangfu and Atkinson 2020; Sagheer et al. 2024). Climate-related events such as wildfires, heatwaves, and hurricanes are increasing in frequency and intensity, disproportionately affecting vulnerable populations including the elderly, children, and low-income communities (Crimmins et al. 2023). The health impacts from wildfire smoke, for example, extend far beyond the burn zone and have been associated with increased hospitalizations and mortality due to respiratory and cardiovascular conditions (Reid et al. 2016). Furthermore, indoor air quality is gaining recognition as a critical issue, with Americans spending approximately 90% of their time indoors, where exposure to volatile organic compounds, mold, and combustion by-products is prevalent and often underregulated (Klepeis et al. 2001).

Legacy pollutants and emerging contaminants also present pressing challenges. Communities across the United States continue to experience the long-term effects of lead exposure, particularly in older housing stock and historically marginalized neighbourhoods (Hanna-Attisha et al. 2016). Widespread contamination with PFAS, often linked to firefighting foams and industrial discharges, has led to bioaccumulation in human populations and contaminated water supplies in nearly all 50 states (Hu et al. 2016). Additionally, microplastics and endocrine-disrupting chemicals like phthalates and bisphenol A are being increasingly detected in environmental samples and human biomonitoring studies, raising concerns about their long-term impacts on reproductive and developmental health (Rochman et al. 2015; Zota et al. 2016). These risks are compounded by stark environmental health disparities, where low-income and racially minoritized communities are often situated near industrial sites, landfills, and highways, increasing both exposure and health burdens (Bullard 2018). Addressing these challenges will require integrated, equity-centered approaches that leverage scientific research, policy, and community engagement.

1.4 Republic of Korea

As a result of rapid industrialization and urbanization, the Republic of Korea is experiencing various environmental health issues. Over the past decade, social awareness of air pollution, particularly PM (2.5 and 10), has increased, leading to continuous efforts to improve the situation. As a result, PM concentrations have been gradually decreasing. However, Korea's annual concentrations of PM_{10} and $PM_{2.5}$ in 2023 were 37 μg/m^3 and 19 μg/m^3, respectively, exceeding the strengthened WHO standards by approximately two times for PM_{10} and four times for $PM_{2.5}$, still posing a significant health risk (NIER 2024). In particular, areas near industrial complexes are more severely affected by air pollution (Bae et al. 2018; Baek et al. 2020). As of 2022, 1,262 industrial complexes are distributed throughout the country, exposing many residents in various regions to air pollutants emitted from these complexes. In fact, government surveys have shown that residents in industrial complex areas have higher incidence rates and mortality ratios for diseases such as diabetes, hypertensive diseases, cerebrovascular diseases, and cancer compared to residents in control areas (Eom et al. 2018; NIER 2017). Residents in these affected areas continue to report health issues, and there is an urgent need for scientific research and social efforts to identify the unique characteristics of air pollution in industrial complex areas and its health impacts.

Recently, concerns have been growing over new pollutants such as microplastics and PFAS. Numerous studies have confirmed the presence of microplastics in Korea's marine and freshwater environments, drinking water, and food (Lee et al. 2023; Pham et al. 2023), and national biomonitoring programs have revealed the detection of endocrine-disrupting chemicals such as PFAS in the human body, highlighting the need for strict management and further research (Hong et al. 2024). Microplastics have also been detected in the Korean atmosphere (Chang et al. 2023; Choi et al. 2022), raising concerns about the possibility of respiratory damage through inhalation exposure, but research on this issue is still in its early stages. It is necessary to consider management strategies for microplastics as a new air pollutant. Like Australia, Koreans spend a high proportion of their time indoors. Therefore, a comprehensive and systematic approach to integrated indoor and outdoor air quality management is necessary to protect public health in the future.

1.5 Malaysia

Malaysia, a low-and medium-income country is facing a range of environmental health challenges stemming from rapid urbanization, industrial growth, and climate change. Air pollution remains a major public health concern, with approximately 21% of death from stroke and ischaemic heart disease in Malaysia are attributed to air pollution exposure (WHO 2025). Besides a growing number of sources from industrial manufacturing, power generation, vehicles, and open burning activities, transboundary haze due to peatland fires from neighbouring country, Indonesia, is one of the main reasons to Malaysia's poor air quality. Malaysia PM2.5 annual average recorded at 17 μg/m ^{3}, over three times the WHO guideline of 5 μg/m^3, which contribute to respiratory and cardiovascular mortality, premature deaths, and dermatological issues (Putra 2025; WHO 2025). While regional efforts exist under the Association of Southeast Asian Nations, political fragmentation has hindered effective enforcement, leading to continued exposure to the hazardous PM2.5 levels (Putra 2025).

Recognizing these challenges, the Ministry of Health Malaysia, through National Environmental Health Action Plan has prioritized most importance environmental health issues. Besides what have been mentioned above, children's environmental health has been identified as the top national priority, due to the vulnerability of children to indoor and ambient environmental hazards. UNICEF Malaysia also reinforces this concern on children, including the impacts from extreme heat from climate change. While cross-sectional studies have explored the effects of traffic-related air pollution (Suhaimi et al. 2024), indoor air contaminants in day care centre (Najib et al. 2024) and preschools (Azam et al. 2025), and biomass fuel combustion (Mohd Shahidin et al. 2021) on Malaysian children, the lack of longitudinal research remains a critical gap. Malaysia urgently needs birth cohort studies to examine early-life exposures and the impacts on children's health. Such evidence is important for designing effective early intervention strategies, besides informing policies to protect the health and development for the next generation.

1.6 Mexico

Mexico is a vast territory with nearly 126 million inhabitants, of which approximately 80% live in cities, such as Mexico City, home of 10 million inhabitants and whose population is expected to exceed 24 million by 2030 (de Lima et al. 2023; INEGI México 2020). This urban growth not only contributes to air pollution but also to solid waste generation, which has been estimated to be around 120,128 tons per day nationwide (México 2020). This situation coexists with the deep-rooted use of biomass fuels for cooking, mainly among the rural population (Juan Rivera Dommarco 2021). Mexico also has a long-standing agricultural tradition marked by the use of pesticides at levels twice the global average (Moreno-Godínez et al. 2024). In addition to this situation, the intensive use of fertilizers and the ongoing discharge of urban and industrial wastewater have led to more than 60% of the fluvial systems showing signs of pollution (Aguilar-Aguilar et al. 2023). Moreover, some regions of Mexico are internationally known for groundwater contamination by arsenic (Podgorski & Berg 2020), and is also a major metal producer and home to one of the largest non-ferrous metal smelting complexes in the world (Mérida-Ortega et al. 2022).

Isolated studies have reported not only the presence of various environmental contaminants in biological samples from Mexicans, but also their association with the development of several diseases such as diabetes (Ugalde-Resano et al. 2024) and breast cancer (Mérida-Ortega et al. 2023). However, Mexico lacks a national biomonitoring program. To date, lead is the only pollutant that has been recently evaluated at the national level and only among children. In 2022, approximately 17% of children aged 1 to 4 were found to have lead poisoning (Bautista-Arredondo et al. 2023), with estimates indicating an average loss of 6.42 IQ points and a corresponding economic impact that could reach up to 7.08% of GDP in certain states (Figueroa et al. 2024). In addition to the limited evidence on the extent of environmental exposure to various contaminants, Mexico has experience increases in droughts, storms, and floods as a consequence of climate change

(Murray-Tortarolo 2021). Moreover, this country generally lags behind in environmental regulation, which often lacks effective enforcement mechanisms (Aguilar-Aguilar et al. 2023; Fisher et al. 2017). This situation might be particularly concerning with regard to emerging contaminants, such as new pesticide formulations, perfluorinated compounds, and micro- and nanomaterials, among others.

1.7 Qatar

Qatar is a small country in the Middle East, one of six Gulf Corporate Countries (GCC). Qatar is an arid country of harsh environment with high summer temperatures (>40°C) and low rainfall (annual average 82 mm) (Darwish and Mohtar 2013). The country has a current population of approximately 3 million compared to 0.744 million in the year 2000, with mostly all living in its capital city Doha. The country has had a rapid increase in economic growth in the last decade, owing to its discovery and production of fuel oil and natural gas (Darwish and Mohtar 2013). Natural gas has become its main source of income leading to an improvement in society's well-being. For example, the income per capita has increased four times in a matter of 8 years, the net enrolment in education has increased by ~ 12% within 20 years, and the life expectancy has increased by 7 years since 1975 (World Bank 2015). Despite these positive outcomes, rapid growth in the economy has had its consequences, mainly environmental challenges such as water security, negative effects on the marine environment, and climate change. These challenges along with the increase in population consequently lead to health issues (Cochrane and Al-Hababi 2023).

Due to its small size, Qatar is the highest emitter of CO_2 globally (Salahuddin and Gow 2019; World Bank 2015). The primary source is the energy sector, particularly the oil and gas industry. Large capacities are available across Qatar's land and ocean, attributing to the release of 96% of greenhouse gasses (GHG), followed by the industrial sector 3%, and waste by 1% (Lanouar et al. 2016). Qatar relies on water desalination as its main source (99%) of water requiring high energy consumption and therefore high emissions (Mannan et al. 2019). Air pollution has become a major concern for Qatar, as it is the main contributor to environmental degradation with levels of PM10 reaching levels of 150 $\mu g/m^3$ for 24 h average exceeding the WHO guidelines by far (50 $\mu g/m^3$) (Lanouar et al. 2016). The main concern with these levels is respiratory diseases such as asthma which has been shown to increase in children in recent years (Janahi et al. 2006; Mannan et al. 2019). Although air pollution is a big cost to economic growth in Qatar, the Qatar National Vision (QNV) report 2030 in conjunction with the Advancing Sustainable Development (ASD) report were launched to find a resolution that allows for the continued economic growth along while conserving the environment, without one having to sacrifice the other (Cochrane and Al-Hababi 2023).

Emergent patterns and shared risks

The Pacific Basin region encompasses diverse countries, populations and geographic characteristics, yet we have shown that across this region there are commonalities in environmental hazards and their drivers. Industrialization, urbanization and population growth and movement remain major drivers of environmental hazards. Additionally, climate and environmental change are driving natural hazards and impacting critical resources. Many of these regions contend with a mix of legacy pollutants and emerging pollutants of concern, revealing the lasting imprint of environmental hazards that continues well beyond their initial release.

While some exposures, such as e-waste, disproportionately affect low to middle income countries, others like air pollution and plastic related pollution are not driven solely by income level. Some of the wealthiest countries discussed (United State and Qatar) still have significant issues with ambient air pollution, driven by the same economic processes that generate their prosperity. It is also worth noting that other countries are bearing the brunt of exposures from which they do not generate any benefit, such as airborne microplastics in Fiji originating from plastic waste in the Pacific Ocean or Malaysia's air quality being affected by agricultural burning in Indonesia. These events remind us that environmental hazards transcend boundaries and require multilateral cooperation to protect vulnerable populations.

2 Conclusion

In this report, we confirm that the various regions and countries within the Pacific Basin—analyzed throughout this work—share several common experiences related to environmental health, including the consequences of climate change, air pollution (both indoor and ambient), widespread exposure to emerging chemicals, and legacy pollutants. However, the magnitude of these environmental treads may vary among the evaluated regions due to their differences in geography, lifestyle, population density, and socio-economic development. While the Pacific Basin might present a distinctive environmental contamination profile we believe that collective action is essential to improving environmental health not only across the region but globally.

We recognize both the significant environmental challenges faced by the Pacific Basin region and the advancement of sophisticated methods that enable the identification of causal relationships, the assessment of coexisting pollutants and the understanding of their role within complex systems. These developments offer an opportunity to improve future investigations to generate more robust scientific evidence regarding the state of environmental health in the region. Crucially, we underscore the importance of effective communication of research findings to promote improved environmental health outcomes. This must always be grounded in an environmental justice framework, prioritizing the needs of communities most vulnerable to environmental exposures or those disproportionately burdened by them. The Pacific Basin Consortium for Environment and Health's Junior Board Advisory Committee is an opportunity for new researchers to come together to address environmental concerns. The Committee formalizes the opportunity for knowledge exchange, supported by senior mentors. Coming together into such a forum fosters a multilateral approach to environmental health challenges. The diversity of perspectives enables the identification of shared priorities, the exchange of innovative solutions, and the development of culturally and regionally appropriate strategies. For example, the Junior Board recognizes the importance of climate change as a driving force of environmental harm, and the fact that environmental hazards do not occur in isolation, but as part of a complex mixture of exposures we encounter daily. We also recognize that children represent both a uniquely vulnerable population and an opportunity for population health improvement across the lifespan. We see these as priority areas for collaboration, with work underway within the board on these topics.

The Junior Board Committee strengthens capacity by pooling expertise, resources, and networks, while also creating opportunities for collaborative research that transcends national boundaries. Such engagement not only accelerates scientific progress but also builds a foundation for long-term partnerships that can influence policy and practice across the Pacific Basin region.

Funding. No funding.

Declaration of Interest. All authors declare have no potential conflict of interest.

Note. Dwan Vilcins is a former Junior Board Advisor, recently appointed as Board Member.

References

Aguilar-Aguilar, A., de León-Martínez, L.D., Forgionny, A., Acelas Soto, N.Y., Mendoza, S.R., Zárate-Guzmán, A.I.: A systematic review on the current situation of emerging pollutants in Mexico: a perspective on policies, regulation, detection, and elimination in water and wastewater. Sci. Total. Environ. **905**, 167426 (2023). https://doi.org/10.1016/j.scitotenv.2023.167426

Al Nabhani, K., Baddock, M., Mani, F., Stuetz, R., Allinson, G., Dansie, A.: Airborne Microplastics in Fiji: Characterisation and contribution of transboundary versus local sources. Science of the Total Environment (in press)

Al Nabhani, K., Salzman, S., Shimeta, J., Dansie, A., Allinson, G.: A temporal assessment of microplastics distribution on the beaches of three remote islands of the Yasawa archipelago, Fiji. Mar. Pollut. Bull. **185**, 114202 (2022). https://doi.org/10.1016/j.marpolbul.2022.114202

Azam, N.S.A., Jalaludin, J., Suhaimi, N.F.: The association between indoor air quality and respiratory health symptoms among preschool children in Penang, Malaysia. Int. J. Environ. Health Res. **35**(3), 606–619 (2025). https://doi.org/10.1080/09603123.2024.2365308

Bae, S., Choi, J., Kim, G., Song, S., Ha, M., Kwon, H.: Evaluation of the exposure to environmental pollutants emanating from national industrial complexes. Environ. Health Toxicol. **33**(2), e2018007 (2018)

Baek, K.-M., Seo, Y.-K., Kim, J.-Y., Baek, S.-O.: Monitoring of particulate hazardous air pollutants and affecting factors in the largest industrial area in South Korea: The Sihwa-Banwol complex. Environ. Eng. Res. **25**(6), 908–923 (2020)

Barraclough, K.A., Carey, M., Winkel, K.D., Humphries, E., Shay, B.A., Foong, Y.C.: Why losing Australia's biodiversity matters for human health: insights from the latest state of the environment assessment. Med. J. Aust. **218**(8), 336–340 (2023)

Bautista-Arredondo, L.F., Trejo-Valdivia, B., Estrada-Sánchez, D., Tamayo-Ortiz, M., Cantoral, A., Figueroa, J. L., Romero-Martínez, M., Gómez-Acosta, L.M., Cuevas-Nasu, L., Rojo, M.M.T.: Childhood lead poisoning in Mexico: other sources of exposure beyond lead-glazed ceramics (Ensanut 2022). Salud pública de México, **65**, S197-S203 (2023). https://doi.org/10.21149/14798

Bonnefoi, M.S., et al.: Human and environmental health challenges for the next decade (2010–2020). Crit. Rev. Toxicol. **40**(10), 893–911 (2010). https://doi.org/10.3109/10408444.2010.506640

Brown, B., Hollins, I., Pickin, J., Donovan, S.: Asbestos stocks and flows legacy in Australia. Sustainability **15**(3), 2282 (2023). https://www.mdpi.com/2071-1050/15/3/2282

Bullard, R.D.: Dumping in Dixie: Race, class, and environmental quality. Routledge (2018)

Chang, D.Y., Jeong, S., Shin, J., Park, J., Park, C.R., Choi, S., Chun, C.-H., Chae, M.-Y., Lim, B.C.: First quantification and chemical characterization of atmospheric microplastics observed in Seoul, South Korea. Environ. Pollut. **327**, 121481 (2023). https://doi.org/10.1016/j.envpol.2023.121481

Choi, H., et al.: Comparison of microplastic characteristics in the indoor and outdoor air of urban areas of South Korea. Water Air Soil Pollut. **233**(5), 169 (2022)

Cochrane, L., Al-Hababi, R.: Sustainable Qatar: Social, political and environmental perspectives. Springer Nature (2023)

Crimmins, A.R., Avery, C.W., Easterling, D.R., Kunkel, K.E., Stewart, B.C., Maycock, T.K.: Fifth national climate assessment (2023)

Darwish, M.A., Mohtar, R.: Qatar water challenges. Desalin. Water Treat. **51**(1–3), 75–86 (2013)

de Lima, G.N., Fonseca-Salazar, M.A., Campo, J.: Urban growth and loss of green spaces in the metropolitan areas of São Paulo and Mexico City: effects of land-cover changes on climate and water flow regulation. Urban ecosystems **26**(6), 1739–1752 (2023). https://doi.org/10.1007/s11252-023-01394-0

Drova, E., Tuqiri, N., Prasad, K., Rosi, A., Gadekilakeba, T., Bai, A., Tunaka, E., Williams, L., Varea, R., Brown, K., Dehm, J.: Microplastic occurrence in 21 coastal marine fish species from fishing communities on Viti Levu, Fiji. Mar. Pollut. Bull., **216**, 118045 (2025). https://doi.org/10.1016/j.marpolbul.2025.118045

Eom, S.-Y., et al.: Health effects of environmental pollution in population living near industrial complex areas in Korea. Environ. Health Toxicol. **33**(1), e2018004 (2018)

Figueroa, J.L., et al.: Loss of cognitive function in Mexican children due to lead exposure and the associated economic costs. Environ. Res. **263**(Pt 1), 120013 (2024). https://doi.org/10.1016/j.envres.2024.120013

Fisher, A.T., López-Carrillo, L., Gamboa-Loira, B., Cebrián, M.E.: Standards for arsenic in drinking water: Implications for policy in Mexico. J. Public Health Policy **38**(4), 395–406 (2017). https://doi.org/10.1057/s41271-017-0087-7

Goldsborough, E., Gopal, M., McEvoy, J.W., Blumenthal, R.S., Jacobsen, A.P.: Pollution and cardiovascular health: A contemporary review of morbidity and implications for planetary health. Am. Hear. J. Plus **25**, 100231 (2023). https://doi.org/10.1016/j.ahjo.2022.100231

Groopman, J.D.: Environmental health in the biology century: transitions from population to personalized prevention. Exp. Biol. Med. **244**(9), 728–733 (2019). https://doi.org/10.1177/1535370219837903

Hagen, T.G., De Nola, G., Lane, A., Woodworth, J.: Australian Case Studies Involving Exposure to PFOS. In: Human and Ecological Risk Assessment (pp. 1051–1090) (2024). https://doi.org/10.1002/9781119742975.ch29

Hanna-Attisha, M., LaChance, J., Sadler, R.C., Champney Schnepp, A.: Elevated blood lead levels in children associated with the Flint drinking water crisis: a spatial analysis of risk and public health response. Am. J. Public Health, **106**(2), 283–290 (2016)

Hong, S., Kim, O.-J., Jung, S.K., Jeon, H.L., Kim, S., Kil, J.: The exposure status of environmental chemicals in south korea: the korean national environmental health survey 2018–2020. Toxics **12**(11), 829 (2024)

Hou, X., Strickland, M.J., Liao, K.-J.: Contributions of regional air pollutant emissions to ozone and fine particulate matter-related mortalities in eastern U.S. urban areas. Environ. Res. **137**, 475–484 (2015). https://doi.org/10.1016/j.envres.2014.10.038

Hu, X.C., et al.: Detection of poly-and perfluoroalkyl substances (PFASs) in US drinking water linked to industrial sites, military fire training areas, and wastewater treatment plants. Environ. Sci. Technol. Lett. **3**(10), 344–350 (2016)

Huangfu, P., Atkinson, R.: Long-term exposure to NO2 and O3 and all-cause and respiratory mortality: a systematic review and meta-analysis. Environ. Int. **144**, 105998 (2020). https://doi.org/10.1016/j.envint.2020.105998

INEGI México.: Población en México in 2020 (2020). https://www.inegi.org.mx/

Jagals, P., Kim, I., Brereton, C., Lau, C.L.: Assessment of environmental impacts on health: examples from the Pacific Basin. Ann. Glob. Health (2022)

Janahi, I.A., Bener, A., Bush, A.: Prevalence of asthma among Qatari schoolchildren: international study of asthma and allergies in childhood Qatar. Pediatr Pulmonol **41**(1), 80–86 (2006). https://doi.org/10.1002/ppul.20331

Juan Rivera Dommarco, T.B.G., Carlos Oropeza.: Síntesis sobre políticas de salud. Propuestas basadas en evidencia (2021). https://www.insp.mx/novedades-editoriales/sintesis-sobre-politicas-de-salud-propuestas-basadas-en-evidencia

Kishi, R., Ikeda, A., Ketema, R.M.: The potential health risks of exposure to environmental chemicals - Global implications for future generations. Proc. Jpn. Acad. Ser. B Phys. Biol. Sci. **101**(4), 197–215 (2025). https://doi.org/10.2183/pjab.101.015

Klepeis, N.E., et al.: The National Human Activity Pattern Survey (NHAPS): a resource for assessing exposure to environmental pollutants. J. Eposure Sci. Environ. Epidemiol. **11**(3), 231–252 (2001)

Lanouar, C., Al-Malk, A.Y., Al Karbi, K.: Air pollution in Qatar : causes and challenges. White Paper, **1**(3), 1–7 (2016). https://doi.org/10.13140/RG.2.1.2269.4004

Lee, Y., Cho, J., Sohn, J., Kim, C.: Health effects of microplastic exposures: current issues and perspectives in South Korea. Yonsei Med. J. **64**(5), 301 (2023)

Mannan, M., Alhaj, M., Mabrouk, A.N., Al-Ghamdi, S.G.: Examining the life-cycle environmental impacts of desalination: a case study in the State of Qatar. Desalination, **452**, 238–246 (2019). https://doi.org/10.1016/j.desal.2018.11.017

Mérida-Ortega, Á., Gennings, C., Rothenberg, S.J., Cebrián, M.E., López-Carrillo, L.: Breast cancer molecular subtypes and supervised analysis of urinary metal mixtures in Mexican women. Expo. Health **15**(4), 903–913 (2023). https://doi.org/10.1007/s12403-022-00532-x

Mérida-Ortega, Á., Rothenberg, S.J., Cebrián, M.E., Arias-Medellín, L.A., Salgado-Salgado, A.L., López-Carrillo, L.: Urinary concentrations of potentially toxic metals and metalloids among women residing in Northern Mexico. Expo. Health **14**(4), 857–870 (2022). https://doi.org/10.1007/s12403-021-00458-w

México, G. d.: Presenta Semarnat el Diagnóstico Básico para la Gestión Integral de Residuos 2020 (2020). https://www.gob.mx/semarnat/prensa/presenta-semarnat-el-diagnostico-basico-para-la-gestion-integral-de-residuos-2020

Mohd Shahidin, F.S., Junaidi, E.S., Jalaludin, J., Tualeka, A.R.: Associations of exposure to PM10 and PM2.5 with respiratory health symptoms and lung function status among children living near palm oil activity in Semenyih, Selangor. J. Sustain. Sci. Manag. **16**(6), 165–179 (2021). https://doi.org/10.46754/jssm.2021.08.015

Moreno-Godínez, M.E., et al.: Trends on deaths from acute pesticide poisoning in Mexico, 2000–2021. Rev. Bras. Epidemiol. **27**, e240001 (2024). https://doi.org/10.1590/1980-549720240001

Najib, S.A.H.M., Jalaludin, J., Feisal, N.A.S., Faruk, T., Khan, M.F.: Interaction of indoor air contaminants and respiratory health among children in the daycare centers. Air Qual. Atmos. Health **17**(8), 1677–1688 (2024). https://doi.org/10.1007/s11869-024-01536-z

[Record #66 is using a reference type undefined in this output style.]

NIER: Comprehensive Evaluation on the Result of four years (2012~2015) Monitoring of Exposure to Environmental Pollutants and Health Effects among Residents Living near Industrial Complex (2017)

NIER: Annual report of air quality in Korea 2023 (2024)

Nilsson, S., Kucharski, N., Orr, J., Bräunig, J., Thompson, K., Jolliet, O., Langguth, D., Kennedy, C., Hobson, P., Thomas, K.V., Mueller, J.F., Toms, L.-M.: Serum concentrations of PFAS across Australian states and territories. Int. J. Hyg. Environ. Health **265**, 114542 (2025). https://doi.org/10.1016/j.ijheh.2025.114542

Pham, D.T., Kim, J., Lee, S.-H., Kim, J., Kim, D., Hong, S., Jung, J., Kwon, J.-H.: Analysis of microplastics in various foods and assessment of aggregate human exposure via food consumption in korea. Environ. Pollut. **322**, 121153 (2023). https://doi.org/10.1016/j.envpol.2023.121153

Podgorski, J., Berg, M.: Global threat of arsenic in groundwater. Sci. (Am. Assoc. Adv. Sci.) **368**(6493), 845–850 (2020). https://doi.org/10.1126/science.aba1510

Poudel, K., Ketema, R.M., Thi Thu Ngo, H., Ikeda, A., Minatoya, M.: E-waste in Vietnam: a narrative review of environmental contaminants and potential health risks. Rev. Environ Health **39**(3), 391–406 (2024). https://doi.org/10.1515/reveh-2022-0233

Putra, B.A.: The politics of environmental policy: haze pollution, ASEAN, and the way forward. Front. Sustain. Cities **6** (2025). https://doi.org/10.3389/frsc.2024.1417746

Reid, C.E., Brauer, M., Johnston, F.H., Jerrett, M., Balmes, J.R., Elliott, C.T.: Critical review of health impacts of wildfire smoke exposure. Environ. Health Perspect. **124**(9), 1334–1343 (2016)

[Record #193 is using a reference type undefined in this output style.]

Sagheer, U., et al.: Environmental pollution and cardiovascular disease: Part 1 of 2: air pollution. JACC. Advances (Online) **3**(2), 100805 (2024). https://doi.org/10.1016/j.jacadv.2023.100805

Salahuddin, M., Gow, J.: Effects of energy consumption and economic growth on environmental quality: evidence from Qatar. Environ. Sci. Pollut. Res. **26**(18), 18124–18142 (2019). https://doi.org/10.1007/s11356-019-05188-w

Scheringer, M., Johansson, J.H., Salter, M.E., Sha, B., Cousins, I.T.: Stories of global chemical pollution: will we ever understand environmental persistence? Environ. Sci. Technol. **56**(24), 17498–17501 (2022). https://doi.org/10.1021/acs.est.2c06611

[Record #65 is using a reference type undefined in this output style.]

Sly, P., Al Nabhani, K., Sripada, K., Kayama, F.: Microplastics in the Asia-Pacific region in the plasticene era: exposures and health risks. Ann. Glob. Health (2024). https://doi.org/10.5334/aogh.4326

Suhaimi, N.F., Jalaludin, J., Roslan, N.I.S.: Traffic-related air pollution (TRAP) in relation to respiratory symptoms and lung function of school-aged children in Kuala Lumpur. Int. J. Environ. Health Res. **34**(3), 1384–1396 (2024). https://doi.org/10.1080/09603123.2023.2211020

Suk, W.A., et al.: Environmental threats to children's health in Southeast Asia and the Western Pacific. Environ. Health Perspect. **111**(10), 1340–1347 (2003). https://doi.org/10.1289/ehp.6059

Ugalde-Resano, R., Mérida-Ortega, Á., Barajas, B., López-Carrillo, L., Cebrián, M.E.: Diabetes mellitus and serum organochlorine pesticides mixtures in Mexican women. Environ. Health **23**(1), 57 (2024). https://doi.org/10.1186/s12940-024-01096-w

United Nations Environment Programme (UNEP) and National Institute for Environmental Studies (NIES). Global Environment Outlook – GEO-5: Environment for the future we want, Asia and the Pacific (2012). https://www.nies.go.jp/whatsnew/2012/20120607/GEO-AsiaPacific.pdf

Vilcins, D., Dangal, P., Lomnicki, S., Cormier, S.A., Lee, W.R., Sly, P.D.: Household dust as a reservoir for environmentally persistent free radicals: a longitudinal study. Int. J. Environ. Health Res., 1–12. https://doi.org/10.1080/09603123.2025.2495202

Vilcins, D., Lee, W.R., Blake, T.L., Wu, W., Cormier, S., Sly, P.D.: Assessing the impact of wildfire smoke exposure on lung health: a pilot feasibility study using prescribed fires as a model. Am. J. Respir. Crit. Care Med. **210**(7), 936–937 (2024). https://doi.org/10.1164/rccm.202403-0625RL

Vilcins, D., Lee, W.R., Blake, T.L., Wu, W., Cormier, S., Sly, P.D.: Reply to myers and rosser: a comment about studying the health effects of smoke produced by prescribed fire. Am. J. Respir. Crit. Care Med. **210**(11), 1384–1385 (2024). https://doi.org/10.1164/rccm.202408-1614LE

WHO, W. H. O.: Malaysia: Country profile of environmental determinants of health 2025 (2025). https://cdn.who.int/media/docs/default-source/country-profiles/environmental-health/environmental-health-mys-2025.pdf

World Bank: World Development Indicators (WDI) (2015). http://www.worldbank.org/

World Health Organization (2024) Report of the Regional Director: the work of WHO in the Western Pacific Region 1 July 2023–30 June 2024. (WPR/RC75/2)

World Health Organization Regional Office for the Western Pacific: Regional agenda on achieving the sustainable development goals in the Western Pacific (2017)

Xu, R., Yu, P., Liu, Y., Chen, G., Yang, Z., Zhang, Y., Wu, Y., Beggs, P. J., Zhang, Y., Boocock, J., Ji, F., Hanigan, I., Jay, O., Bi, P., Vargas, N., Leder, K., Green, D., Quail, K., Huxley, R., Jalaludin, B., Hu, W., Dennekamp, M., Vardoulakis, S., Bone, A., Abrahams, J., Johnston, F.H., Broome, R., Capon, T., Li, S., Guo, Y.: Climate change, environmental extremes, and human health in Australia: challenges, adaptation strategies, and policy gaps. Lancet Reg.Nal Health – West. Pac. **40** (2023). https://doi.org/10.1016/j.lanwpc.2023.100936

Zota, A.R., Phillips, C.A., Mitro, S.D.: Recent fast food consumption and bisphenol A and phthalates exposures among the US population in NHANES, 2003–2010. Environ. Health Perspect. **124**(10), 1521–1528 (2016)

Author Index

L. Vu et al. (Eds.): PBC 2025, 2026.
https://doi.org/10.1007/978-3-032-29579-8

MIX
Papier aus verantwortungsvollen Quellen
Paper from responsible sources
FSC® C105338

If you have any concerns about our products,
you can contact us on
ProductSafety@springernature.com

In case Publisher is established outside the EU,
the EU authorized representative is:
Springer Nature Customer Service Center GmbH
Europaplatz 3, 69115 Heidelberg, Germany

Printed by Libri Plureos GmbH
in Hamburg, Germany